Peptides

Biology and Chemistry

Peptides

Biology and Chemistry

Proceedings of the 1992 Chinese Peptide Symposium
November 3–6, 1992, Hangzhou, China

Edited by

Yu-cang Du
Shanghai Institute of Biochemistry
Chinese Academy of Sciences
Shanghai 200031, China

James P. Tam
Department of Microbiology and Immunology
Vanderbilt University
Nashville, TN 37232, U.S.A.

and

You-shang Zhang
Shanghai Institute of Biochemistry
Chinese Academy of Sciences
Shanghai 200031, China

ESCOM ▪ Leiden ▪ 1993

CIP-data Koninklijke Bibliotheek, Den Haag

Peptides

Peptides: Biology and Chemistry: Proceedings of the 1992 Chinese Peptide Symposium, November 3–6, Hangzhou, China / ed. by Yu-cang Du, James P. Tam and You-shang Zhang – Leiden: ESCOM – Ill.
With index, ref.
Subject headings: Peptides/Proteins.

ISBN-13:978-94-010-9068-1   e-ISBN-13:978-94-010-9066-7
DOI: 10.1007/978-94-010-9066-7

Published by:

ESCOM Science Publishers B.V.
P.O. Box 214
2300 AE Leiden
The Netherlands

Preface

The 1992 Chinese Peptide Symposium was held at Hangzhou, China on November 3–6, 1992 with 152 participants, including 46 from abroad, representing 10 countries. The three days conference was both intense and spiritually rewarding. The Biennial Chinese Peptide Symposium was aimed to provide a forum for advancing the exchange of knowledge, cooperation and friendship between the international and Chinese scientific communities. In order to exchange ideas and review the latest progress in depth, the scientific sessions were held at the mornings and evenings, while afternoons were devoted to social interaction and sightseeing of the unparalleled beauty of Westlake. For realizing the goal to catch up with a fast evolving and fascinating field, the Program Committee arranged eight sessions with 45 oral presentations and 72 posters on topics relating to peptide conformation, design, hormone and bioactive peptides, neuropeptides, recognition, synthesis and methodology in the Symposium and selected 94 articles for publication in this proceedings.

As an organizer I was greatly encouraged by the active response of invited speakers and moved by enthusiastic cooperation and excellent contributions which led to the great success of the Symposium, since their presentations were of high quality and represented the latest advances in all aspects of peptide chemistry and peptide biology.

Many authors concentrated their attention to a great challenge which peptide chemists have nowadays to face, i.e. to discover or invent new chemical structures for practical use as peptide or peptide-like drugs, ligands including agonists and antagonists and synthetic vaccines. Insulin as a typical hormone model is still a very active area and a series analogs of insulin, IGF-I and insulin/IGF-I hybrid were prepared. Beside this, more potent agonists and antagonists of such peptides as glucagon, growth hormone releasing factor, somatostatin, endothelin, neurotensin, gonadotropin-releasing hormone, melanotropin, angiotensin, oxytocin and vasopressin are characterized. Agonist and antagonist ligands for hormone and neurotransmitters receptors are potential drugs, but their potencies on biological functions are remarkably dependent on their space properties and conformational flexibilities. Especially to design and synthesize a potent analog of neuropeptides with the desired long duration and oral availability, topographical considerations and the application of conformational constraint have been proved to be very valuable and effective. Knowledge accumulated from the discoveries of natural bioactive peptides and new functions of known peptides, the interactions between peptide and DNA or lipids and between receptors should be recognized for their intellectual value and potential impact on human welfare.

As the Chairman, it is my pleasure to acknowledge the contributions of those groups, individuals who contributed intellectually, physically, and financially to the successful Symposium. On behalf of the Program Committee, I would like to thank the Shanghai Institute of Biochemistry and the Chinese Biochemical Society for the very helpful assistance of administrators of these two institutions. I am especially indebted to professor James P. Tam and his group for all of their efforts to assist us with communication and organization abroad.

Finally, publication of this Proceedings could not be accomplished without the cooperation of Dr. Schram, ESCOM Science Publishers and the technical assistance of my colleagues in this group especially Ms. Xiaoxia Wang and Ms. Ming-ing Cai.

Yu-cang Du

Chinese Peptide Symposium – 1992

November 3–6, 1992, Hangzhou
Shanghai Institute of Biochemistry, Chinese Academy of Sciences
The Chinese Biochemical Society
The Chinese Pharmaceutical Society

Chairman

Yu-cang Du, *Shanghai Institute of Biochemistry*

Program Committee

James P. Tam, *Vanderbilt University*
You-shang Zhang, *Shanghai Institute of Biochemistry*
Qi-chang Xia, *Shanghai Institute of Biochemistry*
Gui-sheng Lu, *Institute of Materia Medica*
Meng-shen Cai, *Beijing Medical University*
Jie-cheng Xu, *Shanghai Institute of Organic Chemistry*
Jia-cheng Hua, *Shanghai Institute of Materia Medica*
Chong-xi Li, *Peking University*

Awarding Committee of H.H. Liu Peptide Awards

Ching-i Niu, Chairman, *Shanghai Institute of Biochemistry*
Yu-cang Du, *Shanghai Institute of Biochemistry*
Ceng-hu Zhu, Executive Member of *H.H. Liu Education Foundation*

Sponsors

National Natural Science Foundation of China
Chinese Academy of Sciences
H.H. Liu Education Foundation
ESCOM Science Publishers B.V.
Bachem Laboratories Inc.
Hoffmann-La Roche Inc.

Donors

Advanced ChemTech Inc.
Lilly Research Laboratories
Peninsula Laboratories Inc.

Abbreviations

Abbreviations used in the proceedings volume are defined below:

AII	*see* Ang II
AA, aa	amino acids
AAA	amino acid analysis
Aab	3-aminomethyl-4-amino-butanoic acid
Ab	antibody
Aba	2-aminobutyric acid
ABTS	2,2'-azido-bis(3-ethylbenzthiazoline sulfonic acid)
A$_2$bu	2,4-diaminobutyric acid
Abut	4-aminobutanoic acid
ABZ, Abz	aminobutanoic acid
AC	adenylate cyclase
Aca	ε-aminocaproic acid
Acc	1-aminocyclopropane carboxylic acid
Ac$_5$c, Acsc	aminocyclopentane carboxylic acid
ACE	angiotensin-converting enzyme
ACh	acetylcholine
ACHPA	4-amino-3-hydroxy-5-cyclohexylpentanoic acid
AChR	acetylcholine receptor
Acm	acetamidomethyl
ACN,Acn	acetonitrile
ACP,Acp	acyl carrier protein; *see* Aca
ACSA	adenylate cyclase stimulating activity
Acsc, Ac$_5$c	aminocyclopentane carboxylic acid
AcT	N$^{\alpha}$-acetyltransferase
ACTH	corticotropin
AD	Alzheimer's disease
Ada	adamantyl
ADE	atrial peptide degrading enzyme
Adoc	2-adamantyloxycarbonyl
ADR	adriamycin
α-AE	α-amidating enzyme
AEC	3-amino-9-ethylcarbazol
AGSP	atrial granule serine proteinase
AH	amphipathic α-helix
Aha,Ahept	7-aminoheptanoic acid
AHB	acute hepatitis B

Ahept,Aha	7-aminoheptanoic acid
Ahex	6-aminohexanoic acid
AHPBA	3-amino-2-hydroxy-4-phenylbutanoic acid; phenylnorstatine
AHPPA	4-amino-3-hydroxy-5-phenylpentanoic acid
Ahx	aminohexyl
Aib	aminoisobutyric acid
Aic	2-aminoindan-2-carboxylic acid
AIDS	acquired immune deficiency syndrome
Am	amidino
Amc	aminoethylcoumarone
AMD	actinomycin D
Amf	*p*-aminomethylphenylalanine
AMP,Amp	aminomethylpiperidine
AMPA	aminomethylphenylacetic acid
ANF	atrial natriuretic factor
ANG,Ang	angiotensin, angiotensinogen
ANG II	*see* Ang II
Ang II	angiotensin II
ANP	atrial natriuretic peptide
Anq	anthraquinone
AO	antiovulatory
AoA	antiovulatory assay
Aoa	8-aminooctanoic acid
Aoc	1-azabicyclo[3.3.0]-2-carboxylic acid
AOGO	5-amino-4-oxo-8-guanidinooctanoic acid
AP	aminopeptidase
AP,BAP	amyloid protein
APC	antigen presenting cell
Apent	5-aminopentanoic acid
APG	azidophenyl glyoxal
Aph	aminophenylalanine
APM	aminopeptidase M
Apo	apolipoprotein
APP	avian pancreatic polypeptide
A$_2$pr, Dpr	2,3-diaminopropionic acid
APY	anglerfish peptide YG
AR	adrenergic receptor
ARC	AIDS-related complex

AS	ammonium sulfate
Asa	azidosalicylic acid
ASC	antibody-secreting cell
ASF	African swine fever
ASFV	African swine fever virus
Asu	aminosuberic acid
AT	antithrombin
αAT,α₁AT	α-antitrypsin
ATIII	antithrombin III
Atc	2-aminotetralin-2-carboxylic acid
ATEE	acetyl tyrosine ethyl ester
ATR	attenuated total internal reflection
AVP	arginine-8-vasopressin
AZT	3'-azido-3'deoxythymidine; zidovudine
b	bovine
Bab	3,5-bis(2-aminoethyl)benzoic acid
Bal	β-alanine
BAP,AP	β-amyloid protein
BBAL	*t*-butoxycarbonyl-N$^\varepsilon$-(N-bromoacetyl-β-alanyl)-L-lysine
BBB	blood brain barrier
Bct	biocystinyl
BGG	bovine γ-globulin
bGRF	bovine growth hormone releasing factor
BHA	benzhydrylamine
BHAR	benzhydrylamine resin
BHI	biosynthetic human insulin
Biot	biotin
BiP	binding protein
Bipa	biphenylalanine
bipy	bipyridine
BK	bradykinin
BLV	bovine leukemia virus
BME	β-mercaptoethanol
BN, Bn	bombesin
BnPeOH	2,2-[bis(4-nitrophenyl)]-ethanol
Boc	*tert*-butyloxycarbonyl
Boc-ON	2-*tert*-butyloxycarbonyl-amino-2-phenylacetonitrile
BOI	2-(benzotriazol-l-yl-)oxy-1,3-dimethylimidazolinium
Bom	benzyloxymethyl
BOP	benzotriazolyloxy tris-(dimethylamino)phospho-nium hexafluorophosphate
BOP-Cl	bis(2-oxo-3-oxazolidinyl)-phosphinic chloride
BPA	benzylphenoxyacetamido-methyl
Bpa	*p*-benzoylphenylalanine
Bpo	D-α-benzoylpenicilloyl
Bpoc	biphenylpropyloxycar-bonyl
BPTI	bovine trypsin inhibitor
bR	bacterial rhodopsin
Br₂Dmb	2,5-bis(bromomethy)benzoate
BroP	bromo tris(dimethyl-amino)-phosphonium hexafluorophosphate
BSA	bovine serum albumin
Bt	biotinoyl
BTD	bicyclic β-turn dipeptide
BTU	*O*-benzotriazolyl-*N,N,N',N'*-tetramethyl-uronium hexafluorophosphate
BTX	bungarotoxin
Bu	butyl derivative
Bum	*tert*-butyloxymethyl
Butaz	1,2-diphenylpyrazolidine-3,4-dione
BW	body weight
Bzl	benzyl
C,c	cup; cis
c-3-PP	*cis*-3-propyl-L-prolyl
CA	chemical acetylation
Cam	carboxyamidomethyl
cAMP	cyclic adenosyl monophosphate
CAP	core amyloid peptide
CAT	chloramphenicol acetyl-transferase; carboxyl amide terminal
Cbz,Z	carbobenzoxy; benzyloxy-carbonyl
CCD	countercurrent distribution
CCK	cholecystokinin
CD	circular dichroism; complement domain
cDNA	complementary DNA
CE	carbetocin
CE, CZE	capillary zone electrophoresis
CEC	cation exchange chromatography
CFA	complete Freund's adjuvant

CG	chorionic gonadotropin	CTL	cytotoxic T-lymphocytes
CGRP	calcitonin gene-related peptide	CTMS	chlorotrimethylsilane
		Ctp	chloroacetyltryptophan; 6-oxo-3,4,6,7-tetrahydro-1H,5H-azocin[4,5,6-c,d]-indole-4-carboxylic acid
CgTx	conotoxin		
CHA,Cha	cyclohexylamine		
CHAPS	3-[(3-cholamidopropyl)-dimethylammonio]-1-propanesulfonate		
		CVAP	cerebrovascular amyloid peptide
CHB	chronic hepatitis B	CVS	cardiovascular system
cHex,cHx	cyclohexyl	Cya	cysteic acid
CHF	congestive heart failure	Cyp	cyclophilin
Chg	cyclohexylglycine	CZE,CE	capillary zone electro-phoresis
CHO	chinese hamster ovary; aldehyde		
ChTr	chymotrypsin	D	diversity(as with Ig or TCR genes)
ChTX	charybdotoxin		
cHx,cHex	cyclohexyl	DA	D/Ala substitution factor; didemnin A
CID	chemically ionized desorption; collision induced disso-ciation		
		Dab	diaminobutyric acid
		DABCYL	4-dimethylaminoazoben-zene-4'-sulfonyl chlo-ride
CINC	cytokine-induced neutro-phil chemoattractant		
CiTEtOIC	4-chloro-3-(2-isothiureido-ethoxy)isocoumarin	DABITC	4-dimethylaminophenyl-4'-isothiocyanate
CiTPrOIC	4-chloro-3-(3-isothiureido-propoxy)isocoumarin	DADLE	[D-Ala^2, D-Leu^5] enkep-halin
CLA	cyclolinopeptide A	DAGO, DAMGO	[D-Ala^2, N-MePhe^4,Gly^5-ol] enkephalin
CM	chloromethyl; casomorphin		
		Dap	diaminopimelic acid
CMC	carboxymethylcysteine	DAST	diethyl aminosulfur tri-fluoride
C$^\alpha$MePhe	C$^\alpha$-methylphenylalanine		
CNS	central nervous system	Dat	desamino tyrosine
COSY	correlated NMR spectros-copy	DB	didemnin B
		DBF	dibenzofuran
CP	carboxypeptidase	Db$_z$g	dibenzylglycine
Cpa	4-chlorophenylalanine	DBH	dibenzylhydrizide
CPBA	chloroperbenzoic acid	Dbr	α,γ-diaminobutyric acid
CPD	carboxypeptidase	DBU	1,8-diazobicyclo[5.4.0]-undec-7-ene
CPF	caerulin precursor fragment		
		Dcb	2,6-dichlorobenzyl
CPMAS	cross-polarization/magic angle spinning	DCC, DCCI	dicyclohexylcarbodiimide
CP-Y	carboxypeptidase Y	DCHA	dicyclohexylamine
CR	chain recombination	DCI	3.4-dichloroisocoumarin
CRF	corticotropin releasing factor	DCM	dichloromethane
		Dcp	dichlorophenyl
Cro	4-hydroxycrotonic acid	DCU	dicyclohexylurea
CRP	C-reactive protein	DDDA	2,9-diamino-4,7-dioxa-decanedioic acid
CsA	cyclosporin A		
CT	carboxy terminus; calcitonin; chymotrypsin; cholera toxin	DDQ	dichlorodicyanoquinone
		DEAE	diethylaminoethanol
		DEAM	diethylacetamidomalonate
		Deg,Dφg	diphenylglycine
CTAB	cetyl trimethyl ammonium bromide	DG/SA	distance geometry/simu-lated annealing

Dha	dehydroalanine		choline
Dhc	S-(2,3-dihydroxypropyl)-cysteine	DPPG	dipalmitoylphosphatidyl-glycerol
DHO	dihydroorotic acid	Dpr, A_2pr	2,3-diaminopropionic acid
DHP	dihydroxypropyl; dihydropyridine	DQF	double quantum focused
DIBAL	diisobutyl aluminium hydride	DSB	4-(2,5-dimethyl-4-methyl-sulfinylphenyl)-4-hy-droxybutanoic acid
DIC	diisopropylcarbodiimide	DSP	dimethylsulfonium meth-ylsulfate
DIEA	diisopropylethylamine		
DIP	4,7-diphenyl phenan-throline	DTC	dimeric tripeptide chemo-attractants
DIPC	*see* DIC	Dtc	5,5-dimethylthiazolidine-4-carboxylic acid
DIPCDI	*see* DIC		
DIPEA	diisopropylethylamine	DTH	delayed-type hypersen-sitivity
DKP	[Asp^{B10}, Lys^{B28}, Pro^{B29}]-insulin; diketopiperazine	DTNB	dithiobis(2-nitrobenzoic acid)
DLPS	dilauroylphosphatidyl-serine	DTPA	diethylenetriamine penta-acetic acid
DMA	dimethylacetamide	Dts	dithiasuccinoyl
DMAP	dimethylaminopyridine	DTT	dithiothreitol
DMBHA	2',4'-dimethoxybenzhy-dryl amine	Dyn	dynorphin
DMF	dimethylformamide	e	eel; equine
Dmp	dimethylphosphinyl		
DMPC	dimyristoylphosphatidyl-choline	EA	ergotamine
		EACA	*see* Aca
DMPG	dimyristoylphosphatidyl-glycerol	EAE	experimental autoimmune encephalomyelitis
DMPSE	dimethylphenylsilylethyl	EBV	Epstein-Barr virus
DMS	dimethyl sulfide	EDAC, EDC	1-(3-dimethylaminopro-pyl)-3-ethylcarbodiimide hydrochloride
DMSO	dimethyl sulfoxide		
Dmt-OH	2,2-dimethyl-L-thiazoli-dine-4-carboxylic acid	EDC, EDAC	1-(3-dimethylaminopro-pyl)-3-ethylcarbodiimide hydrochoride
Dncp	2,4-dinitro-6-carboxy-phenyl		
DNP,Dnp	dinitrophenyl	EDRF	endothelium-derived relaxing factor
Dns	dansyl		
DOACl	dimethyl dioctadecyl ammonium chloride	EDT	ethanedithiol
		EDTA	ethylenediaminetetra-acetic acid
DOC	deoxycholate		
DOPC	dioleoyl-*sn*-glycerophos-phocholine	EGF, EGFR	epidermal growth factor; epidermal growth factor receptor
Dpa	β,β-diphenylalanine; diphenylalanine	EI	epidermal cell inhibitor
DPBT	diphenylphosphorylbenz-oxazolthione	EIAV	equine infectious anemia virus
DPCDI	*see* DIC	ELAB	enantiomer labeling
DPDPE	cyclo[DPen^2-DPen^5]-enke-phalin	ELISA	enzyme-linked immuno-sorbent assay
Dpg	dipropylglycine	Enk	enkephalin
DPI	despentapeptide insulin	env	envelope protein
DPP	dipeptidyl peptidase	EP	endorphin
DPPA	diphenylphosphorylazide	EPNP	1,2-epoxy-3-(*p*-nitrophen-oxy)propane
DPPC	dipalmitoylphosphatidyl-		

EPR	*see* ESR	GC	gas chromatography
ESR	electron spin resonance	g-CSF	granulocyte-colony stimulating factor
ET, Et	endothelin; etorphine		
EtA	α-ethylalanine	GDA	glutaraldehyde
Etm	ethyloxymethyl	GEMSA	guanidino ethylmercaptosuccinic acid
Et₃N,TEA	triethylamine		
		GH	growth hormone
FAA	fatty amino acid	GHRH, GRF	growth hormone releasing hormone
Fab	antigen binding Ig fragment		
		GHRP	growth hormone releasing peptide
FABMS	fast atom bombardment mass spectrometry		
		GITC	2,3,4,6-tetra-*O*-Ac-β-D-glucopyranosyl isothiocyanate
FACS	fluorescence-activated cell sorter		
Farn	farnesyl	Gla	D-galactopyranosyl; γ-carboxyglutamic acid
Fc	crystallizable Ig fragment		
FeLV	feline leukemia virus	GLP	glucagon-like peptide
FET	fluorescence energy transfer	GM	gramicidin M; [Phe⁹,¹¹,¹³,¹⁵]gramicidin A
Fg	fibrinogen	Gn,Gu	guanidine
FGF	fibroblast growth factor	GnRH	gonadotropin releasing hormone
FI	food intake		
FID	flame ionization detector	GO	8-guanidinooctanoyl
FITC	fluorescein isothiocyanate	gp	glycoprotein
Flg	fluorenylglycine	GPI	guinea pig ileum
Fm, fm	fluorenylmethyl	GRF, GHRH	growth hormone releasing factor
FMDV	foot-and-mouth disease virus		
		GRP	gastrin releasing peptide
FMOC	*see* Fmoc	GS	gramicidin S
Fmoc	fluorenylmethoxycarbonyl	GSH	reduced glutathione
Fn	fibronectin	GSSG	oxidized glutathione
For	formyl	GTP	guanosine triphosphate
Fpa	4-fluorophenylalanine	Gu, Gn	guanidine
FPLC	fast protein liquid chromatography	Gua	guanidino
		GvH	graft vs. host
FPro	5-fluoroproline	GVIA	conotoxin G VIA
FRET	fluorescence resonance energy transfer		
		h	human
FSH	follicle stimulating hormone; follitropin	HA	hemagglutinin
		Hat	6-hydroxy-2-aminotetralin-2-carboxylic acid
FTIR	Fourier transform infrared	HBcAg	hepatitis B core antigen
FXa	blood coagulation factor Xa	HBeAg	hepatitis B e antigen
		hBP	human serum binding protein
G,g	guanine-nucleotide binding; gauche; GTP-binding regulatory		
		HBPyU	2-(1*H*-benzotriazol-1-yl)-1,1,3,3-bis(tetramethylene)-uronium hexafluorophosphate
GA	gramicidin A	HBTU	*O*-benzotriazolyl-*N,N,N',N'*-tetramethyluronium hexafluorophosphate
GABA	γ-aminobutyric acid		
GAL	galanin		
GAP	growth associated protein; gonadotropin-releasing hormone associated protein		
		HBV	hepatitis B virus
		HC	heparin cofactor II
		hCG	human chorionic gonado-

	tropin	HPLC	high pressure liquid chromatography; high performance liquid chromatography	
hCGRP	human calcitonin gene-related peptide			
Hcys,hCys	homocysteine	Hpp	3-(4-hydroxypheny)-propionyl	
HEL	hen egg lysozyme; human erythroleukemia	HPSEC	high performance size exclusion chromatography	
HEPES, Hepes	N-[2-hydroxyethyl]pipera-zine-N'2-ethanesulfonic acid	HR, hr	histamine release; human recombinant	
Hfa, Hfe	homophenylalanine	HSA	human serum albumin	
HFBA	heptafluorobutyric acid	Hse	homoserine	
HFC	human fibroblast collagenase	Hsp	heat shock protein	
HFIP	hexafluoroisopropanol	HSPS	high speed peptide synthesis	
hGH	human growth hormone	HSV	herpes simplex virus	
HHM	humoral hypercalcemia of malignancy	HT	HIV-tachykinin	
HI	hemoregulatory cell inhibitor	Htc	7-hydroxytetrahydroiso-quinoline-3-carboxylic acid	
HIC	hydrophobic interaction chromatography	HTE	hamster trachea epithe-lial cell	
HILIC	hydrophilic interaction chromatography	HTLV	human T-cell leukemia virus	
HIV	human immunodeficiency virus	HUB	hamster urinary bladder	
HIVP, HIV-PR	human immunodeficiency virus protease	HUVEC	human umbilical vein endothelial cell	
hK	homolysine	HYCRAM	hydroxycrotonyl amino-methyl linker	
HLA	human leukocyte antigen	Hyp	hydroxyproline	
HLE	human leukocyte elastase	Hz	hertz	
HMP	hydroxymethylphenoxy-acetic acid; hydroxymercaptopro-pionic acid; 4-hydroxymethylphenoxy-methyl	Ia,I-A	immune-associated anti-gen	
HMPA, HMPT	hexamethylphosphoric triamide	IC	inhibitory concentration	
HNE	human neutrophil elastase	ICAM	intracellular adhesion molecule	
hNP	human neutrophil peptide	ICE	interleukin convertase	
HOBT, HOBt	hydroxybenzotriazole	i.c.v.	intra-cerebro-ventricular	
HODhbt	hydroxyoxodihydrobenzo-triazine	IEC	ion-exchange chromatog-raphy	
HONp	nitrophenol	IEF	isoelectric focusing	
HOOBt	hydroxyoxodihydrobenzo-triazine	IEX	ion exchange	
HOSu	N-hydroxysuccinimide	IFAα	interferon α	
HOTic	7-hydroxy-1,2,3,4-tetra-hydroquinoleic-3-car-boxylic acid	Ig	immunoglobulin	
		IGF	insulin-like growth factor	
		IL	interleukin	
		ILys	lysine(N$^\varepsilon$-isopropyl)	
HPA	hypothalamic-pitituary-adrenal axis	im	imidazole	
		i.m.	intramuscular	
HPCE	high performance capil-lary electrophoresis	in	indole	
		Ind	2-carboxy-indoline	
HPI	human proinsulin	INEPT	insensitive nuclei enhancement by pulse	

	transfer	MARCKS	myristolated alanine-rich C kinase substrate
Ing	indenylglycine	MAS	magic angle spining
IP	inositol phosphate	MAT	mating type as for yeast
IR	infrared;	Mba	2-mercaptobenzoic acid
	insulin receptor	Mbh	methoxybenzhydryl
IRMA	immunoradiometric assay	MBHA	methylbenzhydrylamine
IU	international units	MBHAR	methylbenzhydrylamine resin
IV, i.v.	intravenous		
Iva	2-amino-2-methylbutyric acid;	MBP	myelin basic protein
	isovaline	MBS	*m*-maleimidobenzoyl-*N*-hydroxysuccinimide ester
J	joining (as with Ig or TCR genes)	MBzl,Meb	*p*-methylbenzyl
		MCH	melanin concentrating hormone
KLH	keyhole limpet hemocy-anin	MCPBA	*m*-chloroperbenzoic acid
KP	[Lys^{B28}, Pro^{B29}]-insulin	MD	molecular dynamics
Kpc	ketopipecolyl	ME	mercaptoethanol
		Me	mellitin
LACA	long alkyl chain amino acids	Mea	2-mercaptoethylamine
LAR	leukocyte antigen related	Meb,MBzl	*p*-methylbenzyl
LD	lethal dose	Mel	mellitin; melphalan
LDH	lactate dehydrogenase	MeNTI	*N*-methyl noroxymorphin-dole
LDTOF	laser desorption time-of-flight	MePhg	N^{α}-methylphenylglycine
LEC	ligand-exchange chrom-atography	MHC	major histocompatibility complex
LFA	lymphocyte function-associated antigen	MIC	minimal inhibitory concentration
LH	luteinizing hormone; lutropin	MIR	main immunogenic region
		Mls	minor lymphocyte stimulating gene
LHRH	luteinizing hormone releasing hormone	MMC	migrating motor complex
Lo	oxidized coiled-coil	MMTV	mouse mammary tumor virus
LPC	lauroylphosphorylcholine	Mob	*p*-methoxybenzyl
LPH	lipotropin	Mom	methyloxymethyl
LPS	lipopolysaccharide	MoMuLV	Moloney murine leukemia virus
Lr	reduced coiled-coil		
LSIMS	liquid secondary ion mass spectrometry	MOT,Mot	motilin
LTR	long terminal repeat	MP	mastoparan
LVP	lysine-8-vasopressin	Mpa	mercaptopropionic acid
LZ	Leu-zervamicin	Mpg	3-methoxypropylglycine
		Mpg^p	phosphonic acid analog of 3-methoxyglycine
m	messenger; murine	MPLC	medium pressure liquid chromatography
MAb	monoclonal antibody		
Man	2-mercaptoanaline	Mpr	3-mercaptopropionyl; mercaptopropionic acid
MAP, MAp	membrane anchored protein;	MS	mass spectrometry
	multiple antigen peptide; mean arterial pressure	MSH	melanocyte stimulating hormone; melanotropin
MAPs	macromolecule associated proteins	Msob	methylsulfinylbenzyl

Msc	methylsulfonylethyloxy-carbonyl
Msz	methylsulfinylbenzyloxy-carbonyl
MT	metallothionein
Mtr	methoxytrimethylphenyl-sulphonyl
Mts	mesitylenesulfonyl
MuLV	murine leukemia virus
MVD	mouse vas deferens
MxAn	mixed anhydride
NA	nitroaniline
Nag	naphthylglycine
Nal	3-(2-naphthyl)alanine
1-Nal	3-(1'-naphthyl)alanine
Napa	4-azido-2-nitrophenyl-acetyl
Nase	*Staphylococcus aureas* nuclease
Nbb	*o*-nitrobenzamidobenzyl
NBD	7-nitrobenz-2-oxa-1,3-diazole
NBS	*N*-bromosuccinimide
NCA	*N*-carboxyanhydride
NCp	nucleocapsid protein
NCS	neocarcinostatin
NEM	*N*-ethylmaleimide
NFT	neurofibrillary tangles
Nic	nicotinoyl
NIDD	non-insulin-dependent diabetes
NIS	*N*-iodosuccinimide
NK	neurokinin
NM	neuromedin
NMB	neuromedin B
NMDA	*N*-methyl-D-aspartate
NMM	*N*-methylmorpholine
NMP	*N*-methylphrrolidinone
NMR	nuclear magnetic reso-nance
NMT	*N*-myristoyl transferase
NOE	nuclear Overhauser effect
NOESY	nuclear Overhauser enhanced spectroscopy
NP	neutrophil peptide; neurophysin; neuropeptide
NPE	2-(2-nitrophenyl)ethyl
Npp	nitrophenylpyrazolinone
NPY	neuropeptide Y
Npys	3-nitro-2-pyridylsulfenyl
NT	N terminus; amino terminus; neurotensin
NVOC	nitroveratryloxycarbonyl

NZB	New Zealand black
NZW	New Zealand white
o	ovine
Oic	2,3,4,5,6,7,8-octahydro-indole-2-carboxylic acid
oMePhe	*o*-methylphenylalanine
OMP	outer membrane protein
ONb	*o*-nitrobenzyl
ONp	nitrophenyl ester
OPA	*o*-phthaldialdehyde
OPFp	pentafluorophenyl
ONSu	*N*-hydroxysuccinimide ester
OSu	*O*-succinimide ester
OT	oxytocin
OTf	*O*-triflate
OVA	ovalbumin
OVLT	organum vasculosum laminate terminalis
P	propeller
PA	parent antagonist; phosphatidic acid
PAB	*p*-alkoxybenzyl
PAC, Pac	phenacyl
PAF, Paf	*p*-aminophenylalanine
PAGE	polyacrylamide gel electrophoresis
PAL	photoaffinity labeling; peptide amide linker; tris(alkoxy)benzylamide anchor
Pal	3-(3-pyridyl)alanine
Paloc	3-(3-pyridyl)allyloxy-carbonyl
PAM	phenylacetamidomethyl
Pas	6,6-pentamethylene-2-aminosuberic acid
PBS	phosphate buffered saline
PC	phosphatidyl choline
P_3C	tripalmitoyl-*S*-glyceryl-cysteine
PCP	phencyclidine
P_3CSS	tripalmitoyl-*S*-glyceryl-cysteinylserylserine
PDB	phorbol 12,13-dibutyrate
PDMS	plasma desorption mass spectrometry
PE	phosphatidyl ethanolamine
PEG	polyethylene glycol
Pen	penicillamine
pepy	bipyridine-modified peptide
PFC	plaque forming cell
Pfp	pentafluorophenyl ester

PG	proteoglycan; pharmacophore group
PGF	proteoglycan growth factor
PGL[a]	peptide glycine leucine amide
Pgl	n-pentylglycine
Pgl[p]	phosphonic acid analog of 3-methoxypropylglycine
pGlu	pyroglutamic acid
Ph	phenyl
Phaa,PhAc	phenylacetic acid
PhAc,Phaa	phenylacetic acid
PHBT	polymeric hydroxybenzo-triazole
PHF	paired helical filaments
Phg	phenylglycine
Phi	4-iodophenylalanine
Phol	phenylalaninol
Phpa	3-phenylpropanoic acid
Pht	phthaloyl
PI	phosphoinositide
pI	isoelectric point
Pic	picolinoyl
Pin	β-pineyl
Pip,pip	pipecolinyl; piperidine
Pipes	piperazine-N,N'-bis-[2-ethanesulfonic acid]
Piv	pivaloyl
Piz	piperazic acid
PK	protein kinase
PKA	protein kinase A
PKC	protein kinase C
PLA$_2$	phospholipase A$_2$
PLP	poly-L-proline
PMA	phorbol myristate acetate
PMB	polymyxin B
Pmb	p-methoxybenzyl
Pmc	2,2,5,7,8-pentamethyl-chroman-6-sulfonyl
Pmp	3,3-pentamethylene-3-mercaptopropionic acid; phosphonomethylphenyl-alanine
PMSF	phenylmethylsulfonyl fluoride
pNA	p-nitroaniline
PNb	p-nitrobenzyl
PND	principal neutralizing determinant
PON	periodically oscillating neuron
POPC	1-palmitoyl,2-oleoyl-sn-glycero-3-phospho-choline

POPS	palmitoyl-oleoyl-phos-phatidylserine
PP	pancreatic polypeptide
PPA	n-propylphosphoric anhydride
PPE	porcine pancreatic elastase
PPIase	peptidyl-prolyl cis-trans isomerase
PPL	porcine pancreatic lipase
Ppt	diphenylphosphinothionyl
PQ	paraquat
Pqt	3-(1'-methyl-4,4'-bipy-ridinium-1-yl)propyl
PR	protease
Pra	propargylglycine
PRL	prolactin
PRP	platelet-rich plasma
PS	polystyrene
PSG	pig synovial gelatinase
PT	pertussis toxin
PTH	phenylthiohydantoin; parathyroid hormone
PTHrP	PTH related protein
PTK	protein tyrosine kinase
Ptm	phenyloxymethyl
PTPase	protein-tyrosine phospha-tase
PTZ	phenothiazine
Ptz	3-(10-phenothiazinyl)-propanol
PVA	polyvinyl alcohol
PVDF	polyvinylidene fluoride
PVN	paraventricular nuclei
PyBOP	(benzotriazolyl)-N-oxy-pyrrolidinium phospho-nium hexafluorophos-phate
PYL[a]	peptide tyrosine leucine amide
PYY	peptide tyrosine-tyrosine
QUIS	quisqualate
rDNA recDNA	recombinant DNA
REDOR	rotational echo double resonance
RF	rheumatoid factor; replaceability factor
RGD	Arg-Gly-Asp fibrinogen binding sequence
RIA	radioimmunoassay
RMS	root mean square
RMSD, rmsd	root mean square distance; root mean square

	deviation			combination libraries
RNAP	RNA polymerase II		SPPS	solid phase peptide synthesis
RNase	ribonuclease			
RNP	ribonucleoproteins		SRIF,SS	somatostatin
ROE	rotating frame nuclear Overhauser effect		SS,SRIF	somatostain
			SRP	signal recognition peptide
ROESY	rotating frame nuclear Overhauser enhanced spectroscopy		SS	somatostain
			ss	solid state
			ssDNA	single stranded DNA
ROS	rat osteosarcoma cells		ST	heat stable enterotoxin
RP	reversed-phase		Sta	statin
RPC	reversed-phase chromatography		Su	succinimide
			Sub	substrate
RPHPLC	reversed-phase high pressure liquid chromatography		Suc	succinoyl
			SWM	sperm whale myoglobin
RPIF	relative positional importance factor		t	trans
			t-3-PP	*trans*-3-propyl-L-prolyl
RT	reverse transcriptase; room temperature		Tacm	S-trimethylacetamido-methyl
			TAP	tick anticoagulant peptide
s	salmon; staggered		TASP	template assembled synthetic proteins
SA	symmetrical anhydrides		Tat	transcriptional activator
SAH	*S*-adenosylhomomethionine		TBDMS	*tert*-butyldimethylsilyl
			TBDMSCl	*tert*-butyldimethylsilyl chloride
SAM	*S*-adenosylmethionine		TBTA	*tert*-butyl-2,2,2-trichloroacetamide
SAMBHA	(4-succinylamido-2',2',4'-trimethoxy)benzhydrylamine		TBTU	*O*-benzotriazolyl-*N,N,N',N'*-tetramethyluronium tetrafluoroborate; 2-(1*H*-benzotriazol-1-yl)1,1,3,3-tetramethyluronium tetrafluoroborate
SAP	serum amyloid protein			
SAR	structure-activity relations			
Sar	sarcosyl; sarcosine			
SC	synthetic troponin-C peptide			
sc,s.s.	subcutaneous		Tca	trichloroacetamide
SCC	short circuit current		TCEP	tris(2-carboxyethyl)-phosphine
SCLC	small cell lung carcinoma			
SDS	sodium dodecyl sulfate		TCR	T lymphocyte antigen receptor
SEC	size exclusion chromatography		TCS	trypsin-catalyzed semi-synthesis
SEM	standard error of the mean			
			TCT	tracheal cytotoxin
SH	sulfhydryl		TEA,Et$_3$N	triethylamine
SHMT	serine hydroxymethyl transferase		TEAP	triethylamine phosphate
			TEDOR	transferred echo double resonance
SHR	spontaneous hypertensive rat			
			Teoc	trichloroethyloxycarbonyl
SLE	systemic lupus erythematosus		TEP	triethylphosphite
			TFA	trifluoroacetic acid
SMPS	simultaneous multiple peptide synthesis		TFE	trifluoroethanol
			TFM	trifluoromethyl
sn	small nuclear		TFMSA	trifluoromethanesulfonic acid
SP	substance P			
SPCL	synthetic peptide			

TGF	transforming growth factor	TR-COSY	transferred rotational correlated NMR spectroscopy
THF	tetrahydrofuran		
Thg	2-thienylglycine	TRH	thyrotropin releasing hormone
Thi	tetrahydro-1,4-thiazine-3-carboxylic acid; *see* Dtc	Tris	tris(hydroxymethyl)aminomethane
THIQ	tetrahydroisoquinoline	TRNOE, trNOE	transferred nuclear Overhauser effect
THTP	tetrahydrothiophene		
Thz	thiazolidine carboxylic acid	Trt	trityl
		TSH	thyroid stimulating hormone
Tic,Tiq	1,2,3,4-tetrahydroquinoline-3-carboxylic acid	TT	tetanus toxoid
TicOH	7-hydroxy-1,2,3,4-tetrahydroquinoline-3-carboxylic acid	UDP	uridine diphosphate
		UK	urokinase
TIM,TPI	triosephosphate isomerase	u-PA	urokinase-type plasminogen activator
TLC	thin layer chromatography		
TM	transmembrane	UV	ultraviolet
Tm	melting temperature		
TMBD	N,N,N',N'-tetramethylbenzadine	V	variable(as with Ig or TCR genes)
Tmob	2,4,6-trimethoxybenzyl	VCD	vibrational circular dichroism
TMP	3,4,7,8-tetramethylphenanthroline	VIP	vasoactive intestinal peptide
TMS	trimethylsilyl		
TMSCN	trimethylsilylethyl cyanide	VIS,Vis	visible
TMSE	β-trimethylsilyl ethyl	Vly	valeryl
TMSOTf	trimethylsilyl trifluoromethanesulfonate	VMH	ventro medial hypothalamus
TMTr	4,4',4'-trimethoxytriphenylmethyl	Vn	vitronectin
		VNA	virus neutralizing antibody
Tn	troponin		
TnC	troponin-C	VSMC	vascular smooth muscle cells
TNF	tumor necrosis factor		
TOCSY	total correlation spectroscopy	VSV	vesicular stomatitis virus
Tos	tosyl		
TPA,tPA	12-*O*-tetradecanoylphorbol-13-acetate; tissue plasminogen activator	WSCI	water soluble carbodiimide
		WT,wt	wild type; weight
TPI,TIM	triosephosphate isomerase		
TPK	tyrosine-specific phosphate kinase	XAL	5-(9-aminoxanthen-2-oxy)-valeric acid
TPTU	1,1,3,3-tetramethyl-2-(2-oxo-1(2H)-pyridyl)uronium tetrafluoroborate	Xan	9-xanthenyl
		XPF	xenopus precursor fragment
TPyCIU	1,1,3,3-bis(tetramethylene)chlorouronium tetrafluoroborate	Z,Cbz	carbobenzoxy; benzyloxycarbonyl
Tqu	1,2,3,4-tetrahydroquinoline-2-carboxylic acid	ψ	pseudo
TR	time resolved		
TR$_2$C	tryptic fragment from troponin-C		

Contents

Session I: Peptide hormones

Session II: Neuropeptides

Session III: Immunopeptides

Session IV: Receptor and regulation

Session V: New bioactive peptides

Session VI: Peptide conformation and mimetics

Session VII: Peptide synthesis

Session VIII: Synthetic methodologies

Session I
Peptide hormones

Chairs: Bruce Merrifield
Rockefeller University
New York, New York, U.S.A.

Yue-ting Gong
Shanghai Institute of Biochemistry, CAS
Shanghai, China

Victor J. Hruby
University of Arizona
Tucson, Arizona, U.S.A.

and

Da-cheng Wang
Institute of Biophysics, CAS
Beijing, China

Insulin analogue with substitutions at
the C-terminus of the A-chain

Xin-feng Jie, Ming-hua Xu, Wei Lin and Ching-I Niu
Shanghai Institute of Biochemistry, Chinese Academy of Sciences,
Shanghai 200031, China

Introduction

There are two helical regions in the A-chain: A 2-8 and A 12-18. We started the synthesis of an analogue of the A-chain [A 15-Asn, A 17-Pro, A 21-Ala] $ASSO_3^-$ with alterations at the α-helical region in A 12-18. According to the Chou-Fasman method, proline and asparagine would do great damage to the formation of the α-helix. The α-helix in the C-terminal region of the designed A-chain may be disappeared as predicted by the Chou-Fasman method. So the biological significance of the α-helical region in the C-terminal region of the A-chain can be studied.

As indicated from the crystal structure of insulin, the C-terminus of the A-chain is hydrogen-bonded bridging its α-nitrogen to the carbonyl of B 23 glycine. Hence, this C-terminus A 21 asparagine, although very conserved in the evolutional stage, could still be replaced by other amino acids with simpler side chain functions, including glycine and alanine but not proline, without much abolition of its biological potency.

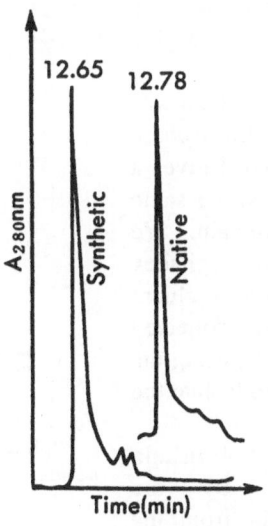

Fig. 1. Synthetic A-chain purified by RP-HPLC. Column: Ultra-sphere Octyl C8 (Beckman) (4.6mm×25cm); Solvent: A=40mM NH₄OAc, 10% CH₃CN, pH 4.01; B=40mM NH₄OAc, 90% CH₃CN, pH 4.01; Flow rate: 1 ml/min.

Table 1 *Mouse convulsion assay*

Sample	Dose (μg)	Convulsion
Insulin	0.5	0/5
	1	2/5
	2	5/5
Insulin analogue	5	0/5
	10	0/5
	15	2/5
	20	5/5

Results and Discussion

The insulin analogue [A15-Asn, A17-Pro, A21-Ala]-Ins. was recombined by a synthetic A-chain and a native B-chain. The A-chain analogue was prepared by solid phase method and started by coupling Nα-Fmoc and side chain-tBu protected amino acids with DCC-HOBt to the p-alkoxy benzylated resin. It was deprotected by M TMSOTf / thioanisale / TFA[1]. After sulfitolysis, the analogue was separated by sephadex G-15 and purified by HPLC. Then it was recombined with the native $BSSO_3^-$ to give the insulin analogue[2].

Fig. 1 shows that the synthetic A-chain was purified by RP-HPLC on an Ultra-sphere Octyl C8 column (Beckman), and it gives a peak with a retention time almost the same as the peak of a native A-chain. We separated the insulin analogue by sephadex G-50, using 10% HOAc solution as eluent (Fig. 2). The peak 1 in Fig 2 was collected and purified by RP-HPLC (Fig. 3), using the salt NH_4OAc in the procedures to isolate the peptide (peak 1).

The receptor binding activity of insulin analogue was determined by the displacement of labelled insulin from the

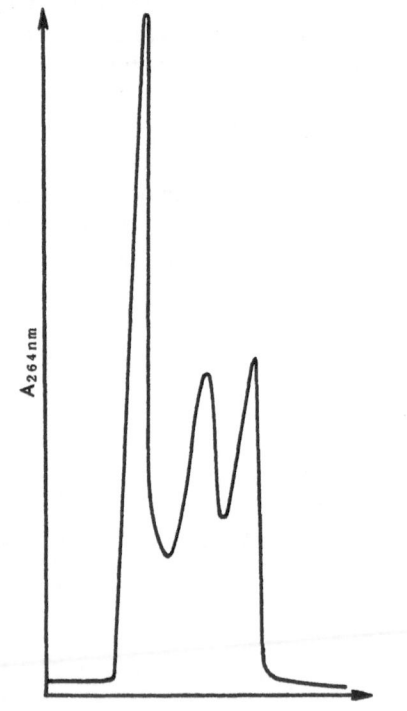

Fig. 2. *Insulin analogue separated by sephadex G-50. Solvent: 0.01 M NH_4HCO_3; Flow rate: 0.8 ml/min.*

insulin receptor on human placenta membrane (Fig. 4). Table 1 shows the biological activity of the insulin analogue tested by mouse convulsion assay. Insulin analogue exhibited *in vitro* (receptor binding) activity of 10.4%, and *in vivo* (mouse convulsion)

Table 2 *Secondary structure*

Sample	Sequence
A-chain of insulin	G I V E Q C C T S I C S L Y Q L E N Y C N α α α α α α α α α α α α α β β β β β β β β β β β β t t t t t t t t
A-chain of insulin analogue	G I V E Q C C T S I C S L Y N L P N Y C A α α α α α α α β β β β β β β β β β β β t t t t t t t t

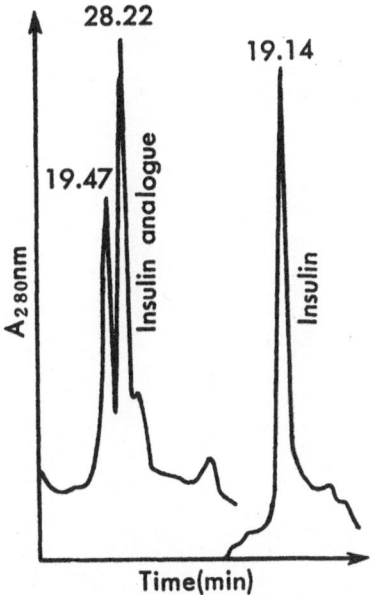

Fig. 3. *RP-HPLC analysis of insulin and insulin analogue; Column: Ultra-sphere Octyl C8 (Beckman) (4.6mm×25cm); Solvent: A=40mM NH₄OAc, 10% CH₃CN, pH 4.01; B=40mM NH₄OAc, 80% CH₃CN, pH 4.01; Flow rate: 1 ml/min.*

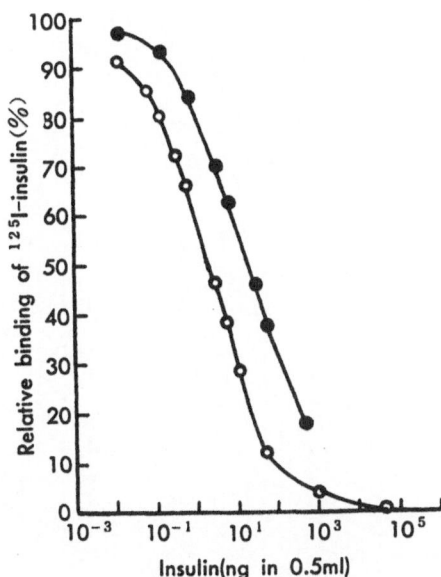

Fig. 4. *Displacement of labelled insulin from insulin receptor of human placental membrane by insulin analogue.*

activity of 10% natural porcine insulin. The secondary structure of the insulin analogue predicted by the Chou-fasman method is shown in Table 2.

References

1. Fujii, N., Otaka, A., Ikemura, S., Halana, M., Okamachi, A., Funakoshi, S., Sakurai, M., Shioiri, T. and Yajima, H., Chem. Pharm. Bull., 35(1987)3447.
2. Paynovich, R.C. and Carpenter F.H., Int. J. Peptide Protein Res., 13(1977)121.

Insulin and IGF-I analogs: novel approaches to improved insulin pharmacokinetics

L.J. Slieker, G.S. Brooke, R.E. Chance, R.D. DiMarchi, L. Fan, J.A. Hoffman, D.C. Howey, H. Long, J. Mayer, W.N. Shaw, J.E. Shields, and K.L. Sundell
Lilly Research Laboratories, Lilly Corporate Center, Indianapolis, IN 46285 U.S.A.

Introduction

Current insulin formulations do not mimic the normal glucose-induced release of insulin by the pancreas in a physiologic manner. Regular insulin(currently the most rapid acting formulation) is too slow and has too long a duration of action. Long acting human ultralente insulins are too short, show a pronounced peak in activity and are suspensoins, resulting in variability in administration.

Because of the structural homology between insulin and insulin-like growth factor-1(IGF-I), we have investigated specific IGF-like modifications in the insulin sequence to determine if these will transfer to pharmacokinetic differences in insulin absorption and clearance. Inversion of the 28-29 position of the insulin B chain to that of the homologous region of IGF-I generated $[B_{28}$-Lys, B_{29}-Pro]-insulin. This molecule self associates much less than does insulin, and has been demonstrated clinically to be more rapid acting than regular insulin. In this report we have assessed the IGF-like activity of the entire $[B_{28}$-Xaa, B_{29}-Pro]-insulin series to determine if these analogs have increased mitogenic potential.

Since IGF-I and -II circulate bound to binding proteins (IGFBPs)which do not bind insulin, we attempted to engineer IGFBP affinity into insulin to determine if this would result in decreased clearance and delayed time action. Modifications of both the insulin and IGF-I nucleus were investigated.

Results and Discussion

The insulin analogs were made by recombinant methods, by trypsin-asisted semi-synthesis from des octapeptide insulin (or des octapeptide B_{10}-Asp-insulin) and the corresponding octapeptide modified in the 28-29 position, or by chain combination. IGF-I analogs were prepared by solid phase synthsis and chain combination.

Approximately 30 insulin analogs in the $[B_{28}$-Xaa, B_{29}-Pro]-insulin series and structurally related series were prepared. Previous reports have demonstrated the monomeric nature of these analogs and the clinical candidate $[B_{28}$-Lys, B_{29}-Pro]-insulin has been demonstrated to be more rapid acting than Humulin R[1]. Because of the slight increased homology of $[B_{28}$-Lys, B_{29}-Pro]-insulin to IGF-I, we examined the binding affinity of these analogs for both the human insulin and IGF-I receptors. Substitutions in the B_{28} position, with the exception of Phe, Trp, Ile, Leu and Gly, had relatively little effect on insulin receptor affinity. However, the IGF-I receptor affinity was more

7

sensitive to charge differences, as well as steric bulk, in this region. Specifically, basic residues(Lys, Arg and Orn) increased relative IGF-I receptor affinity approximately 1.5 to 2 fold, whlie acidic residues decreased it another two fold, relative to insulin(Fig.1).

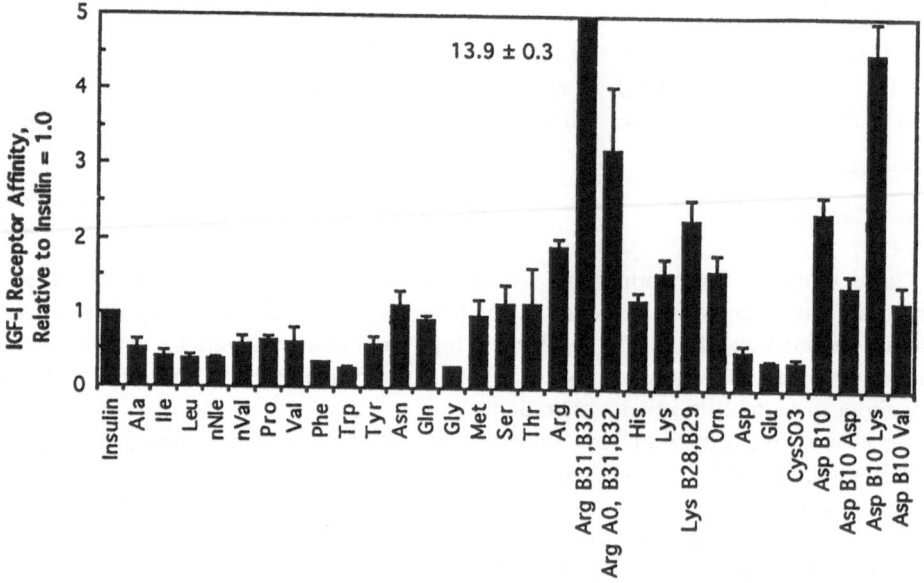

Fig. 1. *Affinity of insulin analogs for the human IGF-I receptor. Single amino acids refer to the B_{28} position of $[B_{28}$-Xaa, B_{29}-Pro]-insulin. Asp B_{10} Xaa refers to the corresponding B_{28}-Xaa, $[B_{29}$-Pro]-insulin with an additional substitution of Asp at B_{10}. Human placental membranes were used as a source of IGF-I receptor.*

This effect of charge alteration in the C terminal region of the B chain was particularly evident in analogs having an increased number of basic residues, such as $[B_{28}$-, B_{29}-Lys]- and $[B_{30}$-, B_{31}-Arg]-insulin ("Diarg insulin"). These effects were also additive to the enhancement of IGF-I receptor affinity induced through the Asp B_{10} substitution. Mitogenic assays employing normal human mammary epithelial cells have confirmed that analogs with particularly high affinity at the IGF-I receptor (such as $[B_{10}$-Asp]-insulin and $[B_{30}$-, B_{31}-Arg]-insulin) are more potent than insulin at stimulating cell growth (3 and 7 x, respectively), but that $[B_{28}$-Lys, B_{29}-Pro]-insulin is equipotent to insulin.

IGF-I and -II circulate bound to members of a family of IGF binding proteins (IGFBPs). The physiological role of these proteins is complex, and may involve targeting IGFs to particular locations, potentiating IGF action or transporting IGFs across capillary surfaces[2]. IGFBPs are characterized by generally having greater affinity for IGF-II over IGF-I, and having no measurable affinity for insulin. IGF-I analogs that do not bind IGFBPs have shorter half lives *in vivo*, suggesting that binding of IGFs to BPs might increase clearance times[3]. Residues 3,4,15,16 and 49-51 of IGF-I have been reported to be important in mediating affinity for several of the IGFBPs [4], so we substituted these residues into the homologous regions of insulin. The following insulin analogs

Table 1 *Receptor and IGFBP affinities for insuln/IGF-I hybrids*

Analog	%Ins Rec	%IGF-I Rec	%IGFBP[a]
Insulin	100	0.7	---
IGF-I	nd	100	100
A[Phe8,Arg9,Ser10]:B	32	0.3	---
A:B[Glu4,Gln16,Phe17]	5	0.1	---
A[Phe8,Arg9,Ser10]:B[Glu4,Gln16,Phe17]	2	0.3	0.01

[a] Acid stable binding protein from human serum (largely IGFBP3). Purified IGFBP1 gave similar results.

Table 2 *Receptor and IGFBP binding affinity of IGF-I analogs*

Analog	%IGF-I Rec	%Ins Rec	%IGFBP[a]
IGF-I	100	1.2	100
Insulin	0.37	100	---
A:B	7.7	17.8	151
A:BC	127	9.7	226
CA:B	2.7	1.4	154
DA:B	7.8	7.2	69

[a] Acid stable binding protein from human serum (largely IGFBP3).

were prepared by solid phase synthesis and chain combination: A[Phe8, Arg9, Ser10]:B insulin, A:B[Glu4, Gln16, Phe17] insulin and A[Phe8, Arg9, Ser10]: B[Glu4, Gln16, Phe17] insulin. Table 1 shows the affinity of these analogs for the insulin and IGF-I receptors, and IGFBP1 & 3.

A[Phe8, Arg9, Ser10]:B[Glu4, Gln16, Phe17] insulin demonstrated low, but measurable affinity for both IGFBPs (EC$_{50}$ approx 1-2 μM), at the expense of a dramatic reduction in insulin receptor affinity and, interestingly, a 2 fold reduction in IGF-I receptor affinity. Neither of the singly modified A or B chain analogs had measurable IGFBP affinity.

Because modification of the insulin nucleus generated analogs with low affinity for both the insulin receptor and the IGFBPs, another approach was taken: start with the IGF-I nucleus and increase the affinity for the insulin receptor. The following two-chain IGF-I analogs were prepared by solid phase synthesis and chain combination: A:B, A:BC, CA:B and DA:B, where A, B, C and D refer to domains of IGF-I. Table 2 shows the affinity of these analogs for the insulin and IGF-I receptors, and IGFBP3.

A:B IGF-I has high affinity for the IGFBP, moderate affinity for the insulin receptor and significantly reduced IGF-I receptor affinity. This latter point confirms the importance of the C domain of IGF-I for receptor selectivity [4]. The moderate affinity of A:B IGF-I for the insulin receptor combined with high affinity for IGFBPs suggests that this analog might be useful as a foundation for the development of long acting basal insulin agonists.

References

1. DiMarchi, R.D., Mayer, J.P., Fan, L., Brems, D.N., Frank, B.H., Green, L.K., Hoffman, J.A., Howey, D.C., Long, H.B., Shaw, W.N., Shields, J.E., Slieker, L.J., Su, K.S.E., Sundll, K.L. and Chance, R.E., In Smith, J.A. and Rivier, J.E. (Eds.) Peptides: Chemistry and Biology (Proceedings of the 12th American Peptide Symposium), ESCOM, Leiden, 1992, pp.26–28.
2. Bar, R.S., Clemmons, D.R., Boes, M., Busby, W.H., Booth, B.A., Dake, B.L. and Sandra, A., Endocrinology, 127(1990)1078.
3. Cascieri, M.A., Saperstein, R., Hayes, N., Green, B.G., Chicchi, G.G., Applebaum, J. and Bayne, M.L., Endocrinology, 123(1988)373.
4. Clemmons, D.R., Dehoff, D.L., Busby, W.H., Bayne, M.L. and Cascieri, M.A., Endocrinology, 131(1992)890.
5. Bayne, M.L., Applebaum, J., Underwood, D., Chicchi, G.G., Green, B.G., Hayes, N.S. and Cascieri, M.A., J. Biol. Chem., 264(1988)11004.

Preparation and properties of [B_2,Lys]-insulin and [B_3,Lys]-insulin

Ming-hua Xu, Ying Ye, Xin-tang Zhang, Ying-gao Xu and Shang-quan Zhu
Shanghai Institute of Biochemistry Chinese Academy of Sciences,
Shanghai 200031, China

Introduction

As we know, the C-terminal region of insulin B-chain is very important in its receptor binding activity and the putative receptor binding region includes this part[1]. The N-terminal of insulin B-chain has been thought to be crucial in its immunogenicity, but not so important in the receptor binding process. But some evidences showed that the gradual removal of residues from the N-terminal of insulin B-chain resulted in a simultaneous decrease in its receptor binding capacity[2]. This phenomenon suggests that the N-terminal of insulin B-chain does play a certain role in the process of binding with its receptor. In order to obtain fruther information about whether the mutation in the N-terminal of B-chain would influence its receptor binding capacity, we prepared two insulin analogues, with B_2-Val or B_3-Asn replaced by Lys respectively through chemical semisynthesis. Preliminary study of their *in vivo* as well as *in vitro* activities have been carried out and the results are rather interesting.

Results and Discussion

Methylsulfonylethoxycarbonyl(MSC) was used as an acid-stable amino protecting group. Its hydroxysuccinimide ester(MSC.ONSu) reacted selectively with the A1, B29 amino groups of insulin. [MSC]$_2^{(A1, B29)}$insulin was degraded two and three steps successively by Edman method. After coupling with active esters Boc-Phe-Lys(BOC)-ONSu and BOC-Phe-Val-Lys(Boc)-ONSu, the deprotected products were purified by Sephades G-25, followed by further purification with SP-Sephadex C-25 and DEAE-Sepharose CL-6B. The final products were identified by amino acid analysis, cellulose acetate electrophoresis and polyacrylamide gel electrophoresis(Figure 1).

The in vivo activity
According to the methods described in pharmacopoeia, the biological activities were measured by mouse convulsion tests using native insulin as control. The *in vivo* activity for [B_2,Lys]-insulin and [B_3,Lys]-insulin were 85% and 100% of that of native porcine insulin, that is 22 IU/mg and 26 IU/mg respectively.

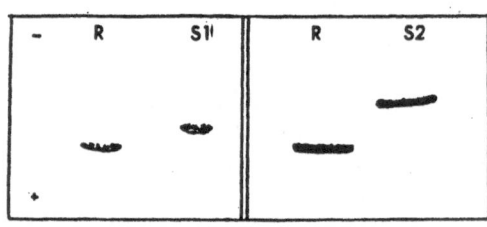

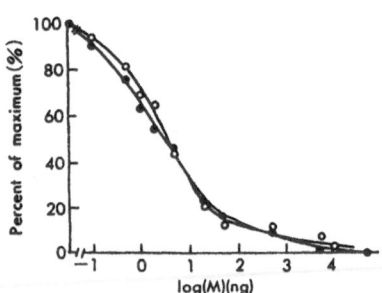

Fig. 1. Polyacrylamide gel electrophoresis of insulin analogues. R, MC insulin; S1, [B₃,Lys]-insulin; S2, [B₂,Lys]-Insulin.

Fig. 2. Displacement curve of [B₂,Lys]-insulin in the binding of insulin to human placentl membrane. ●--● insulin; O--O [B₂,Lys]-insulin.

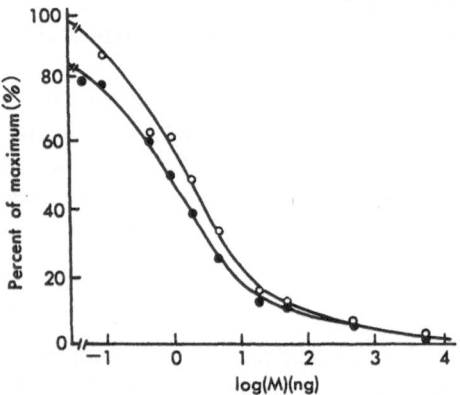

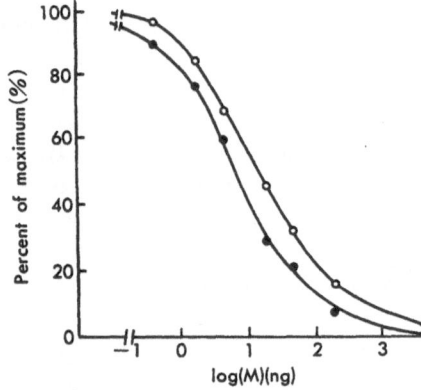

Fig. 3. Displacement curve of [B₃,Lys]-insulin in the bnding of ¹²⁵I-insuling to human placental membrane. O--O insulin; ●--● [B3,Lys]-insulin.

Fig. 4. Displacement curve of [B₃,Lys]-insulin in the binding of ¹²⁵I-insulin to rat adipocytes. O--O insulin; ●--● [B₃,Lys]-insulin.

Insulin receptor binding capacity

The receptor binding assay for [B$_2$,Lys]-Insulin and [B$_3$,Lys]-insulin in human placental membrane(HPM) were done at 4°C. Figures 2 and 3 represent the displacement curves for these two analogues. According to the IC$_{50}$ values of samples and native insulin, the receptor binding capacity for [B$_2$,Lys]-insulin and [B$_3$,Lys]-insulin in HPM were determined to be 80% and 200% of that of native porcine insulin respectively.

The receptor binding assay for [B$_3$,Lys]-insulin was also carried out in rat adipocyted at 37°C. Figure 4 shows the displacement curve. Calculated from the IC$_{50}$ values, the receptor binding capacity for [B$_3$,Lys]-insilin in rat adipocytes was 220% of that of

Table 1   *Dissociation constants of [B₃,Lys]-insulin and insulin in binding with HPM insulin receptor*

	$K_{d1}(M)$	$K_{d2}(M)$
(B₃,Lys)-Insuin	5.8×10^{-10}	4.4×10^{-7}
Iusulin	7.9×10^{-10}	6.8×10^{-7}

Note: K_{d1}--high affinity sites; K_{d2}--low affinity sites.

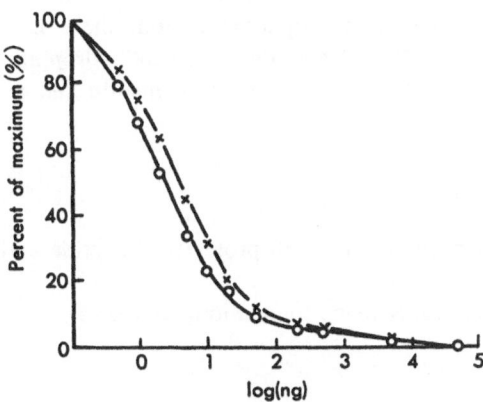

Fig. 5. *Displacement curve of [B₃,Lys]-insulin in the binding of ¹²⁵I-[B₃,Lys]-insulin in human placental membrane.* ×--× insulin; O--O [B₃,Lys]-insulin.

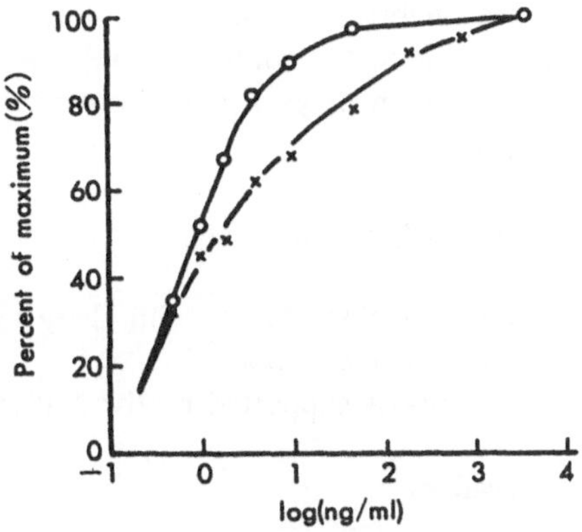

Fig. 6. *Does dependent curve of lipogeneses stimulated by [B₃,Lys]-insulin and insulin.* O--O [B₃,Lys]-insulin; ×--× insulin.

native porcine insulin.

In order to estimate the affinity of [B₃,Lys]-insulin with insulin reciptor in HPM, [¹²⁵I-B₃,Lys]-insulin was uase as tracer in the receptor binding assay in HPM. Figure 5 is the displacemet curve. By method of Scatchard analysis, the dissociation constants for [B₃,Lys]-insulin and native procinee insulin were calculated and listed in Table 1. It clearly shows that the affinity between [B₃,Lys]-insulin and insulin receptor is stronger than that between native porcine insulin and insulin receptor.

In vitro activity

We measured the *in vitro* activity by method of lipogenesis-stimulating ¹⁴C-Glucose incorporating into lipids in adipocytes. Figure 6 is the dose dependent curve. From the EC_{50} values we estimated the activity of [B₃,Lys]-insulin was 1.7 times as high as that of native porcine insulin.

It has generally been agreed that the receptor binding region of insulin is a large hydrophobic part composed of Gly(A1), Glu(A5), Tyr(A19), Asn(A21) and Val(B12), Phe(B24), Phe(B25), Tyr(B26). N-terminal of B-chain is an extended part far from this region, but the analogues with this part shortened have lower receptor biding capacity[2]. When B2-Val was replaced by Lys, the receptor binding capacity was also decreased, that is 80% of that of native porcine insulin. However, when B3-Asn was replaced by Lys, the receptor binding capacity increased remarkably, that is about 2 times as high as native porcine insulin. These results may suggest that the N-terminal of B-chain plays an important part not only in stabilizing insulin conformation but also in the interaction with its receptor either directly or indirectly. As we know, chicken insulin and fish insulins are more active than porcine insulin[3,4]. The A8-A10 are His Asn Thr (chicken), His Arg Pro (cod), His Lys Pro (tuna) and His Lys Arg (silver carp), which are more hydrophilic than that of porcine insulin(Thr Ser Val). A8-A10 and B3-B5 are close to each other in insulin molecule and may form a hydrophilic surface. So it may be reasonable to believe that besides the putative hydrophobic region, there is a hydrophilic region composed of A8, A9, A10, B3, B4, B5 and other adjacent hydrophilic groups which plays a rather important part in the interation between insulin and its receptor.

Acknowledgements

We are indebted to Ms. Cui Heng-ran for providing us with protected dipeptide and tripeptide active esters.

This work is supported by the National Natural Science foundation(39070244).

References

1. Gammeltolft, S., Physiol. Rev., 64(1984)1321.
2. Lei, K.J., Dong, B., Xie, D.L. and Yuan, H.S., Acta Biochimi. Biophysi. Sin., 15(1983)457.
3. Simon, J., Freychet, P., Rosselin, G. and Demeyts, P., Endocrin., 100(1977)115.
4. Wang, S. and Liang, D., Chinese Biochem. J., 1(1986)9.

Biological activities of thymopentin, tuftsin and neurotensin(8-13) are profoundly influenced by metabolites resulting from enzymatic degradation of their retro-inverso analogues

Antonio S. Verdini

Italfarmaco S.p.A., Via dei Lavoratori, 54 20092 Cinisello Balsamo, Milano, Italy

Introduction

Resistance to enzymatic hydrolysis is critically important for the development of peptide-based drugs. The short half-life of small unmodified peptides can be dramatically prolonged *in vitro* and *in vivo* by insertion into peptide chains of the "reversed amide" peptide bond surrogate (retro-inverso peptide analogues)[1,2]. End-modified retro-inverso peptides resist progressive digestion by exopeptidases while peptides modified at internal sites are stable to specialized endopeptidases that cleave the chain to turn off their activity. In spite of unavoidable backbone modifications and structural perturbations, partially modified retro-inverso analogues retain their biological activity [2,3]. In particular, an isolated reversed bond, besides stabilizating the corresponding CONH bond, could also prevent adjacent as well as remote peptide bond metabolization. In partially modified retro-inverso peptides, hydrolysis at the remaining peptidase-labile sites may lead to stable active fragments which, being present in significant concentrations at the site of action, may control biotransformation and perhaps the activity of the parent pseudopeptide. Oddly enough, information on the metabolization in biological fluids of partially modified retro-inverso peptides is still fragmentary, as are attempts to bring to light the biological activities of small pseudopeptide metabolites. We have addressed these issues by studying some retro-inverso analogues of thymopentin(TP5), tuftsin (TUFT) and neurotensin(8-13) [NT-(8-13)] *in vitro* and *in vivo*.

Results and Discussion

The TUFT analogue H-g Thr-(R,S)m Lys-Pro-Arg-OH was completely resistant to human plasma peptidases. When administered to mice, either i.v. or p.o., it was significantly more active than TUFT as an immunostimulant and about 10 fold more potent in reducing rat adjuvant arthritis [2]. The activity increase has been attributed to the preservation of TUFT chain integrity which, in turn, prevents generation of the primary metabolites H-Lys-Pro-Arg-OH and H-Thr-Lys-Pro-OH are thus eliminates their antagonism by negative feedback effects. These results demonstrated once again how effectively a single backbone modification at the N-terminus of a tetrapeptide chain might influence the enzymatic susceptibility of a "remote" peptide bond. The TP5 analogue H-g Arg-(R,S)m Lys-Asp-Val-Tyr-OH, with a reversed amide group placed at the major enzymatic cleavage site, showed: a) increased half-life in human plasma ($t_{1/2}$: min vs 1.5 min) and b) immunostimulatory activity in mice much higher than that of TP5 and comparable to that of thymosin $\alpha 1$, a natural 28-aminoacid polypeptide(Fig. 1 and 2).

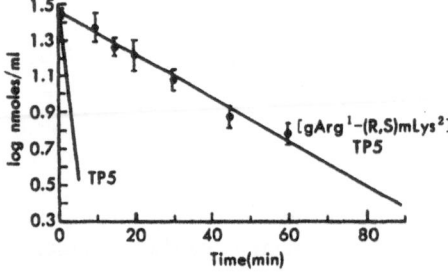
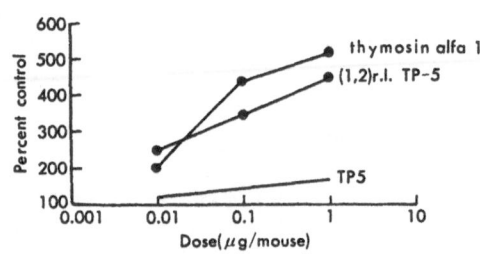

Fig. 1. Thymopentin and [gArg¹-(R,S)ₘLys²]TP5: degradation in human plasma.

Fig. 2. In vivo immunostimulatory activity of TP5, (1,2)r,i TP5 and thymosin alfa 1.

This single backbone modification profoundly altered the peptide degradative pattern by human plasma enzymes affording H-Val-Tyr-OH and H-g Arg-(R,S)m Lys-Asp-OH as the only primary metabolites(Fig. 3).

The pseudotripeptide fragment was still active, although less potent than the parent peptide, and totally resistant to further enzymatic breakdown. This stable fragment, which is released during enzymatic cleavage, should be present in significant amounts at the site of action of its parent peptide and hence may contribute to some extent to the rise in TP5 retro-inverso analogue potency. The N-terminal Arg-Lys and adjacent Lys-Asp bonds were refractory to cleavage in the pentapeptide chain but Asp-Val turned out to be an additional site of enzymatic cleavage. The previously undetected lability of Asp-Val to endopeptidase(s) suggested a demolition pattern for TP5 more complex than that previously proposed by G. Goldstein and his colleagues [4]. In fact, the analogue with a reversed Asp-Val bond, normally cleaved at Arg-Lys by aminopeptidase(s), lasted longer than TP5 in human plasma ($t_{1/2}$:5 min vs 1.5 min, Fig. 4).

It was also equipotent to TP5, confirming the Arg-Lys bond as the TP5 major site of enzymatic cleavage and further suggesting that immunostimulatory potency could be increased through the stabilization of both Arg-Lys and Asp-Val bonds. The analogue bearing these two isolated retro-inverso peptide bonds has been found to be almost completely stable to human plasma enzymes and two hundred fold more potent than TP5. Backbone modifications properly placed at some sites of enzymatic cleavage may allow modulation of peptide susceptibility to enzymes and reveal short peptide fragments with inherent biological activity. By introducing a reversed amide bond at the C-terminus of NT(8-13) we have also attempted to correlate the effects of the perturbation of the pseudopeptide processing *in vivo* with blood pressure changes induced by i.v. injection of the NT (8-13) analogue H-Arg-Arg-Pro-Tyr-gIle-(R,S)m Leu-OH into pentobarbital

16

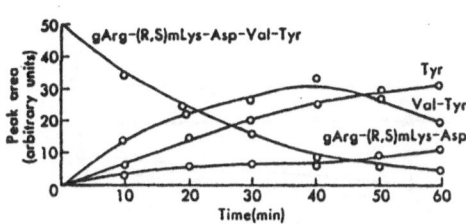

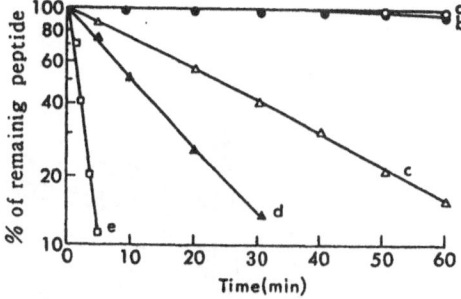

Fig. 3. Time-curve of breakdown of the thymopentin analogue retro-inverted at Arg¹-Lys² in human plasma.

Fig. 4. Kinetics of degradation in human plasma of retro-inverso thymopentin analogues. a, pseudo tripeptide; b, peptide retro-inverted at Arg¹Lys² and Asp³Val⁴; c, peptide retro-inverted at Arg¹Lys²; d, peptide retro-inverted at Asp³Val⁴; e, TP5.

anaesthetized rats [5]. Several short peptide metabolites are produced through a concerted chain cleavage of NT(8-13) by a number of exo- and endopeptidases. According to a previous suggestion such small peptides may play a role in inducing the characteristic triphasic blood pressure change (1st and 3rd hypotensive phases; 2nd hypertensive phase, about 60 sec after injection). Degradation studies in human plasma indicated that incorporation of a reversed Ile-Leu bond into NT(8-13) did, as expected, protect from carboxypeptidase(s) while the cleavage of adjacent Tyr-Ile by plasma dipeptidylcarboxypeptidases was almost abolished. While the degradation rate was identical for both the natural and retro-inverso peptides, indicating that breakdown of the remote unmodified scissile bonds is barely influenced by modification at Ile-Leu, the degradation pattern of the retro-inverso analogue was greatly simplified. Only two primary pseudopeptide metabolites H-Arg-Pro-Tyr-gIle-(R,S)m Leu-OH and H-Tyr-gIle-(R,S)m Leu-OH, and Arg-Pro were observed upon protracted incubation of the pseudohexapeptide in human plasma. (12,13) Retro-inverso NT(8-13), administered i.v. to anaesthetized rats as a mixture of the two diastereomers, induced a single-dose depressor response. The second phase pressor response followed by a subsequent depressor response was never observed (Fig. 5).

These results suggest that the reversed peptide bond may prevent generation of short metabolites with pressor activity. The degradation pattern in human plasma and blood pressure results, taken together, seem to indicate H-Arg-Pro-Tyr-OH as the secondary metabolite which may determine the hypertensive phase of blood pressure change.

In conclusion, the proper incorporation of "reversed amide" surrogates of enzyme-labile CONH bonds into bioactive peptides not only increases resistance against enzymatic breakdown but may also lead to detection of biologically active short fragments generated during biotransformation of the parent peptide. The peptide metabolites, either natural or retro-inverted, may in turn influence the activity of the parent peptides at their site of action.

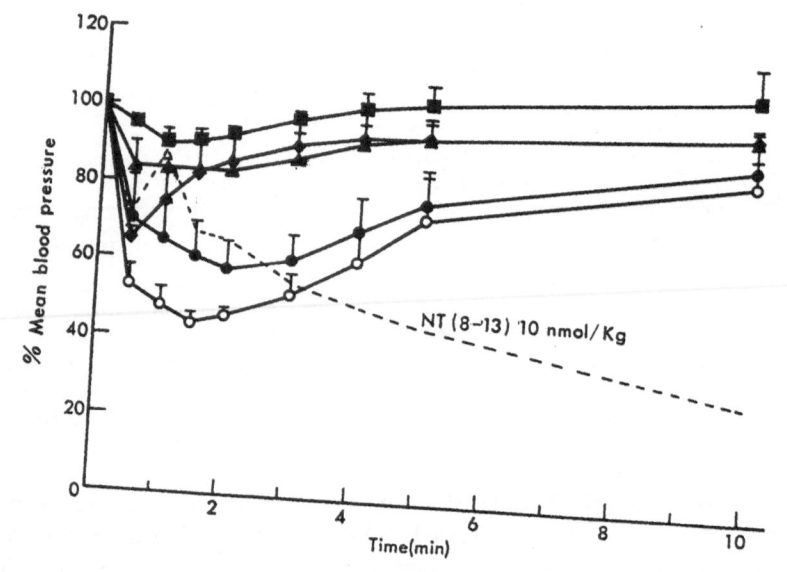

Fig. 5. Hypotensive effect in the anaesthetized rat. ■, [12-13]r.i.NT(8-13), 10 nmol/kg; ▲, [12-13]r.i.NT(8-13), 30 nmol/kg; ♦, [12-13]r.i.NT(8-13), 100 nmol/kg; ●, [12-13]r.i.NT(8-13), 183 nmol/kg; O, [12-13]r.i.NT(8-13), 300 nmol/kg.

Acknowledgements

I wish to thank for their contributions: G.P. Viscomi, A. Sisto, M. Pinori, S. Mariotti, L. Gazerro, A. Pessi, F. Bonelli, M.G. Longobardi, F. Cardinali, F. Centini, S. Cappelletti, G. Di Gregorio and A. Biffi (synthesis); G. Esposito, A. Mancini, P. Filtri and M. Barbini (analysis); A. Groggia and G. Marcozzi (degradation studies); P. Kitabgi, L. Nencioni, R. Cereda, L. Villa, L. Parente, S. Peppoloni, M. Perretti, P. Pileri, C. Becherucci (biological studies).

References

1) Goodman, M. and Chorev, M., Acc. Chem. Res., 12(1979)1.
2) Verdini, A.S., Silvestri, S., Becherucci, C. et al., J. Med. Chem., 34(1991)3372.
3) Pallai, P., Richman, S., Struthers, R.S. and Goodman, M., Int. J. Peptide Protein Res., 21(1983)84.
4) Tischio, T.P., Patrick, J.E., Weintraub, H.S., et al., Int. J. Peptide Protein Res., 14(1983)179.
5) Di Gregorio, G., Pinori M. and Verdini A.S., in Innovation and Perspectives in Solid Phase Synthesis, Epton, R. Ed., Intercept Ltd, Andover, 1992, pp.311.

Production of inhibin by human luteal cells

Yue-ting Gong[a], Han-zheng Wang[b], Zhi-da Sun[b], Wei-xiong Sheng[b], Pei-feng Ren[a] and Min-yi Wang[b]

[a]*Shanghai Institute of Biochemistry, Chinese Academy of Sciences, Shanghai 200031, China*

[b]*Shanghai Institute of Planned Parenthood Research, Shanghai 200032, China*

Introduction

Inhibin is a glycoprotein synthesized by the gonad to inhibit preferentially the secretion of follicle-stimulation hormone (FSH) from the pituitary[1]. It is a heterodimeric molecule composed of α-and β-subunits(β_A or β_B) which are linked up through disulfide linkages and expressed by separate genes.

The production of inhibin and progesterone by the human corpus luteum (CL) was under the control of luteinizing hormone (LH)[2] and the regression of CL resulted in a decreased inhibin production. Dynamic changes in circulating inhibin levels were correlated with the peripheral concentrations of progesterone and estradiol during the human menstrual cycle[3]. Ovarian inhibin has been considered to be secreted by granulosa cells while testicular inhibin is originated from Sertoli cells. By Northern blot analysis, messenger RNAs for the α-and β_A were detectable only in ovine and bovine follicles but not in corpora lutea[4]. Recently, immunostaining of the inhibin-α subunit was also observed within the granulosa-lutein cells of the human CL[5]. However, the direct effects of gonadotropins and steroids on inhibin production by primate or human luteal cells have not been reported. The present study was designed: 1) to examine the *in vitro* production of immunoreative inhibin and mRNA of inhibin-α by dispersed human luteal cells obtained in the early to mid (days 16-19) and late stage (day 23) of the menstrual cycle; 2) to determine the effects of hCG, FSH, estradiol and testosterone on inhibin production *in vitro*; and 3) to examine the immunocytochemical localization of inhibin in human luteal cells. Estradiol and progesterone rates were also measured to confirm the activity of the cell preparations and their responsiveness to hCG.

The N-terminal peptide segment of inhibin-α (1-26)-Tyr-Gly-OH, i.e., H-Ser-Thr-Pro-Leu-Met-Ser-Trp-Pro-Trp-Ser-Pro-Ser-Ala-Leu-Arg-Leu-Leu-Glu-Arg-Pro-Pro-Glu-Glu-Pro-Ala-Ala-Tyr-Gly-OH, was synthesized in HPLC pure form by stepwise elongation solid-phase procedure, then conjugated to thyroglobulin by coupling with bisdiazotized benzidine[6]. The antiserum obtained from rabbits immunized by this peptide complex having a titer of 1:32,000 and Ka=0.23 $\times$ 10^{10} M^{-1}, could exhibit an obvious immune cross reaction with human pregnant serum, human and bovine follicular fluids, human placental tissue homogenates and porcine semen, but did not cross react with hCG, FSH, PRL, oxytocin, vasproessin on the basis of displacement method in RIA for inhibin.

Isolation and culture of human luteal cells were conducted as described in our previous work[7]. Radioimmunoassays of progesterone and estradiol were made by using validated RIA kits provided by the WHO Matched Reagent Programme. Immuno-reactive inhibin was measured by Dr. Findlay's laboratory. Immunocytochemical localization of inhibin was examined with our antiserum in human luteal cells by means of the Avidin-Biotin-Complex method, the immunostaining of inhibin in the Sertoli cells of adult rat seminiferous tubule was used for comparison[8]. Northern blot analysis was conducted on the basis of hybridization of mRNA extracted from human corpus luteum with inhibin-α cDNA as the probe.

Results and Discussion

The N-terminal region of inhibin-α (1-26) was considered to be the main antigenic site of the glycoprotein. The conjugation of this synthetic peptide with thyroglobulin could be used as an immunogen for raising antibody which was proved to be specific for immunoreaction towards human and bovine follicular fluids, human placental tissue homogenates and other tissue fluids containing inhibin, but non-reactive with other reproductive hormones such as hCG, FSH, PRL, oxytocin, vasotocin and vassopressin (Fig.1).

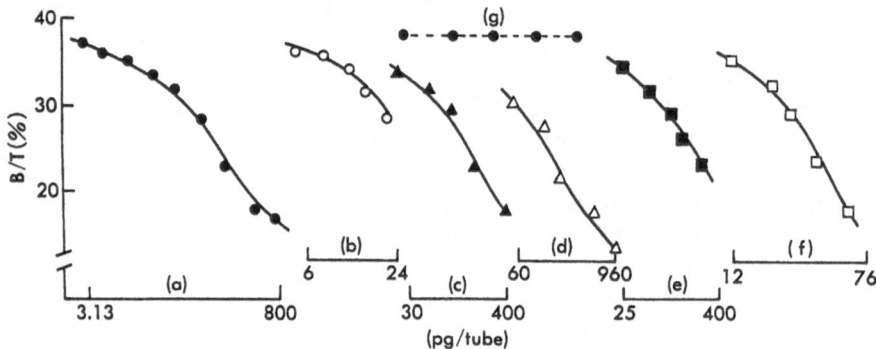

Fig. 1. Antibody against hla(1-26)TyrGlyOH. (a) hla(1-26)TyrGlyOH standard; (b)pregnant human serum; (c) porcine semen; (d) homogenates of human term placenta; (e) bovine follicular fluid; (f) human follicular fluid; (g) hCG, FSH, PRL, oxytocin, vasotocin, vasopressin.

The antiserum was not sufficiently sensitive for determination of peripheral inhibin in circulation. Furthermore, an intense immunostaining with this antiserum could be specifically identified in the localization of inhibin within the granulosa-lutein cells mainly but not found in the thecal and other cells of human corpus luteum, thus supporting the view that the large cells are believed to be derived from granulosa cells and could also be a major source of inhibin. There were no significant effects of either FSH, estradiol or testosterone on inhibin production by dispersed human luteal cells under either basal or hCG-stimulated condition. Neither FSH nor testosterone had any significant effects on progesterone production. However, estradiol induced a significant dose-related decrease in both basal and hCG-stimulated progesterone production (Fig.2). It was found that only the early/mid CL cells responded to hCG with an increase of inhibin production (Fig.3), but not in cells from late stage CL. Levels of mRNA of inhibin-α in different stages detected by the Northern blot analysis were consistent with the above results (to be published). It was concluded on the basis of these results that inhibin production by dispersed human luteal cells *in vitro* is influenced by the age of the corpus luteum and LH(hCG) but not FSH of sex steroids dependent.

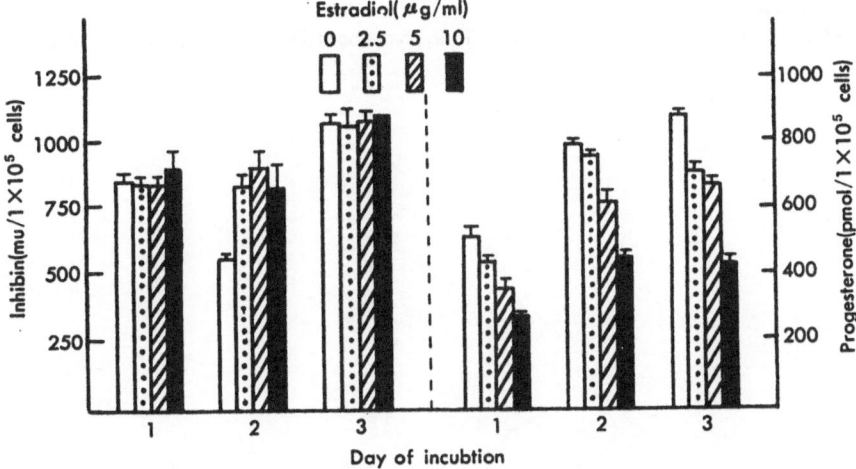

Fig. 2. *Effect of estradiol on the inhibin and progesterone production by human luteal cells.*

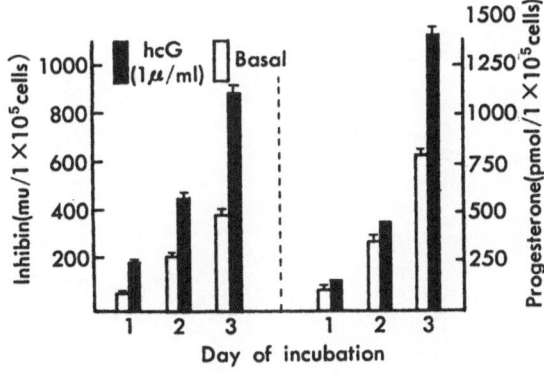

Fig. 3. *Effect of treatment with hCG on inhibin and progesterone production by human luteal cells (early stage, triplicate incubation).*

References

1. Ying, S.Y., Endocrine Review, 9(1988)267.
2. McLachlan, R.I., Cohen, N.L., Vale, W., Rivier, J., Burger, H.G., Bremner, W.J. and Soules, M.R., J. Clin. Endocrinol. Metab., 68(1989)1078.

3. Roseff, S.J., Bangah, M.L., Kettle, L.M., Vale, W., Rivier, J., Burger, H.G. and Yen, S.S.C., J. Clin. Endocrinol. Metab., 69(1989)1033.
4. Rodgers, R.J., Stuchbery, S.J. and Findlay, J.K., Mol. Cell. Endocrinol., 62(1989)95.
5. Smith, K.B., Millar, M.R., McNeilly, A.S., Illingworth, P.J., Fraser, H.M. and Baird, D.T., J. Endocrinol., 129(1991)155.
6. Shen, W.X., Wei, W., Cui, D.F., Sun, Z.D., Zhou, W., Wang, H.Z. and Gong Y.T., Reproduction and Contraception, 13(1993)in press.
7. Wang, H.Z., Lu,S.H., Han, X.J., Sun, Z.D, Shen, W.X., Zhou, W., Bangan, M.L. and Findlay, J.K., Reprod. Fert. Dev., 4(1992)67.
8. Wang, M.Y., He, Z. Y., Wang, H.Z. and Gong Y,T., Reprodution and Contraception, 13(1993) in press.

Synthesis of B22 Asp human insulin and its biological activity

Yue-hua Tang[a], Da-fu Cui[a], You-shang Zhang[a] and Richard J. Simpson[b]

[a]*Shanghai Institute of Biochemistry, Academia Sinica, Shanghai 200031, China*
[b]*Joint Protein Structure Laboratory, Ludwig Institute for Cancer Research and Walter & Eliza Hall Institute for Medical Research Victoria, Australia 3050*

Introduction

In previous studies, it was found that B22 Arg of B23 D-Ala deshexapeptide insulin could be replaced by Asp and the product was still active[1]. Here, B22 Asp human insulin B-chain was made by solid phase synthesis and combined with natural A-chain to obtain B22 Asp human insulin and its biological activity was studied.

Results and Discussion

Preparation of natural S-sulfonated A-chain [2]

1.5g of crystalline porcine insulin were dissolved in 70 ml of pH 7.5, 0.1 M Tris buffer containing 7 M urea. 2.8 g of sodium sulifite were added and dissolved at 37°C, 1.4 g of sodium tetrathionate were added. After standing at room temperature overnight, the mixture was dialysed against water for 1.5 h, adjusted to pH 6.4 with 1 M acetic acid and centrifuged at 10,000 rpm for 5 min. The supernatant was desalted with Sephadex G-15 column (3.5×44cm) which was eluted with 1 M acetic acid. The eluent was lyophilized to obtain 400 mg of dry powder which was HPLC pure and its amino acid composition was consistent with that of A-chain.

Solid phase synthesis of B22 Asp SH B-chain

Applied Biosystems 430A peptide synthesis and 0.5 mole of Boc-threonylmethyloxy-PAM resin were used. The N-terminal amino group was protected by Boc group. The protecting groups of side chains were as follows: chlorocarbobenzoxy for ε-amino group, O-benzyl for β-carboxyl group, O-cyclohexyl for α-carboxyl group, benzyl for hydroxyl group, benzoxymethyl for imidazole group, 4-methylbenzyl for SH group and bromocarbobenzoxy for phenolic group. The carboxyl group of protected amino acid was activated by hydroxybenzotriazole and the peptide was elongated from the C-terminal to N-terminal. The peptide was cleaved from the resin by treating with liquid HF (10% paracresol as scavenger) at −5°C for 75 min. The residue was washed with ether to remove the scavenger and extracted with a solvent mixture (15% acetic acid/acetonitrile/0.1% trifluoroacetic acid=1:1:1) and lyophilized to obtain 1.33g of dry powder (yield 87%). 800 mg of the dry powder were dissolved by adding 1 ml of acetic acid, 1 ml of 0.1% trifluoroacetic acid (TFA) and 13 ml of 0.1 M HCl. 10 volumes of acetone were added to obtain 530 mg of acetone powder (yield 66%), its amino acid composition was consistent with the theoretical values.

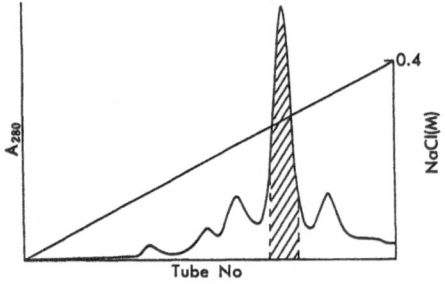

Fig. 1. *Purification of B22 Asp S-sulfonated B-chain on DEAE-Sepharose Cl-6B column (2.2×20 cm) eluent–pH 8.0, 0.05 M Tris buffer containing 40% isopropanol with salt gradient (0-0.4 M NaCl); main peak–B22 Asp S-sulfonated B-chain.*

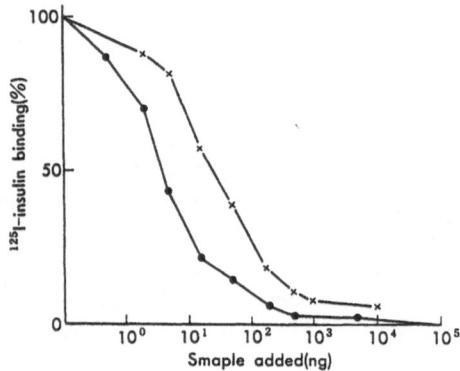

Fig. 3. *Binding of B22 Asp human insulin with insulin receptor on human placental membrane left–porcine insulin, right–B22 Asp human insulin.*

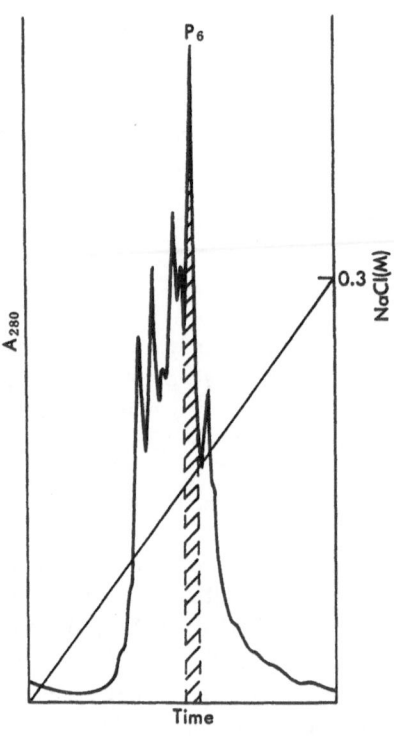

Fig. 2. *Purification of B22 Asp human insulin on FPLC Mono Q HR 5/5 column (0.5x5 cm), eluent A: pH 7.8, 0.05 M Tris buffer containing 40% isopropanol; eluent B: eluent A containing 0.5 M NaCl; gradient: 0-60% B in 45 min; flow rate: 0.5 ml/min.*

Preparation of S-sulfonated B22 Asp B-chain

137 mg of B22 Asp SH B-chain were dissolved in 5 ml of pH 7.5, 0.1M Tris buffer containing 7 M urea. 0.2 g of sodium sulfite and 0.1g of sodium tetrathionate were added. After standing overnight at room temperature, the mixture was filtered through Sephadex G-15 column (3.5×44cm) which was eluted with 0.01 M ammonium bicarbonate to obtain 108 mg of dry powder. The product was further purified with DEAE-Sepharose C1-6B column chromatography (Fig.1) to obtain 30 mg of B22 Asp S-sulfonated B-chain.

Combination of synthetic B22 Asp B-chain and natural A-chain

20 mg of B22 Asp S-sulfonated B-chain and 40 mg of natural S-sulfonated A-chain were dissolved in 7 ml of pH 10.5, 0.1 M glycine buffer. The mixture was cooled in an

ice bath and 6.55 mg of DTT dissolved in the above buffer were added. After reacting at 4°C for 24 h, the mixture was filtered through G-50 fine column. The peak with insulin activity (mouse convulsion assay) was collected, lyophilized and further purified by FPLC on Mono Q column (Fig.2). Peak 6 with insulin activity was lyophilized to obtain 0.89 mg of B22 Asp human insulin.

Characterization of B22 Asp human insulin

The final product was proved to be B22 Asp human insulin by determining the amino acid sequences of A- and B-chains simultaneously with ABI 477A sequencer.

Biological activity of B22 Asp human insulin

The biological activity of B22 Asp human insulin determined by mouse convulsion assay was 50% (13 IU/mg) compared with porcine insulin.

Receptor-binding capacity of B22 Asp human insulin

The binding of B22 Asp human insulin with insulin receptor on human placental membrane was 15% compared with porcine insulin (Fig.3).

Previous work demonstrated that B22 Arg of B23 D-Ala deshexapeptide insulin could be replaced by Asp with no influence on its biological activity. However, the product obtained by early semisynthetic method was not homogeneous because of side reactions. Now, we have prepared B22 Asp human insulin which is homogeneous in HPLC by combining natural A-chain and synthetic B22 Asp B-chain. B22 Asp human insulin is still biologically active but its activity is reduced to 50%. In the three-dimensional structure of native insulin, the molecular conformation is stabilized by a salt bridge between the (quanidino group) of B22 Arg and the carboxyl group of A21 Asn. When B22 Arg is replaced by Asp, this salt bridge is no longer present but the biological activity is only half reduced instead of totally lost. Probably, the side chain of Asp introduced to B22 of insulin may form another kind of non-covalent linkage to stabilizing the molecular conformation.

Acknowledgements

We thank Xin-tang Zhang for doing the receptor-binding assay.

References

1. Zhu, S.Q., Li, T.F., Cao, Q.P. & Zhang, Y.S., Sci. Sin., 24(1981)264.
2. Du, Y.C., Zhang, Y.S., Lu, Z.X. & Tsou, C.L., Sci. Sin., 10(1961)84.
3. Chance, R.E., Hoffmann, J.A., Kroeff, E.P., Johnson, M.G., Schirmer, E.W., Bromer, W.W., Ross, M.J. & Wetzed, R., In: Peptides, Proc. Am. Peptide Symp. 7th, Rich, D.H. and Gross, E.(eds), Pierce Chem. Co., Rockford, Illinois, 1981, p.721.

High level expression of human insulin genes in *Escherichia coli*

Jian-guo Tang, Ying-zi Xue, Xian-bin Fan, Yi-xin Fu, Ge Wei, Yan Feng, Huai-yu Sun, Bin Xie, Xiong-biao Li, Hong-tao Zhang and Mei-hao Hu
Department of Biology, Peking University, Beijing 100871, China

Introduction

A major problem in the production of low molecular weight proteins or peptides by recombinant DNA technique in E. coli is their short half-life in bacterial hosts[1,2,3]. The half-life for rat proinsulin in E.coli has been reported to be just two minutes. In order to increase the stability of the target product, different methods have been employed. (a) In frame fusion of a large gene with the coding sequence of the desired peptides results in a fused protein. Usually the target peptide just constitutes a small part of the fusion protein, resulting in reduced yield and increased difficulties in purification. (b) A multimeric form of proinsulin gene has been reported to express in high level of the multi-proinsulin peptide. The product could then be cut into proinsulin-like form, but need more protein-processing work. (c)Fusion of a bacterial secretary signal sequence to the target peptide leads to the transport of the fusion protein into the periplasmic region of the bacteria to prevent its degradation. The yield is also limited. We have constructed a series of high level expressin vectors containing Tac or $P_R P_L$ promoters and fused insulin precursor genes with limited length of the fusion part, limited CNBr cleavage site and shortened induction time. The expressed fused insulin precusor proteins (proinsulin and mini-proinsulin) could amount up to 50% (sometimes near 80%) of the total cellular protein whereas the insulin precursor parts constitute more than half of the fusion proteins. After sonication to break down the cells, isolating the inclusion bodies, cleavage with CNBr to release the target peptide, sulfitolysis to break down the disulfide bonds, refolding of proinsulin single chain and transpeptidation, the human insulin could then be obatained.

Construction of the expression plasmid

The plasmid pBV220 containing a $P_R P_L$ promoter is a direct expression vector. The plasmid pBCA containing a synthetic human proinsulin gene was constructed by insertion of human proinsulin gene into pBR332 by EcoR I and BamH I. The plasmid pJG104 was then constructed by insertion of proinsulin gene (BCA) fused with part of proinsulin gene(BC') at the polylinker site of pBV220 as shown in Table 1. pJG105 and pJG107 were the same except that the inserted proinsulin gene was changed by a mutant with the connetion peptide shortened to 6 and 2 amino acids respectively. The constructed plasmid was transformed E.coli strain DH5α. The plasmid pTcaC containing Tac promoter is an expression vector for calf prochymosin[4]. Different length of the 5'-terminus of the prochymosin gene was used to construct the plasmid pJG202 and

Table 1 *General structure of the expression vectors with fused genes for human insulin*

Construct	General structure	Amino acid
pJG202	<u>1 Prochy 1 63 BCA</u>	249
pJG204	<u>1 Prochy 3 5 BCA</u>	121
pJG104	<u>1 BC' 6 9 BCA</u>	155
pJG105	<u>1 BC' 6 9 BAC(6)</u>	126
pJG107	<u>1 BC' 6 9 BCA(2)</u>	122

pJG202 and pJG204 are Tac promoter system, while pJG104, pJG105 and pJG107 belong to P_RP_L promoter system. Prochy stands for prochymosin, BCA for proinsulin, BCA(6) for mini-proinsulin with ArgArgGlySerArgLys as mini-C peptide, BCA(2) for mini-proinsulin with ArgArg as mini-C peptide, BC' for B chain and part of C peptide.

pJG204 also shown in Table 1. The plasmid was transform E.coli strain JM105.

Cell-culture

For temperature inducible constructs, one bacterial colony was transfered into 300μl LB medium containing ampicillin at 30°C overnight with vigorous shaking. After adding 3ml LB the bacterial culture was kept still at 30°C for 4 h. Then the temperature was raised to 42°C quickly for another 4 h for induction. For high density fermentation in a 5 liter fermentor, a single bacterial colony was transfered into 3ml of LB mediun containing ampicillin at 30°C with vigorous shaking. 12 h later it was transfered into 300ml of LB. Another 12 h later the bacterial culture was put into the fermentor containing 3 liter M9 minimal medium except with the adding of 18g tryptone and 24g yeast extract. 150g yeast extract and 150g of glucose in 750ml of water was added. When OD_{600} reached 40, the temperature was raised to 42°C for induction for one hour. The cells(80g per liter)were harvested with and kept at −20°C.

For Tac promoter system, one bacterial colony was transferred into 300μl of LB medium containing ampicillin at 37°C overnight with vigorous shaking. After adding 3ml of LB the bacterial culture was kept at 37°C for 2 h. Then the temperature was raised to 42°C quickly for 4 hours with the addition of 20μM IPTG for induction. Lactose could also be used for induction with HB101 as bacterial host.

Purification

80g wet cell was thawed in 5 volumes Lysis buffer (0.05M Tris, 5% Triton, 8% sucrose, 0.05M EDTA, pH8.0) and lysed by sonication. The lysis pellet was collected by centrifugation and suspended by stirring overnight at 10°C with 4 volumes of 8M urea or 6M guanidine-HCl, 1mM DTT, 0.1M Tris/HCl, pH8.0. After centrifugation 3g Na_2SO_3, 2g $Na_2S_4O_6$ were added to the supernatant and the mixture was kept at 37°C for one hour. Then the mixture was diluted 5 times with cold water with pH adjusted to 4.5 and kept at 4°C for 2 h. The precipitate was collected by centrifugation and dissolved in 50ml 0.05M Gly-NaOH buffer with pH adjusted to pH 10.8. The sample was centrifuged and the supernatant loaded onto a Sephadex G75 column (4×100)eluted with the same buffer. The second peak was pooled, ultrafiltrated and lyophilized. The lyophilized protein was cleaved by CNBr and refolded. The human proinsulin was then obtained by DEAE-Sephadex A25 separation.

Results and Discussion

We have shown two high level expression systems of fused human proinsulin and its analogs. The purpose of our work is to find out a suitable system with high level expression, high proportion of insulin-like material, few methionine residue, quick induction of the expression of our target product and simple purification procedures. Much previous work was done with Lac, Trp or Tac promoters requiring a long period of time for induction, usually more than ten hours' culturing time[1,2,3]. The existance of expressed protein in living cells for long hours may be the crucial reason for their failure to obtain the product . The gene under the control of Tac promoter could be induced by IPTG and increased temperature in a short time. But it can not meet the requirement of industrial purpose. Lactose could also be used as both carbon resource and inducer, nevertheless it needs a long time for induction, usually more that 15 h. The construction of pJG204 seems better than pJG202 because the former contains only small part of the prochymosin gene while the latter a large one. Moreover the fused protein of pJG202 has two methionine residues, while that of pJG204, just has one at the connection site of the fused part and proinsulin as shown in Table 2. The temparature inducible plasmids, pJG104, pJG105 and pJG107, containing P_RP_L promoter are better than pJG202 and pJG204 because it could be induced in a short time, with no addition of inducer except increasing temperature to 42°C, it also contains only one methionine residue at the conception site.

Large amount of fusion protein existed in pellet after sonication suggests that the expressed product is in insoluble form. The formation of inclusion bodies is a major factor for the stability of the product. The formation of inclusion bodies also helps us to simplify the purification steps by just separating inclusion bodies from soluble proteins and other soluble cellular materials through centrifugation to give a purity of

Table 2 *Comparision of the fusion proteins among previous and present work*

Construct	Previous(%)		Present(%)				
	Wetzel[3]	Guo[2]	202	204	104	105	107
Expression	20	30	35	38	54	80	58
Proinsulin/fused protein	12.7	35	71	55			
Insulin/fused protein	4.6	7.5	20	42	33	40	42
Insulin/total protein	0.9	2.2	7.0	15.9	17.8	32	24.4
Methionine	23	15	2	1	1	1	1
Connection	6	35	35	35	35	6	2

80% as determined by densitometric scanning of the product on SDS-PAGE gel.

Under appropriate conditions the target products could constitute up to 50% of the total cellular protein(sometimes near 80%). The high density fermentation gives the expression level of 20% which is as expected a little lower than that of the fermentation in a shaker. The present results also show that proinsulin analogs with very small part of the connection C peptide and superior to proinsulin in refolding and transpeptidation yield, could be expressed in high level. The purified proinsulin can be directly changed into human insulin with correct amino acid composition and native biological activity by trypsin catalyzed transpeptidation.

The distance between the SD sequence and ATG is very important for the starting of strong translation. There are two SD sequences in our Tac promoter system, 3 and 10 bases up-stream form ATG, so are there in our $P_R P_L$ promoter system 7 and 10 bases ahead. We have constructed some other systems. Although they have the appropriate distances, the expressions are not satisfactory. Other factors such as the stable secondary structure of mRNA should be considered.

Acknowledgements

We thank Prof. Yunte Hou of the Institute of Virology for a gift of the expression vector pBV220, Prof. Kaiyu Yang of the institute of Microbiology, Acadimia Sinica. for the expression vector pTcaC and Prof. Tongjian Shen of the Institute of Biophysics, Acadimia Sinica. for the plasmid pBCA.

References

1. Goeddel, D.V., Kleid, D.G., Bolivar, F., Heyneker, H.L., Yansura, D.G., Crea, R., Hirose, T., Kraszewski, A., Itakvra, K. and Riggs, A.D., Proc. Natl. Acad. Sci. U.S.A., 76(1979)106.
2. Guo, L.H., Stepien, P.P., Tso, J.Y., Brousseau, R., Narang, S., Thomas, D.Y. and Wu, R., Chinese J. of Biotechnology, 1(1985)14.
3. Wetzel, R., Kleid, D.G., Crea, R., Heynader, H.L., Yansura, D.G., Hirose, T., Kroszewski, A., Riggs, A.D., Itakura, K. and Goeddel, D.V., Gene, 16(1981)63.
4. Zhang, Y. Y., Zhow, W., Liu, N.J. and Yang, K.Y., Chinese J. of Biotechnology, 7(1991)195.

Preparation and properties of [des(B1-3),B4-pGlu]-insulin and [des(B1-4)]-insulin

Ying-gao Xu, Xin-tang Zhang, Ying Ye, Ming-hua Xu and Shang-quan Zhu
Shanghai Institute of Biochemistry, Chinese Academy of Sciences,
Shanghai 200031, China

Introduction

According to the structure of insulin, N-terminal of B-chain resides in the surface of the dimer in an extended form. So it has been excluded from the putative receptor binding region and considered to be important only to its immunogenisity. To get further information about whether the changes in N-terminal of insulin B-chain really affects its activity, we prepared two analogues with shortened N-terminal of B-chain -- [des(B1-3),B4-pGlu]-insulin and [des (B1-4)]-insulin. Our results demonstrated that the gradual removal of N-terminal of insulin B-chain does cause a remarkable decrease in its biological activity simultaneously.

$[MSC]_2^{A1,B29}$-insulin was prepared as reported[1], followed by three and four steps of Edman degradation and then deprotected by treatment of 2N NaOH, followed by Sephadex G-25 chromatography to obtain crude products of [des(B1-3), B4-pGlu]-insulin and [Des(B1-4)]-insulin.

Results and Discussion

These two analogues were purified on Beckman semipreparative column RP C-3 (1.0×25cm). Fig.1 and Fig.2 are the HPLC profiles of [des(B1-3), B4-pGlu-insulin and [des(B14)]-insulin respectively. Fig.3 shows the retention time for insulin, [des(B1-3)]-insulin and [des(B1-4)]-insulin on HPLC RP C-3 column (0.46×7.5cm) are 9.40, 9.74 and 10.35 minutes respectively. [Des(B1-3),B4-pGlu]-insulin and [des(B1-4)]-insulin were identified by protein sequencing on 477A Protein Sequencer (Applied Biosystem Inc.). The N-terminal amino acid sequence of [des(B1-3),B4-pGlu]-insulin is Gly Ile Val Glu Gln and the sequence of des (B1-4)-insulin is

    A: Gly Ile Val Glu Gln
    B: His Leu --- Gly Ser

Compared with the sequence of native porcine insulin, it was determined that the N-terminal residues of A and B-chain of [des(B1-4)]-insulin are Gly and His respectively. In [des(B1-3),B4-pGlu]-insulin case only A chain is sequenced, because the N-terminal residue of B-chain is pGlu, which can not be further degraded. The receptor binding assay for these two analogues were carried out in human placental membrane at 4°C. Fig.4 is the displacement curve. From the IC_{50}, the binding capacities of [des(B1-3),-B4-pGlu]-insulin and [des(B1-4)]-insulin were calculate to be 50% and 33% of that of native porcine insulin.

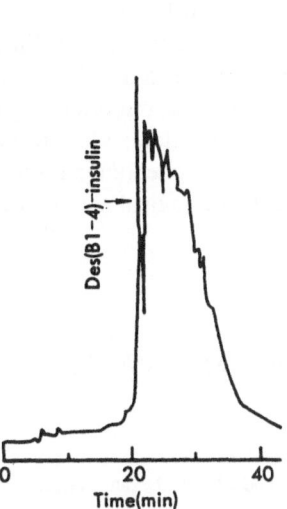

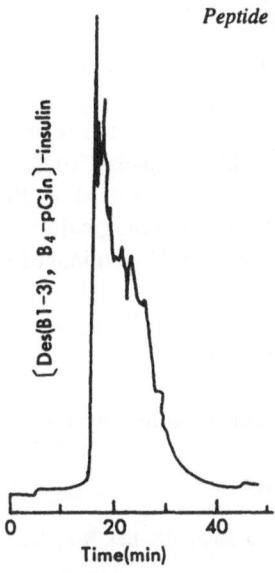

Fig. 1. Preparation of [des(B1-3),B4-pGlu]-insulin on HPLC RP C-3(1.0×25cm), 230nm, at 15°C. Solvent system: A: 10% CH₃CN, 90% H₂O, 0.1% TFA. B: 100% CH₃CN, 0.1% TFA. Flow rate: 2.35ml/min with a linear gradient 30 min from 10% to 20% solution B.

Fig. 2. Preparation of [des(B1-4)]-insulin on HPLC RP C-3 1.0×25cm, 230nm, at 15°C. Solvent system: A: 10%CH₃CN, 90% H₂O, 0.1% TFA. B: 100% CH₃CN, 0.1% TFA. flow rate: 2.35 ml/min with a linear gradient 30 min from 10% to 20% solution B.

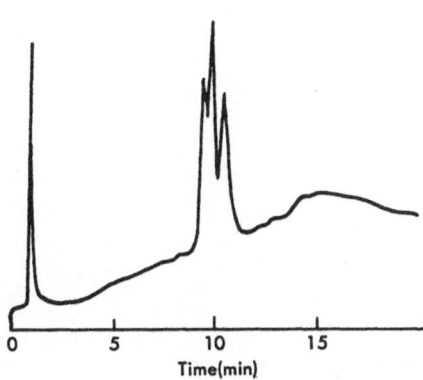

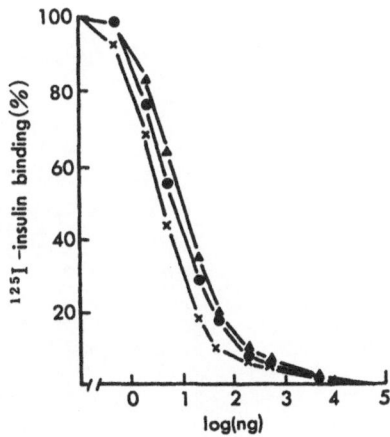

Fig. 3. Profile of insulin, [des(B1-3),B4-pGlu]-insulin, and [des(B1-4)]-insulin(from left to right) on HPLC RP C-3(0.46×7.5cm), 230nm, at 15°C. Solvent system: A: 10% CH₃CN, 90% H₂O, 0.1% TFA; B:100% CH₃CN, 0.1% TFA. Flow rate: 1ml/min with a linear gradient 15 min from 0 to 20% solution B.

Fig. 4. Displacement curve of [des(B1-3),B4-pGlu]-insulin, [des(B1-4)]-insulin analogue in human placental membrane. ×--× insulin, ●--● [des(B1-3),B4-pGlu]-insulin, ▲--▲ [des(B1-4)]-insulin.

31

It has been generally believed that the N-terminal of insulin B-chain is far from the putative receptor binding region, and was not possible to interact with insulin receptor directly. In addition, the residues in N-terminal of B-chain are quite different among species, which suggests that residues in this part were not so important in receptor binding. But our results show that analogues with gradually shortented N-terminal of B-chain had lower biological activity than native insulin. These phenomena may suggest that the length in N-terminal of B-chain is necessary for the biological activity of insulin.

Acknowledgement

This work is supported by the National Sciences foundation(39070244).

Reference

1.   Brandenburg, D., Lei, K.J., Wang, Z.Z., Dong, B., Ru, B.G. and Zhu, S.Q., Sci. Sin., 23(1980)1443.

Semisynthesis of insulin/IGF-I hybrid from desalanylinsulin

Xiao-hong Lin, Da-fu Cui, Ling Qin and You-shang Zhang

Shanghai Institute of Biochemistry, Chinese Academy of Sciences,
Shanghai 200031, China

Introduction

Insulin-like growh factor I (IGF-I) is structurally related to human insulin in amino acid sequence and three dimensional structure[1-3]. Insulin exhibits its biological activity only after the removal of the C peptide from proinsulin, whereas IGF-I is active when the the C domain is present. While comparing the A and B chains of insulin with A and B domains of IGF-I, one will find a high homology in amino acid sequences, the invariant residues and the disulfide bonds.

IGF-I exhibits much higher mitogenic activity but much lower metabolic activity than insulin. It has been reported that the binding of IGF-I to type 1 receptor requires multiple structural determinants in IGF-I and that IGF-I C-domain with no homology to insulin C-peptide may be important for the recognition of type 1 receptor[4].

In order to gain more information of the structure and function relationships of IGF-I and insulin, we have synthesized a hybrid consisting of insulin and the C domain of IGF-I by means of enzymatic formation of phenylhydrazide and subsequent conpling of unprotected IGF-I C-domain to desalanylinsulin(DAI). Preliminary mitogenic assays showed that this hybrid displayed an increased mitogenic potency (compared to insulin) in the range of 1 to 10 nM concentration.

Results and Discussion

Zinc free porcine insulin and phenylhydrazine in 1000 fold in excess were dissolved in pH 7.4, 72% 1,4-butanediol. Trypsin was added and the mixture was incubated at 25°C for 17 h. Over 85% of the starting material was converted to desalanylinsulin phenylhydrazine as shown by PAGE (Fig. 1). The solution was then dialyzed against 1M acetic acid and the product was purified by gel filtration through Sephadex G-50. The crude DAI-phenylhydrazine was directly oxidized in dark by 2[(2-nitrophenyl) sulfenyl] - 3-methyl-3 '-bromoindolinine (BNPS-skatole) in DMF/DMSO at pH 6.5 for 0.5 h and IGF-I C-domain (TGYGSSSRRAPQT) was added in dark. After adjusting the pH to 8.5 by N-methylmorpholine, the mixture was incubated at 25°C for 24 h. The product was precipitated by acetone and purified through DEAE-sepharose CL-6B column (Fig. 2).

The product was homogeneous in PAGE (Fig. 3) and its amino acid composition was in accordance with the theoretical values.

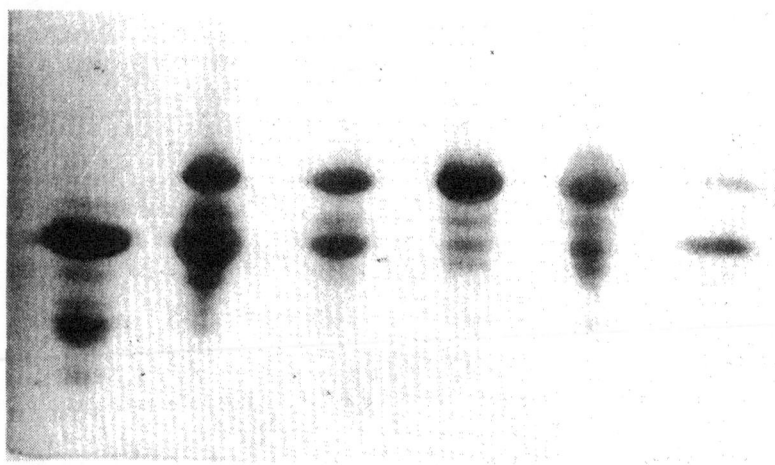

Fig. 1. Time-course of DAI-phenylhydrazide formation analysed by PAGE from left to right: 2.5 h, 6 h, 17 h, 20 h and Insulin.

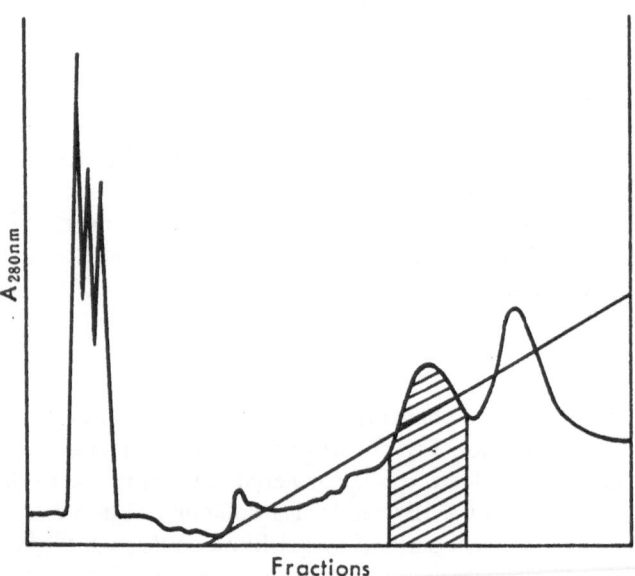

Fig. 2. Using DEAE-Sepharose CL-6B (1.1 × 12 cm) to purify Insulin/IGF -I hybrid, eluent: 0.05 M Tris buffer containing 40% isopropanol (pH 8.3) with salt gradient (0-0.1 M NaCl). The peak 4 (shaded) is the Insulin/IGF-1 hybrid.

In the three-dimensional structure of insulin, B30 Ala is exposed to the surface of the molecule. Therefore, by controlling the proper reaction conditions, it is possible to react at B29 Lys instead of B22 Arg. The yield of phenylhydrazide is high (85%, see Fig.1) but the yield of oxidation and couping is not satisfactory because of the instability of

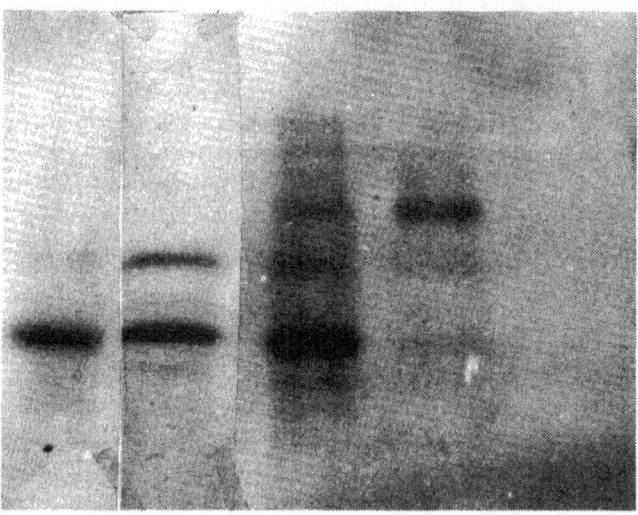

Fig. 3. Ployacrylamide gel electrophoresis at pH 8.30, from left to right: Insulin, DAI-phenyl-hydrazide, crude Insulin/IGF-I hybrid and purified Insulin/IGF-I hybrid.

DAI-phenylhydrazide and DAI-phenyldiimide.

In the hybrid, the C-domain of IGF-I is not covalently joined to the insulin A-chain. It remains to be seen whether this hybrid is similar to IGF-I in their three-dimensional structures.

References

1. Rinderknecht, E. and Humbel, R.E., J. Biol. Chem., 253(1978)2763.
2. Zapf, J., Schoene, E. and Fooesch, E.R., Eur. J. Biochem., 87(1978)285.
3. Fooesch, E.R., Scamid, C., Schwande, J. and Zapf, J., Annu. Rev. Physiol., 47(1985)443.
4. Caia, J.F., and Tager, H.S., J. Biol. Chem., 265(1990)17820.

Insulin analogues with alteration at two regions of the A chain

Shi-zhen Yang, Wei Lin, Yi-ding Huang, Xin-feng Jie, Ming-hua Xu and Ching-I Niu

Shanghai Institute of Biochemistry, Chinese Academy of Science, Shanghai 200031, China

Introduction

Hundreds of insulin analogues with changes at the various position of insulin molecule have been prepared by either chemical modification, protein engineering or peptide synthesis durin the past several decades. Most of those studies involve changes of biolgical activity with changes of the nature of individual amino acids. What interested us here is the effect on biological activities by the change of certain regions of the A chain. Although it is difficult to demarcate accurately regions in the molecule for specific function, we still think it feasible that a certain region be more important for certain function while other region for another. In this communication, a series of 7 porcine insulin analogues were prepared (Table 1) with changes in two regions of the A chain: C-terminal helical region and the region inside the disulfide bond of A6 to A10. From sequences as listed in Table 1: analogue I was made for scrutinizing the essentiality of A21 asparagine; analogues II,III and IV for studying the importance of the α helical region at A12-18 by either shortening the chain[1] or by substituting with some helix-abolishing amino acid residues; and analogues V, VI and VII for exploring the biological function of the least conserved, species different sequences of A8-10, in order to search for more potent or less immunogenic insulin analogues.

Results and Discussion

The fully protected peptide chains with the various alterations of the A chain of

Table 1 *Sequences of altered A chain*

	1				5					10					15					20	
A (poroine)	G	I	V	E	Q	C	C	T	S	I	C	S	L	Y	Q	L	E	N	Y	C	N
I.	G	I	V	E	Q	C	C	T	S	I	C	S	L	Y	Q	L	E	N	Y	C	A
II.	G	I	V	E	Q	C	C	T	S	I	C	S	.	.	Q	L	E	N	Y	C	A
III.	G	I	V	E	Q	C	C	T	S	I	C	.	.	.	.	L	E	N	Y	C	A
IV.	G	I	V	E	Q	C	C	T	S	I	C	S	L	Y	N	L	P	N	Y	C	A
V.	G	I	V	E	Q	C	C	H	K	P	C	S	L	Y	Q	L	E	N	Y	C	A
VI.	G	I	V	E	Q	C	C	H	R	P	C	S	L	Y	Q	L	E	N	Y	C	A
VII.	G	I	V	E	Q	C	C	H	K	R	C	S	L	Y	Q	L	E	N	Y	C	A

porcine insulin was assembled by stepwise elongation of Fmoc amino acids with DCC-HOBt to the 2% cross linked p-alkoxybenzyl alcohol resin. The functional side chain groups were protected by tBu for Cys, Glu, Tyr and Ser; Trt for His; Mtr for Arg and Boc for Lys. The final protected peptide was detached from the resin support and deprived of all protecting groups in one step using M TMSOTf-thioanisole-TFA system, and followed by sulfitolysis. The S-sulfonate of individual chain was purified by HPLC on a Ultrasphere C8 column. The semisynthetic insulin analogues obtained by reconstituting the synthetic $Asso_3^-$ analogues with the natural $Bsso_3^-$ from porcine insulin by modified Lilly procedure was purified by HPLC. Table 2 showed the biological activities of the analogues in three criteria. C-terminal A21 asparagine, although very conserved in evolutional stage, could be replaced by other amino acids with simpler side chain function, including glycine and alanine but not proline without much abolishing of its biological potency. The C-terminal helical region of the A chain is important for full activity but still may tolerate shortening of up to 4 amino acid residues (Fig. 1) or substituion of helix-abolishing amino acid residues with retention of substantial biological activity (Fig. 2). Analogues VB, VIB and VII B with alteration at A8-10 exhibit parallel reduction of potencies with a decreasing order from VB to sample VIIB. The differences in biological activities although not very large are significant enough when they are compared with porcine insulin (Fig. 3). It is noteworthy that the immunological activity of sample VIIB, in which A10(Arg) is different from sample VB and VIB(Pro), is comparatively low suggesting that A10 position is the immunodominant group[2] of this antigenically important region of the insulin molecule.

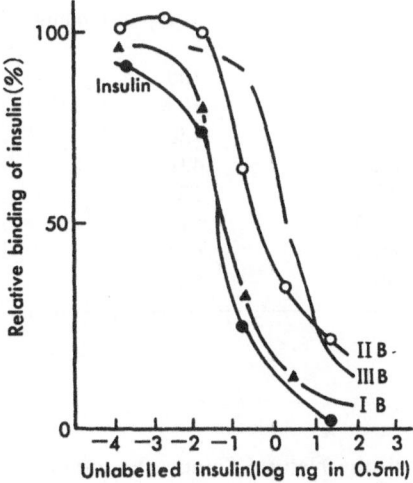

Fig. 1. Displacement of labelled insulin from insulin receptor on human placental membrane by insulin analogue.

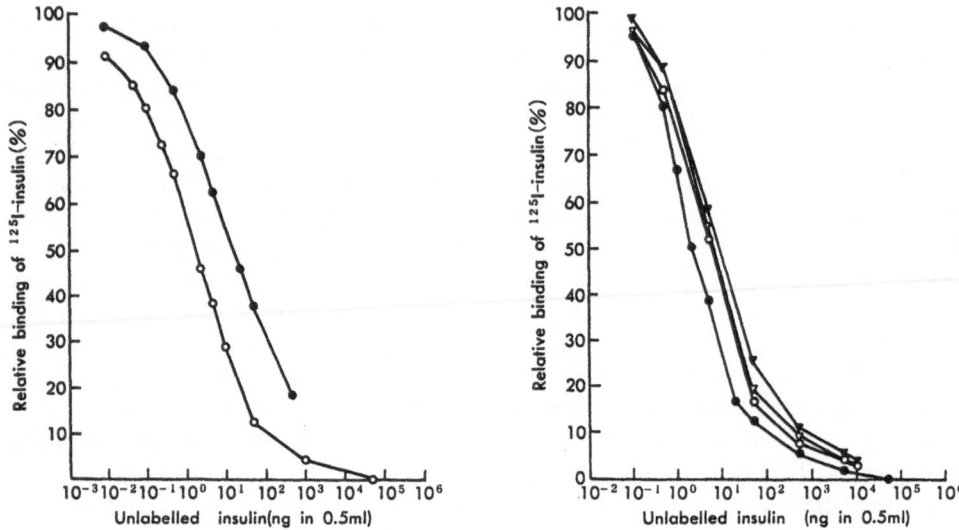

Figs. 2,3. *Displacement of labelled insulin from insulin receptor on human placental membrane by insulin analogues.*

Table 2 *Biological activities of insulin analogues*

Sample	in vivo(%)	Receptor binding(%)	Antibody binding(%)
Poroine Insulin	100	100	100
IB.	60.0	70.8	
IIB.	6.2	7.9	
IIIB.	3.3	4.0	
IVB.	10.0	10.4	
VB.	50.0	51.9	28.8
VIB.	40.0	44.3	29.6
VIIB.	30.0	32.1	15.4

Acknowledgement

This work is supported by the Natural Science Foundation. (3880193).

References

1.   Yang, S., Huang, Y., Feng, Y., and Niu, C., Acta Biochim. Biophys. Sin., 24(1992)503.
2.   Berson, S., and Yallow, R., J. Clin. Invest., 38(1959)2017.

Angiotensin II in the myocardium and aging

**Shao-jun Wen, Jia-rui Wang, Shi-da He, Li-jun Huang, De-hong Lu,
Bao-gang Xu and Zhong-yuan Yu**

Xuan Wu hospital, Capital Institute of Medicine, Beijing 100053, China

Introduction

Forty-one female Wistar rats were divided into 5 groups (1,3,6,10,21 months old). The 21-month group was taken as aged rats. With radioimmunoassay and immuno-histochemistry methods[1], we have recently studied the local renin-angiotensin system(LRAS) of the myocardium and the relations between cardiac endocrine functions and physical aging by means of the characteristic distribution of local myocardial angiotensin (LM Ang) levels and their changes with aging, the correlation between Ang I and II concentrations in the heart, and comparison of the plasma and myocardial Ang levels.

Results and Discussion

There were Ang I and Ang II like materials in the myocardium of rats. The immuno-reactive granules were principally located in the cytoplasm of myocardial cells. Ang had a characteristic distribution in different parts of the heart. The level of Ang II was highest in the atria and lowest in the ventricle (Figure 1). The order (from high to low)

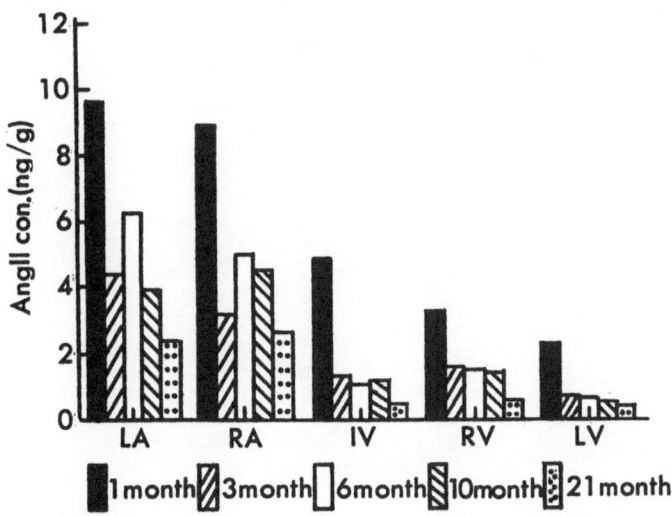

Fig. 1 Ang II had a characteristic distribution in different parts of the heart and there was a negative correlation between the level of Ang II and the age in month.

was the left (LA) and right atria (RA) > right ventricle (RV) and inter ventricular septum (IV) > left ventricle (LV). The experiments showed that the endocrine functions of LM Ang in the atria was stronger than in the ventricles. The level of Ang II in the 1-month group was the highest, that in the 21-month group being the lowest. As age increased (beyond six month), the levels of LM Ang II droped gradually. There was a positive correlation between the levels of Ang I and Ang II in the same part of the heart and a negative correlation between the level of Ang II and the age in month. There was no correlation between the plasma and myocardial Ang levels. The secretion of LM Ang II may be a paracrine, or autocrine-paracrine system. The experiments done by others showed that LM Ang II could regulate the inotropic and chronotropic activity and LM Ang II had a positive action on the myocardium and the coronary arteries[2]. The experiment suggested that as age increased, the hormonal function of local Ang in the myocardium gradually declined.

References

1.   Wen, S.J. and He, S.D., Radiommunoassay, 3(1990)81.
2.   Dzau, V.J., Circulation (suppl I), 77(1988)I-4.

Preparation and identification of [$B_{1,2}$-Ala, Ala]-insulin

Ying Ye, Ying-gao Xu, Ming-hua Xu, Xin-tang Zhang and Shang-quan Zhu
Shanghai institute of biochemistry, Chinese Academy of Sciences,
Shanghai 200031, China

Introduction

The crystal structure of insulin shows that the N-terminal of B-chain is on the surface of the dimmer in the form of extended structure. It has not generally been regarded as a region involved in the interaction with its receptor because it is variable among different species. But some evidences have suggested that the integrity in this terminal is important in maintaining activity. In order to study the influences on activities of the mutation in N-terminal of B-chain, an analogue, [$B_{1,2}$-Ala, Ala]-insulin in which B1 and B2 were both replaced by Ala was prepared from native porcine insulin. The preliminary study on its receptor binding capacity has been done.

The following scheme is used in the preparation of [$B_{1,2}$-Ala,Ala]-insulin from porcine insulin.

Crystalline insulin(porcine)
$\downarrow$ MSC • ONSu, DMSO, pH 8.5, 25°C, 25min.
[MSC] × insulin
$\downarrow$ SP-Sephadex C-25 at pH 3.0
[MSC]$_2$^{A1,B29}-insulin
$\downarrow$ Edman degradation for 2 times
[MSC]$_2$^{A1,B29}des(B1-2)-insulin
$\downarrow$ (1) Coupling with Boc•Ala•Ala•OSu
$\downarrow$ (2) TFA
[MSC]$_2$$^{A1, B29}$[$B_{1,2}$-Ala, Ala]-insulin
$\downarrow$ treatment with NaOH
crude [$B_{1,2}$-Ala, Ala]-insulin
$\downarrow$ HPLC
[$B_{1,2}$-Ala,Ala]-insulin

Results and Discussion

[$B_{1,2}$-Ala, Ala]-insulin was purified by RP HPLC on Beckman C-8 column (0.46×10cm) according to the conditions listed in Fig.1. The shadowed peak was collected and lyophilized. HPLC purification of this analogue is indispensable, because the remaining des(B1-2)-insulin is only two residues (Ala-Ala) less than the product and difficult to remove.

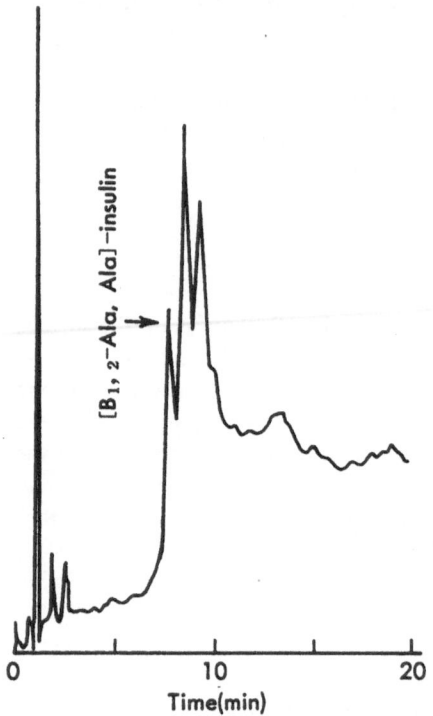

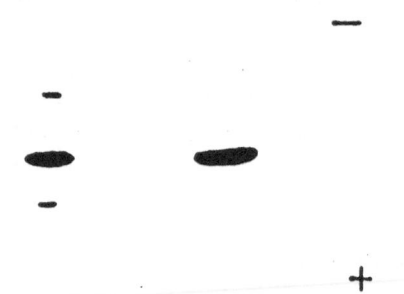

Fig. 2. *Polyacrylamide gel electrophoresis of (B$_{1,2}$ - Ala, Ala]-insulin at pH 8.3. From left to right: Porcine insulin and [B$_{1,2}$-Ala, Ala]-insulin.*

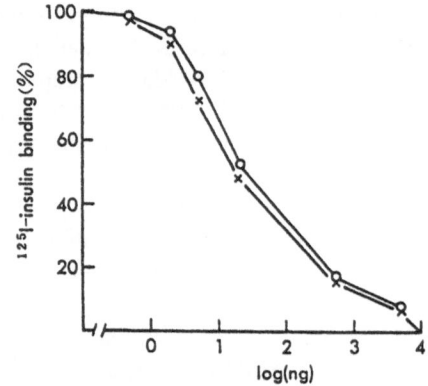

Fig. 1. *Isolation of [B$_{1,2}$-Ala, Ala]-insulin on HPLC RP C-8 (0.46×10cm), 230nm, at 18°C, 230nm. Solvent system: A: 10% acetonitrile, 90% H$_2$O, 0.1%TFA. B: 100% acetonitrile (CH$_3$CN), flow rate: 1ml/min, with a linear gradient from 25% to 35% solution B in 20 min.*

Fig. 3. *Displacement curve of [B$_{1,2}$-Ala, Ala]-insulin binding in human placental membrane. ×--× insulin; ○--○ [B$_{1,2}$-Ala, Ala]-insulin.*

Fig. 2 shows the polyacrylamide gel electrophoresis(pH 8.3) of the purified [B$_{1,2}$ -Ala, Ala]-insulin.

The purified product was homogeneous on electrophoresis. The amino acid composition analysis results were also consistent with the theoretical values.

The receptor binding assay was done in human placental membrane at 4°C, Fig.3 is the displacement curve of [B$_{1,2}$ -Ala, Ala]-insulin. The receptor binding capacity of [B$_{1,2}$ -Ala, Ala]-insulin was estimated to be 73.6% of that of native porcine insulin. This result shows that the replacement of B1-Phe and B2-Val by two Ala will result in a decrease in its receptor binding capacity.

Acknowledgement

This work is supported by the National Natural Sciences Foundation (39070244).

Session II
Neuropeptides

Chairs: Gui-sheng Lu
Institute of Materia Medica, CAMS
Beijing, China

and

Geoffrey W. Tregear
Howard Florey Institute, University of Melbourne
Melbourne, Australia

Structure-activity studies of LHRH antagonists with side-chain modified D-Lysine in position 6

Zhen-ping Tian, Yong-liang Zhang, Maria Kowalczuk, Tanya Hrinyo-Pavlina, Patrick Edwards and Roger Roeske
Department of Biochemistry and Molecular Biology, Indiana University School of Medicine, Indianapolis, Indiana 46202-5122, U.S.A.

Introduction

Luteinizing hormone-releasing hormone (LHRH), also called gonadotropin-releasing hormone (GnRH), with the sequence of pGlu-His-Trp-Ser-Tyr-Gly-Leu-Arg-Pro-Gly-NH_2 plays a very important role in the control of mammalian reproduction by regulation the secretion of two gonadotropins: luteinizing hormone (LH) and follicle-stimulation hormone (FSH) from the anterior pituitary gland. After its isolation [1,2] in 1971, thousands of agonists and antagonists of this native hormone have been synthesized in a search for better therapeutic agents for the treatment of GnRH-related endocrine diseases (e.g. prostate cancer and breast cancer) and for contraception [3].

Continuing studies have led to the development of several very potent antagonists of LHRH. For example, "Nal-Arg" (Ac-D-Nal[1]-4-F-D-Phe[2]-D-Trp[3]-Ser-Tyr-D-Arg[6]-Leu-Arg[8]-Pro-Gly-NH_2) completely inhibits ovulation in rats of proesterus at a dose of 1.0 µg/rat [4]. Unfortunately, when administered subcutaneously (sc), this type of potent antagonist causes transient edema of the face and extremities [5a] and a cutaneous anaphylactoid-like reaction [5b] in rats.Structural analysis of the LHRH antagonists and other peptides which have histamine-release (HR) activities indicated that the combination of a positively charged D-amino acid residue (D-Arg or D-Lys) at position 6 in close proximity to Arg[8] and a cluster of hydrophobic aromatic residues at the N-terminus (positions 1,2,3) are the chemical basis for triggering the mast cells to release histamine [6].

Progress has been made in the design of LHRH antagonists with low histamine-releasing toxicity but maintaining high antiovulatory (AO) activity. Transposing residues 5 and 6 to separate the two strong cationic charges (D-Arg[6] and Arg[8]) reduces histamine releasing activity. Switching the residues Tyr[5] and D-Arg[6] of LHRH antagonists [(Ac-D-Nal[1]-4-Cl-D-Phe[2]-D-Trp[3]-Ser-Tyr-D-Arg[6]-Leu-Arg[8]-Pro-D-Ala-NH_2) with HR ED_{50} = 0.10 µg/ml] to give (Ac-D-Nal[1]-4-Cl-D-Phe[2]-D-Trp[3]-Ser-Arg[5]-D-Tyr[6]-Leu-Arg[8]-Pro-D-Ala-NH^2) reduces the HR activity to about 3% that of its parent compound. The AO activity is lowered by approximately 50% [7]. Another approach to lower HR activity is to shield the two positive charges [8]. Replacement of D-Arg[6] with N-isopropyl-D-lysine [D-Lys(iPr)] and Arg[8] with N-isopropyl-lysine [Lys(iPr)] to give (Ac-D-Nal[1]-4-Cl-D-Phe[2]-D-Trp[3]-Ser-Tyr[5]-D-Lys(iPr)[6]-Leu-Lys(iPr)[8]-Pro-D-Ala-NH_2) lowers the HR activity from ED_{50} of 0.10 µg/ml to 6.6 µg/ml [9]. A dramatic advancement in terms of lowering HR toxicity was made in the dasign of Antide[10] (Ac-D-Nal^1-4-Cl-D-Phe^2-D-Pal^3-Ser-Lys(Nic)-D-Lys(Nic)^6-Leu-Lys(iPr)^8-Pro-D-Ala-NH$_2$), which has a HR toxicity of ED$_{50}$ of 261 µg/ml [11]. The exact explanation for this low histamine release is still unknown, but is probably due to removal of the positive charge in position 6. However, Antide has a tendency to form a gel at the subcutaneous injection site, which discourages its further study for clinical application.

The effect of N$^\varepsilon$-alkylation on D-lysine^6 of LHRH antagonists has been studied [9,12]. It was found that as long as the basic characteristic of the D-lysine side chain was maintained, the HR activity did not decrease and the AO activity did not change very much either. We carried out an extensive study on the side chain modification of D-lysine in position 6 with complete elimination of the basicity by acylation, and wish to report our results here.

Results and Discussion

A series of LHRH antagonists was synthesized with a side-chain modified D-lysine in position 6, having the general sequence Ac-D-Nal^1-4-Cl-D-Phe^2-D-Pal^3-Ser-Tyr-D-Lys(X)^6-Leu-Lys(iPr)^8-Pro-D-Ala-NH$_2$). A divergent solid phase modification strategy was used for the syntheses of these analogs. The parent peptide-resin was synthesized on a large scale (Boc/Bzl chemistry) using 4-methylbenzhydrylamine (MBHA) resin (Fig. 1) in which D-lysine was incorporated with its ε-amino group protected by an Fmoc group. After the peptide chain was assembled on the resin, the Fmoc group was selectively removed by using 50% piperidine-DMF. To avoid any undesired oxidation, the free amino group was converted to its trifluoroacetate salt form. Then the peptide-resin was divided into portions and stored for later modification. Acylation was carried

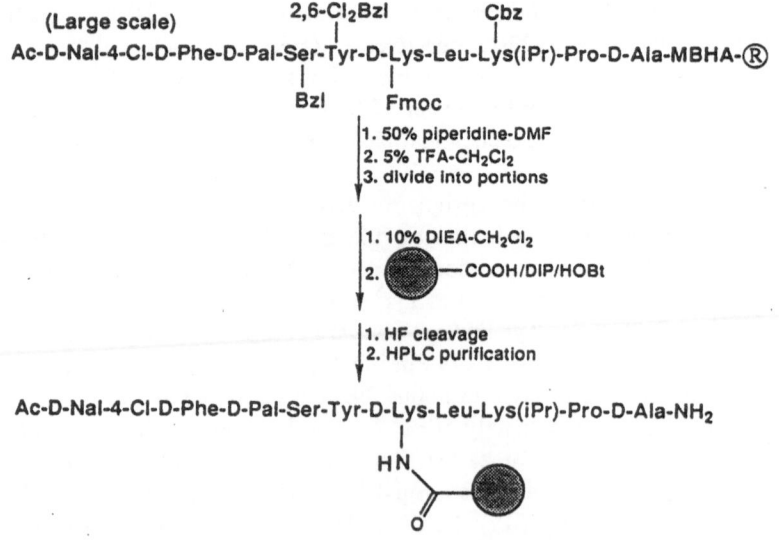

Fig. 1. General scheme of the side-chain modification.

Table 1 *Biological characterization of LHRH antagonists with side-chain modified D-Lys in position 6 [Ac-D-Nal-4-Cl-D-Phe-D-Pal-Ser-Tyr-D-Lys(X)-Leu-Lys(iPr)-Pro-D-Ala-NH₂]*

Entry	Modifying moieties(X)	AO Activity[a]	Histamine Release ED_{50} $(\mu g/ml)$[b]
1	dansyl	5/10 @ 10,9/10 @ 2.0	219.213
2	2-furoyl	4/10 @ 3.0	61
3	4-dimethylaminobenzoyl	10/10 @ 1.0	127
4	3-dimethylaminobenzoyl	10/10 @ 1.0	-
5	(3-pyridyl)acetyl	7/8 @ 1.0	-
6	2-chloronicotinoyl	4/8 @ 2.0, 8/8 @ 1.0	-
7	pyrazinoyl	7/8 @ 1.0	-
8	N-methylnicotinoyl	2/10 @ 0.5, 0/10 @ 1.0	5.4
9	N-isopropylpicolinoyl	2/8 @ 1.0	-
10	N-n-butylnicotinoyl	1/8 @ 1.0	-
11	nicotinoyl N-oxide	1/8 @ 1.0	-
12	histidyl	2/10 @ 1.0	22
13	pipecolinoyl	2/8 @ 1.0	-
14	N-diisopropyl-ε-aminocaproyl	6/10 @ 0.5, 1/10 @ 1.0	21
15	acetyl	2/10 @ 1.0, 0/10 @ 2.0	144,146
16	hydroxyacetyl	7/8 @ 1.0	-
17	cyanoacetyl	8/8 @ 1.0	-
18	trifluoroacetyl	6/8 @ 1.0	-
19	oxamyl	4/8 @ 1.0	-
20	gulonoyl	4/8 @ 1.0	-
21	pyroglutamyl	5/10 @ 0.5, 1/10 @ 1.0	61
22	2-oxo-4-thiazolidinoyl	6/8 @ 0.5, 0/8 @ 1.0	43

a, AO =antiovulatory: rats ovulating/total rats @ dosage in micrograms; b, ED_{50} for ''Nal-Arg''was 0.17 µg/ml and for ''Antide'' was 261 µg/ml.

out by coupling the acid to the resin-bound ε-amino group using diisopropyl-carbodiimide/HOBt. With the coupling of N-alkylated pyridyl carboxylic acids, a catalytic amount of 4-dimethylaminopyridine (DMAP) is needed in order to have a complete reaction. In some cases where the acid is not soluble in DMF or CH_2Cl_2, DMSO is added to increase its solubility. The peptide was deprotected and released from the resin by HF treatment, purified by medium pressure C18 column chromatography, and characterized by FAB-MS and amino acid analysis.

The biological data of 22 LHRH antagonists are shown in Table 1. It can be seen that modification Of D-lysine side chain ε-amino group in position 6 with aromatic moieties give less active compounds (1 to 7). Incorporation of an aromatic heterocyclic moiety with a positive charge on the ring increases the AO activity dramatically in comparison with their uncharged counterparts (8 to 11), although the HR toxicity is also increased. This observation is in good agreement with the working model of the structural basis for histamine release. Replacement of the ε-amino group of D-lysine with either another primary amino group (12), a secondary amino group (13) or a tertiary amino group (14)

47

does provide good AO activity, but the HR ED_{50} is not improved very much. It is interesting that a simple acetylation (15) gives a very active compound with low HR toxicity. In comparison, modifications (16 to 19) with related acids all give compunds with somewhat lower AO activity. We have made one antagonist (20) with a carbohydrate residue attached to the ε-amino group of D-lysine[6] which increased the water solubility of the peptide and the AO activity is good.

Compounds 21 and 22 showed the best combination of high AO activity and low HR toxicity. Compound 21 with pyroglutamic acid attached to D-lysine sidechain shows 90% inhibition of ovulation at 1 µg; with the thio-analog of pyroglutamic acid , the AO activity was further increase to give complete inhibition of ovulation at 1 µg and 25% inhibition at 0.5µg. Compound 22 (named Otac) is one of the most active LHRH antagonists with good water-solubility and reasonably low HR toxicity to date.

References

1.   Schally, A.V., Arimura, A., Kastin, A.J., Matsuo, H., Baba, Y., Redding, T.W., Nair, R. M.G., Debeljuk, L. and White, W.F., Science, 173(1971)1036.
2.   (a) Monahan, M., Rivier, J., Burgus, R., Amoss, M., Blackwell, R., Vale, W. and Guillemin, R.C.R.H., Acad. Sci. Ser., D273(1971)508; (b) Burgus, R., Butcher, M., Amoss, M., Ling, N., Monahan, M., Rivier, J., Fellows , R., Blackwell, R., Vale, W. and Guillemin, R., Proc. Natl. Acad. Sci. USA, 69(1972)278.
3.   Karten, M.J. and Rivier, J.E., Endocr. Rev., 7(1986)44.
4.   Rivier, J.E., Rivier, C., Perrin, M., Porter, J. and Vale, W.W., in LHRH and Its Analogs-Contraceptive and Therapeutic Applications, Vickery, B.H., Nestor, J.J., Jr. and Hafez, E.S.E., Eds., MTP Press: Lancaster, UK, 1984, pp.11-22.
5.   Horvath, A., Coy, D.H., Nekola, M.V., Coy, E.J., Schally, A.V. and Teplan, I., Peptides, 3(1982)969.
6.   (a) Schmidt, F., Sundaram, K., Thau, R.B. and Bardin, C.W., Contraception, 29(1984)283; (b) Hahn, D.W., McGuire, J.L., Vale, W.W. and Rivier, J., Life Sci., 37(1985)505.
7.   Roeske, R.W. and Chaturvedi, N.C., in Peptides: Structure and Function, Poceedings of the Ninth American Peptide Symposium, Deber, C. M., Hruby, V.J. and Kopple, K.D., Eds., Pierce Chemical Co., Rockford, IL, 1985, pp.561-564.
8.   Karten, M.J., Hook, W.A., Siraganian, R.P., Coy, D.H., Folkers, K., Rivier, J.E. and Roeske, R.W., in LHRH and Its Analogs Part. 2, Nestor, J.J. Jr. and Vickery, B.H., Eds., MPT: Lancaster, UK, 1987, pp.179-190.
9.   Hocart, S.J., Nekola, M.V. and Coy, D.H., J. Med. Chem., 30(1987)1910.
10. Roeske, R.W., Chaturvedi, N.C., Hrinyo-Pavlina, T. and Kowalczuk, M., in LHRH and Its Analogs Part.2, Nestor, J.J. and Jr. and Vickery, B.H., Eds., MPT: Lancaster, UK, 1987, pp.17-24.
11. Ljungqvist, A., Feng, D., Tang, P.L., Kubota, M., Okamoto, T., Zhang, Y., Bowers, C.Y., Hook, W.A. and Folkers, K., Biochem. Biophys. Res. Commun., 148(1987)849.
12. Hocart, S.J., Nekola, M.V. and Coy, D.H., J. Med. Chem., 30(1987)739.

Effect of GABA on LHRH release in median eminence of rats

Xin-zu Zhu[a], Lu-guang Luo[a], Bai-ge Zhao[b] and Wei Zhao[b]

[a]*Shanghai Institute of Materia Medica, Chinese Academy of Sciences, Shanghai 200031, China*
[b]*Shanghai Institute of Planned Parenthood Research, Shanghai 200032, China*

Introduction

Growing evidence suggests an important role played by r-aminobutyric acid (GABA) in the control of gonadotropin secretion. GABA injected into the third ventricle of male rats promotes the release of pituitary luteinizing hormone (LH). When GABA was injected directly into the pituitary it was ineffective in promoting LH release. It was suggested that GABA may modulate LH release at the hypothalamic level, most likely by influencing the luteinizing hormone-releasing hormone (LHRH) system[1]. The present *in vitro* study was designed to evaluate the effect of GABA on LHRH release and the possible interactions between GABAergic and noradrenergic system involved in LHRH release.

Preparation of tissue for incubation

Adult male Sprague-Dawley rats obtained from Shanghai Institute of Planned Parenthood Research were used in this study. The animals were housed in a temperature-controlled room with a 12-h light, 12-h dark cycle for at least 2 weeks before the experiment. Food and water were provided ad libitum. On the day of the experiment, the animals were killed, and median eminence (ME) was removed as previously described[2].

In vitro incubation system

ME fragments were incubated in a polypropylene tube in a Dubnoff shaker at 37°C with constant shaking (60 cycles/min). Each tube contained two ME fragments and 0.5 ml incubation medium, consisting of Hepes-Yamamoto-BSA buffer (HYB, NaCl 145 mM, KCL 5 mM, $MgSO_4$ 2 mM, $CaCL_2$ 2 mM, $NaHCO_3$ 5 mM, Hepes 10 mM, D-glucose 1 mg/ml, BSA 0.5 mg/ml, pH 7.4). The tissues were preincubated for a period of 60 min, at the end of which time, medium was replaced by fresh medium containing the test substances. Incubations were then carried out for a period of 10 min. After the incubation, medium was immediately transferred to microtubes and centrifuged at low speed for 15 min at 4°C. The medium were collected for LHRH assay and NE measurement[3].

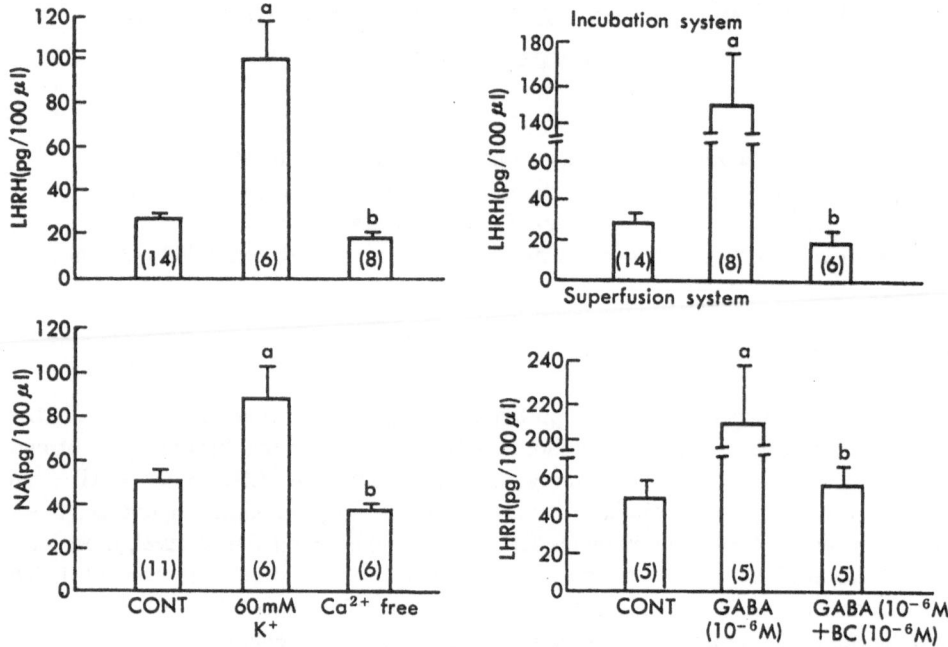

Fig. 1. Effect of K⁺-depolarization on LHRH (top panel) and NA (lower panel) release from ME. CONT=control. Each column represents mean ± SE of the number of observations shown at its base. a: p<0.01 vs control group; b: p<0.01 vs 60 mmol/L K⁺ group.

Fig. 2. Effect of GABA on LHRH release from ME. CONT=control. Each column represents mean ±SE of the number of observations shown at its base. a: P<0.01 vs control group; b: P<0.01 vs GABA group.

Results and Discussion

When ME was incubated in vitro, significant amounts of LHRH were released to the medium during incubation. The release of LHRH was 27.3 ± 2.5 pg/100 μl (n=14). High concentration of potassium (HYB buffer containing 60 mM) induced an increment in LHRH release from the basal level to 91.2 ± 19.2 pg/100 μl. The treatment also induced an increase in the NE release from 50.9 ± 4.2 to 89.6 ± 14.9 pg/100 μl (Fig 1). GABA (10^{-6} M) significantly increased LHRH and NE release from ME from 27.3 ± 2.5 to 150.4 ± 27.9 pg/100 μl (Fig 2) and from 50.9 ± 4.2 to 105.5 ± 19.1 pg/100 μl (Fig 3), respectively. The effect of GABA promoting the LHRH and NE release was blocked by bicuculline (10^{-6} M). When NE (10^{-6} M) was added to the medium during incubation, the release of LHRH also increased from 27.3 ± 2.5 to 56.1 ± 6.6 pg/100 μl suggesting that NE could induce LHRH release from ME (Fig 4). When rats were pretreated with reserpine (8mg/kg), GABA increased LHRH only by 26.5% instead of 451.9% in the normal animal (Fig 5). It has been suggested that GABA may modulate LH release at the hypothalamic level, most likely by influencing the LHRH system. The

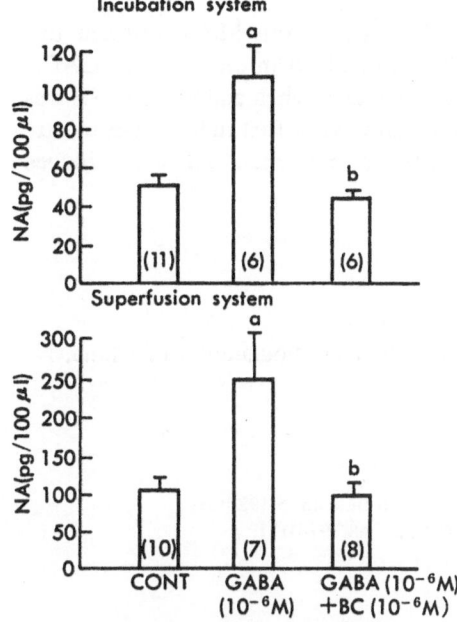

Fig. 3. *Effect of GABA on NA release from ME. CONT=control. Each column represents mean ±SE. a: P<0.01 vs control group; b: P<0.01 vs GABA group.*

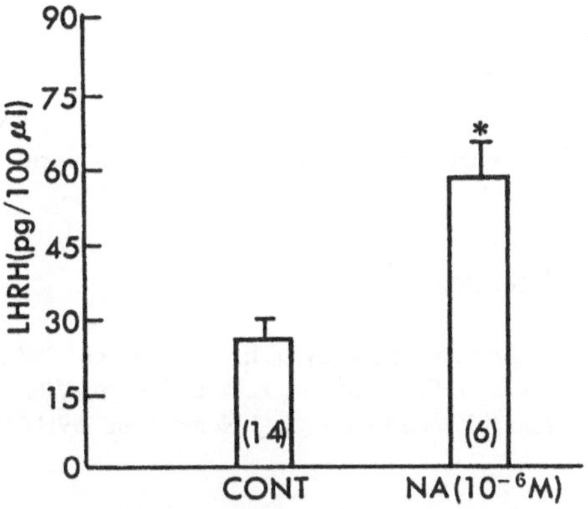

Fig. 4. *Effect of exogenous NA on LHRH release from ME. CONT=control. Each column represents mean ±SE of the number of observations shown at its base. P<0.01 vs control group.*

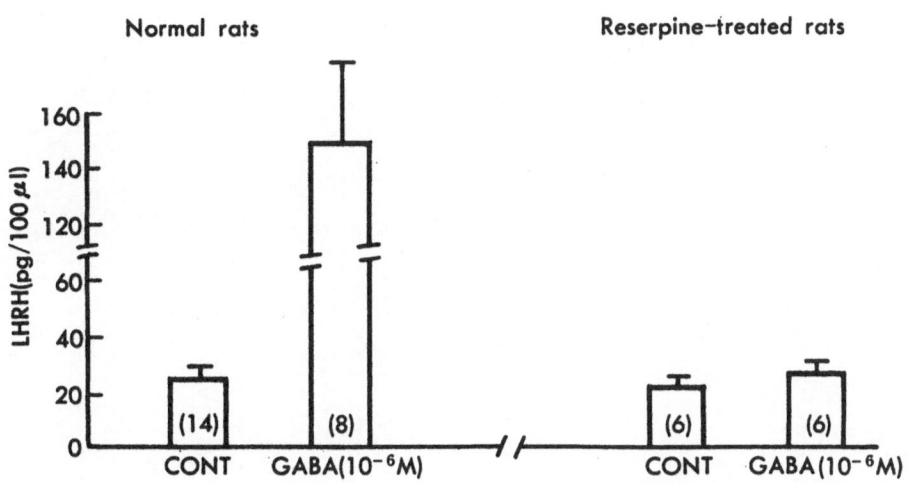

Fig. 5. *Effect of GABA on LHRH release from ME from the reserpine- treated rats. CONT=control. Each column represents mean ±SE of the number of observations shown at its base.*

present results that GABA could induced LHRH release from ME confirmed this hypothesis. Our results also demonstrated that NE is involved in the effect of GABA promoting the LHRH release since this effect was attenuated when endogenous NE was depleted with reserpine pretreatment. It is possible that GABA first induces NE release and NE further induces LHRH release. The results that exogenous NE could induce LHRH release support this hypothesis.

Acknowledgement

This work was supported by the National Natural Science Foundation of China, No 9389007.

References

1. Negro-vilar, A., Vijayan, E. and McCann, S.M., Brain Research Bulletin, 5(1980)239.
2. Gallardo, F., Voloschin, L.M. and Negro-Vilar, A., Brain Res., 148(1978)121.
3. Zhu X.Z. and Luo, L.G., J Neurochem., 59(1992)932.

Melanotropic peptides for the identification, localization (imaging) and chemotherapy of melanoma

Mac E. Hadley[a], Shubh D. Sharma[b] and Victor J. Hruby[b]

[a]Departments of Anatomy and [b]Chemistry, University of Arizona, Tucson, AZ 85721, U.S.A.

Introduction

α-Melanocyte stimulating hormone (α-MSH,α-melanotropin) is a tridecapeptide (Ac-Ser-Tyr-Ser-Met-Glu-His-Phe-Arg-Trp-Gly-Lys-Pro-Val-NH$_2$) synthesized and secreted by the pituitary of most vertebrates. The hormone is also synthesized within melanocorticotropic neurons of the brain wherein it may function as a neurotransmitter and/or neuromodulator to regulate one or more physiological functions. For example, in its role as a neurohormone the peptide may regulate body temperature and enhance cognitive skills and memory retention. A role in neuroplasticity has also been suggested. Our goal is to determine the nature and distribution of melanotropin receptors in normal and abnormal (melanoma) tissues and to ascertain whether or not these receptors could serve as a membrane marker for melanoma.

Results and Discussion

Melanotropin design, synthesis, and biological characterization

We have designed hundreds of α-MSH analogs for structure-function studies and have determined the minimal message sequence of the hormone for biological activity. Utilizing the classical frog and lizard skin bioassays it has been determined that the minimal sequence required for agonistic activity is the central sequence, His^6-Phe^7-Arg^8-Trp^9 (Ac-α-MSH$_{6-9}$ -NH$_2$). The minimal sequence for equipotency to α-MSH was found to be the 4-11 sequence in the lizard [1] and the 4-12 sequence in the frog [2]. These results have important implications for the design of melanotropin analogs for clinical medicine, for they suggest that the nonessential sequences of the hormone might be modified for the attachment of diagnostic ligands or cytotoxic agents [3,4].

A series of α-MSH competitive antagonists of the generic structure H-His^6-Xaa^7-Yaa^8-Trp^9-D-Phe^{10}-Lys-NH$_2$ have been designed which possess moderate antagonist potency (pA$_2$ values of 5-6) in the frog (*R. pipiens*) or lizard (*A. carolinensis*) bioassays [5]. These and other antagonists that we have designed may prove useful for physiological and clinical studies.

Melanotropin analogs possessing superpotency, prolonged (residual) activity and resistance to proteolytic inactivation have also been designed and biologically characterized [6]. Two such analogs, [Nle^4,D-Phe^7]α-MSH (MT-1) and Ac-[Nle^4,Asp^5,D-Phe^7,Lys^{10}] α-MSH$_{4-10}$ -NH$_2$ (MT-II) will be discussed in this paper.

Fluorescent melanotropins for receptor identification and visualization

We have prepared biocytin derivatives of MT-I: [N$^\alpha$-Bct-Ser^1,Nle^4,D-Phe^7]α-MSH and [12-Bct-N$^\alpha$-dodecanoyl-Ser^1,Nle^4,D-Phe^7]α-MSH. These melanotropins possessed almost identical potency and prolonged activity to the parent analog, MT-I, as determined in several bioassays [7]. We also synthesized a fluorescein-labeled MT-I analog using dichlorotriazinylamino-fluorescein (DTAF). The fluorescent conjugate was a superagonist, was nonbiodegradable, and possessed ultraprolonged biological activity similar to that of the parent analog, MT-I [3].

We have designed a new class of multivalent peptide hormone macromolecular composites that may serve as powerful diagnostic, imaging, and therapeutic tools. Composites have been synthesized in which multiple copies of a biospecific ligand (e.g., a hormone) were covalently attached to a biologically compatible but inert polyfunctional macromolecule. In addition, the conjugation of multiple copies of a fluorophore directly to the macromolecule provided an enhanced visual means of detecting ligand-macromolecular conjugates bound to target cells in *in vitro* binding assays. A fluorescent melanotropin macromolecular conjugate has been synthesized and used to demonstrate the presence of specific melanotropin receptors on various human melanoma cell lines [8].

Most importantly, every cell of every melanoma cell line possessed melanotropin receptors as visualized by fluorescence microscopy. Cells of nonmelanocyte origin did not exhibit such receptors. These cell-specific melanotropin receptors may serve as cell surface markers for melanoma. Fluorescent melanotropin conjugates should be useful in determining whether all (primary and metastatic) tumors possess such receptors. These receptors may provide targets for the identification, localization, and chemotherapy of melanoma. For example, a fluorescent melanotropin conjugate could be used to identify visually metastatic melanoma cells in lymph nodes before elective lymph node dissection.

Radiolabeled melanotropins

Radiolabeled antibodies specific for tumor-associated antigens promise to improve diagnostic imaging of tumors. Unfortunately, the antigenic profile of melanoma cells changes with time and labeled antibodies have proven ineffective as diagnostic or chemotherapeutic tools. Hormones also provide vehicles for the tumor-specific delivery of radioisotopes. Some melanotropin analogs irreversibly interact with melanoma cells; attached to a radioisotope (^{125}I,^{131}I,^{111}In) they should provide a method for melanoma tumor localization by external scintigraphy.

α-MSH activity is lost due to oxidative/reductive effects during radioiodination. [Nle^4]-substituted analogs are less susceptible to such inactivation. Iodinated MT-I has been utilized successfully by several investigators [9]. A chelating derivative of one of our analogs, Ac-[Nle^4,Asp^5,D-Phe^7,Lys^{10}]α-MSH$_{4-10}$, has also been used as a potential imaging agent for malignant melanoma [10]. The [^{111}In]-DTPA-labeled MSH analog preferentially localized to melanotic tumors in mice relative to other tissues.

Melanotropic peptides for chemotherapy of melanoma

The results described above indicate that rather bulky ligands can be attached to melanotropin analogues without any appreciable loss of biological activities. It should be possible therefore to attach cytotoxic drugs to melanotropic peptides. These melanotropins would then act as vehicles for the cell-specific delivery of the cancer drugs to melanoma cells. The specificity of such delivery would be assured by a receptor-directed mechanism. The melanotropin drug complex might then manifest its cytotoxic actions at the level of the plasma membrane, or, if internalized, its actions might be exerted within the cell.

The demonstration that radiolabeled melanotropins can bind specifically to melanoma tumors in mice and humans suggests that radioactive conjugates of melanotropin analogs can be used for the radiotherapy of melanoma. In addition, melanotropin drug conjugates might similarly be targeted to melanoma tumors. We synthesized Ac-[Nle^4,Glu^5(gamma-4'-hydroxyanlide),D-Phe^7]α-MSH$_{4-10}$-NH$_2$. Although the potency of the peptide was identical to that of peptide substrate to which the drug was attached, the conjugate failed to exhibit cytotoxic activity [11]. More recently, a recombinant α-MSH-diptheria toxin fusion protein was developed as a potential specific cytotoxic agent for melanoma [13]. A more reasonable tactic might be to conjugate a melanotropin such as MT-I or MT-II to the numerous lysine moieties of a toxin through a peptide bond.

Melanotropic peptides for protection against skin cancer

We have utilized melanotropic peptides to induce an authentic tanning (melanin production) of the skin that does not require ultra violet light [13]. Both MT-I and MT-II are presently in phase-1 clinical trials. In both preclinical animal studies and clinical studies on human volunteers no undue side effects have been noted. Additional studies are directed at determining whether a melanotropin induced-pigmentation of human skin will provide some degree of protection against the deleterious effects of solar radiation.

Melanotropin delivery

Following intramuscular injection of MT-I into human volunteers melanotropic activity can be detected in the urine. Data suggests that the activity is due to a metabolite other than the authentic MT-I. Following injection of MT-II, the intact peptide can be found within the serum and urine.

Melanotropic peptides injected into humans increase melanin pigmentation of the skin [12]. Alternate routes of delivery would, however, be desirable. We have determined that MT-II can be delivered to the systemic circulation following oral delivery of 2 to 20 mg of the peptide. The presence of melanotropic activity in the urine indicates that the peptide has been delivered from the lumen of the gut to the systemic circulation and is then excreted by the kidneys.

MT-I, when applied topically to mice, induces darkening of follicular melanocytes throughout the skin. The melanotropin was also applied to the surface of human skin samples through a permeation apparatus and allowed to penetrate for 24 h at 36°C. Passage of the analog was shown by both bioassay and redioimmunoassay. These assays correlated well and demonstrated both the presence and the biologic integrity of the peptide after transdermal passage. This study is the first to show that a melanotropic peptide can be delivered transdermally through human skin *in vitro* [14].

We have also determined that MT-I can be delivered to the systemic circulation by an iontophoretic device. Melanotropic activity within the serum and urine can be detected following one millivolt delivered over a one-hour period of time.

References

1. Castrucci, A.M.L., Hadley, M.E., Sawyer, T.K., Wilkes, B.C., Al-Obeidi, F., Staples, D.J., de Vaux, A.E., Dym, O., Hintz, F.M., Riehm, J.P., Rao, K.R. and Hruby, V.J., Gen. Comp. Endocrinol., 73(1989)157.
2. Hruby, V.J., Wilkes, B.C., Hadley, M.E., Al-Obeidi, F., Sawyer, T.K., Staples, D.J., deVaux, A.E., Dym, O., Castrucci, A.M.L., Hintz, M.F., Riehm, J.P. and Ranga Rao, K., J. Med. Chem., 30(1987)2126.
3. Chaturvedi, D.N., Hruby, V.J., Castrucci, A.M.L., Kretuzfeld, K.L. and Hadley, M.E., J. Pharm. Sci., 74(1985)237.
4. Hadley, M.E. and Dawson, B.V., Pigment Cell Res., 1(Suppl)(1988)69.
5. Sawyer, T.K., Staples, D.J., Castrucci, A.M.L. and Hadley, M.E., Pept. Res., 2(1989)140.
6. Al-Obeidi, F., Hruby, V.J., Castrucci, A.M.L. and Hadley, M.E., J. Med. Chem., 32(1989)174.
7. Chaturvedi, D.N., Knittel, J.J., Hruby, V.J., Castrucci, A.M.L. and Hadley, M.E., J. Med. Chem., 27(1984)1406.
8. Sharma, S.D., Hruby, V.J., Hadley, M.E., Granberry, M.E. and Leong, S.P., In Smith, J.A. and Rivier, J.E. (Eds.) Peptides: Chemistry and Biology (Proceedings of the 12th American Peptide Symposium), ESCOM, Leiden, 1992, p. 599.
9. Tatro, J.B., Atkins, M., Mier, J.W. et al., J. Clin. Invest., 85(1990)1825.
10. Bard, D.R., Knight, C.G. and Page-Thomas, D.P., Biochem. Soc. Trans., 18(1990)882.
11. Al-Obeidi, F., Mulcahy, M., Pitt, V.S., Begay, V., Hadley, M.E. and Hruby, V.J., J. Pharm. Sci., 79(1990)500.
12. Tatro, J.B., Wen, Z., Entwistle, M.L., Atkins, M.B., Smith, T.J., Reichlin, S. and Murphy, J.R., Cancer Res., 52(1992)2545.
13. Levine, N., Sheftel, S.N., Eytan, T., Dorr, R.T., Hadley, M.E., Weinrach, J.C., Ertl, G.A., Toth, K., McGee, D.L. and Hruby, V.J., J. Amer. Med. Assoc., 266(1991)2730.
14. Dawson, B.V., Hadley, M.E., Kretuzfeld, K.L., Don, S., Levine, N., Eytan, T. and Hruby, V.J., J. Invest. Dermatol., 94(1990)432.

Binding capability of ZNC(C)PR and its effects on potentiation of synaptic transmission in rat hyppocampus

Yu-cang Du[a], Xin-wei Rong[b], Jian-hua Wu[a], Tong Tang[a] and Xiu-fang Chen[b]

[a]Shanghai Institute of Biochemistry,
[b]Shanghai Institute of Physiology, Chinese Academy of Sciences,
Shanghai 200031, China

Introduction

ZNC(C)PR (AVP$_{4-8}$) is a pentapeptide found in rat brain, as a enzymatic product of arginine-vasopressin. As showing more potent on facilitating acquesition and maintenance of learning and memory than AVP did and no substantial press nor antidiuretic activities[1], ZNC(C)PR has regarded as a new neuropeptide in mammalian brains. We have reported that ZNC(C)PR not only *in vivo* enhanced significantly behavioral responses in rats but also *in vitro* induced a series of biochemical changes in rat hippocampus[2], e.g. enhancing accumulation of second messenger IP$_3$ in hippocampus slices, accelerating the maturation of phosphorylated B50/GAP-43 and inducing a couple of proto-oncogenes transcription[2]. Currently we found that ZNC(C)PR not only specifically bound to rat brain, but also remarkably stimulated long-term potentiation(LTP) in rat hippocampus slices. These pharmaceutical evidences indicated that ZNC(C)PR could be an important regulator in rat memory processes.

Results and Discussion

^{35}S-labelled ZNC(C)PR with high specific activity (>280 Ci/mmol) was synthesized by solid-phase method in our lab. By using this radioactive peptide we investigated ZNC(C)PR specific binding site in rat brain. As we have memtioned previously[2], density-scanning on radiophotogram of rat brain slices showed a tight binding to pyramidal cell and dentate granular cell in hippocampus and a few binding sites distributed in other brain regions. The results as shown in Table 1 generated from radio-receptor assay (RRA) showed that they are amygdala, hypothalamus, cortex, septum and hippocampus etc. It is worthy to note that the biological functions of these regions are somewhat related to the learning and memory.

In accordance with radiophotogram, RRA of ZNC(C)PR resulted in, in the presence of 10 mM or MgCl$_2$, a specific binding to the synaptic plasma membranes of rat hippocampus. The characters of this binding is given by Fig. 1 and 2: there is a typical saturation curve and a straight line on the scatchard polt. It indicates that only one binding site of ZNC(C)PR with high affinity could be found on this synaptic membrane. From Scatchard analysis, Bmax can be estimated as 19 fmol /mg protein and dissociation constant of ZNC(C)PR to this membrane preparation is 1.69 nM. obviously there is a

Table 1 *The amount of* ^{35}*S-ZNC(C)PR binding site in different brain regions*

Brain region	Binding site (fmol/mg protein)
amygdala	29.3
hypothalamus and hypophysis	29.0
cortex	24.1
nucleus caudatus	21.6
nuclei tractus and solitarii	21.3
septum	19.6
superior and inferior colliculus	17.7
hippocampus	15.7
medulla oblongata	10.2
celebellum	5.5

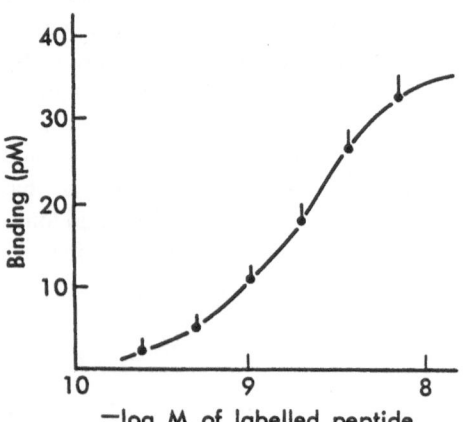

Fig. 1. *Dose-dependent binding of* ^{35}*S-ZNC(C)PR to hippocampal synaptic plasma membranes.*

Fig. 2. *Schatchard analysis of binding curve showed in figure 1.*

real difference between AVP and ZNC(C)PR binding in rat brain. In AVP case, there were two kinds of binding site: high and low affinity and the K_D for the former was round 11 nM to 100 nM [3]. The reversibility of ZNC(C)PR binding was confirmed by dilution. Based on Kinetic parameters K_D calculated from the first order velocity constants was $k_2/k_1 = 1.02$ nM. It is in the same range with the figures generated from binding curve (1.69 nM) and from concentration of half saturation (1.65 nM). Results obtained from rat cortex was similar to rat hippocampus. In general, they reveals a tight (nM) and specific binding of ZNC(C)PR in rat brain and this receptor may not be the same with AVP'S.

In order to answer the question:through which way ZNC(C)PR could affect memory responses after they have bound to rat hippocampus and cortex? Did any cellular events

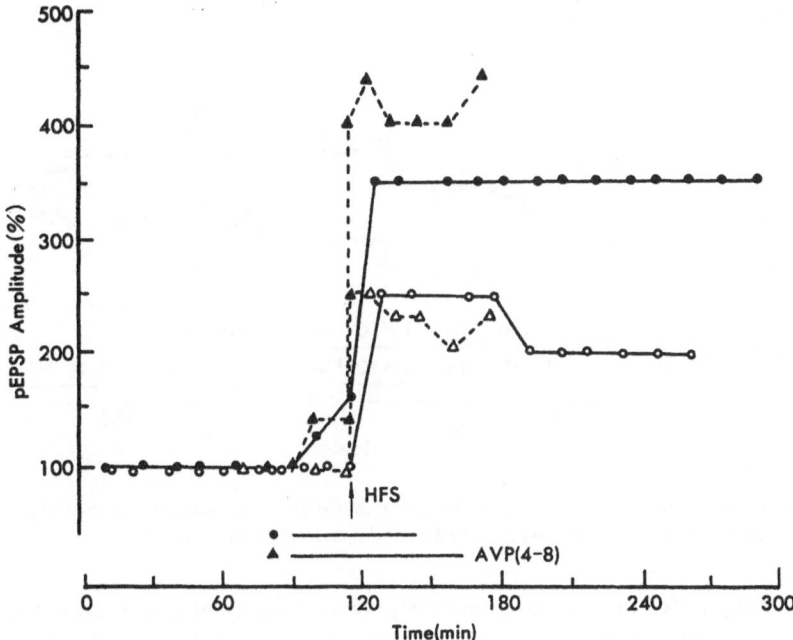

Fig. 3. *ZNC(C)PR enhances LTP induced by brief high frequency stimulation. Two experiments showed the amplitude of pEPSP in slices perfused with (solid line) and without 5×10⁻⁷M of peptide.*

subsequently happened? As LTP are thought to be underlie certain forms of learning and memory, we studied the effects of ZNC(C)PR on the synaptic transmission in rat hippocampus slices, at both extracellular and intracellular levels.

As we have memtioned previously, intracellular recording of several CA1 neuron showed that 0.5 μM - 10 nM of ZNC(C)PR dramatically stimulates the amplitude of excitatory post-synaptic potential(EPSP) in CA1 region of Schaffer collateral/ commissural fibers. After incubation for 15 minutes or more the EPSP become action potential. On the other hand, similar stimulation was found in extracellular level: 2×10^{-7} M of ZNC(C)PR increased the amplitude of field EPSP evoked in CA1 region by low frequency stimulation (0.1 HZ) and after washing the potentiation effect was irreversible (n=6), at least for as long as the recording lasted (3.5 h).

Furthermore, we observed the effect of ZNC(C)PR on the long-term potentiation (LTP) of synaptic transmission by brief high frequency stimulating of afferent fibers in two slices of the same hippocampus simultaneously in which slice perfused with standard medium but containing or not containing the peptide as sample or control. Fig. 3 shows the enhancing effects of 0.2 μM of ZNC(C)PR on the amplitude of LTP of CA1 region of the slice. It is significantly larger than that of control slice by 50-100% (n=6) and the increase in amplitude was long lasting after washing.

Dose-dependent effect of ZNC(C)PR on the amplitude of pEPSP in CA1 region was

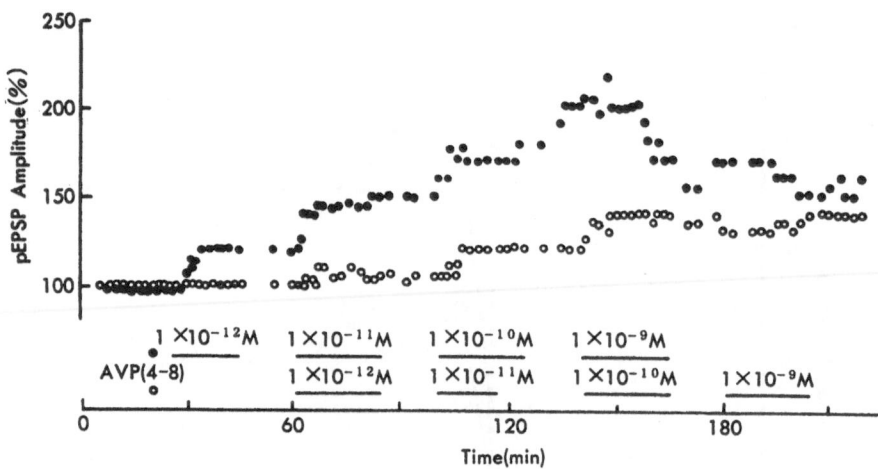

Fig. 4. *Potenciation effects of different concentration of ZNC(C)PR. Two experiments showed the effects of different concentration of ZNC(C)PR on the pEPSP amplitude in the same slice.*

tested by using different concentration of peptide and the result is showed in Fig. 4. The threshold concentration for generating the effect was 1×10^{-12} M (n=5). We compared the effect of arginine-vasopressin (AVP) and found in five experiments that the lowest effective concentration of AVP was usually in the range of 1×10^{-8} M to 10^{-9} M. It means that the potency of ZNC(C)PR on enhancing effect is thousands times higher then AVP. It should be pointed out that there existed a clear parallelism in comparison with potencys of these two peptides, as we have found in passive avoidance trials on adilt rats and in the assays of facilitating PIP_2 metabolism of rat hippocampus slices that ZNC(C)PR is hundrads or thousand times effective than its parent molecule-AVP[4,5].

The potentiation of synaptic transmission by ZNC(C)PR in mossy fiber-CA3 pyramidal neuron synapse was similar to the potentiation in Schaffer collateral/Commissural fiber synapse. It is very interesting to note that as a well established concept, NMDA-receptor is not included in the transmission of mossy fiber--CA3 but it is in that of CA1--Sch/com fibers. Obviously this fact indicates that the way for ZNC(C)PR stimulating LTP may not be through NMDA-receptor. Additional evidence we have found is that: DL-APV, an NMDA-receptor antagonist, completely block the LTP following tetanic stimulation, but did not block the potentiation effect of ZNC(C)PR.

In general, our results presented here postulates that ZNC(C)PR receptor in rat brain is a new neuropeptide receptor, as it was much different to the AVP receptor V_1 or V_2 and after binding ZNC(C)PR potentiates the synaptic transmission in hippocampus long-lastingly and this potentiation is mainly due to presynaptic mechanism.

Acknowledgements

The study was supported by grants from the National Natural Science Foundation (9389007) and Chinese Academy of Sciences (87-50-01).

References

1. Lin, C., Liu, R.Y., and Du, Y.C., Peptides, 11(1990)633.
2. Du, Y.C., Gu, B.X., Tang, T., Rong, X.W., Chen, X.F. and Chen, Z.F., In Peptides: Biology and Chemistry: Proceedings of the Chinese Peptide Symposium 1990, Du, Y.C. et al.(Eds.) Science Press, Beijing, 1991, pp. 71.
3. Pearlmutter, A.F., Costantini, M. and Loeser, B., Peptides, 4(1983)335.
4. Gu, B.X. and Du, Y.C., Acta Biochem. et Biophy. Sin., 23(1991)331.

Novel oxytocin antagonists

Daniel. F. Veber[a], Mark G. Bock[a], Bradley V. Clineschmidt[b], Robert M. DiPardo[a], Jill M. Erb[a], Ben E. Evans[a], Roger M. Freidinger[a], Kevin F. Gilbert[a], Norman P. Gould[a], James B. Hoffman[a], James L. Leighton[a], George F. Lundell[a], Debra S. Perlow[a], Douglas J. Pettibone[b], Roger D. Tung[a], Willie L. Whitter[a] and Peter D. Williams[a]

[a]Departments of Medicinal Chemistry, [b]New Lead Pharmacology,
Merck Research Laboratories, West Point, PA 19486, U.S.A.

Introduction

A major objective in the field of medicinal chemistry has become peptide mimetic design and discovery directed toward both agonist and antagonist properties. In general, design has not yielded substances with the desired long duration and oral availability. The search for ligands for peptide receptors has increasingly turned to receptor based screening. Agonist and antagonist ligands for peptide receptors have been discovered from both natural product sources and chemical sample collections in which the structural relationship to the peptide is difficult to discern. The term "limetic" or ligand mimetic[1] has been proposed to describe these substances which often show the desired oral availability and long duration that has been sought in designed mimetics.

Results and Discussion

We have been searching for antagonists of oxytocin for prevention of preterm labor. Cyclic hexapeptide antagonists were discovered from microbial fermentation sources[2]. These supplied a new view of the nature of antagonist binding to oxytocin receptors. Structure-activity studies of L-365,209 have revealed the importance of the D-phenylalanine and isoleucine for receptor binding. This dipeptide unit may relate to the D-Tyr(OEt)-Ile present in previously studied antagonists such as atosiban[3]. The high potency of a thioamide analog involving the bond between these two amino acids points to another important interaction which led to a thirty-fold enhancement of antagonist potency[4]. In addition, a hydrogen bond acceptor group appears important in the D-dehydropiperazic acid(Δ-Piz) residue adjacent to isoleucine[5,6]. This residue may relate to the glutamine of oxytocin or threonine of atosiban.

Basic amino acids in place of the second dehydropiperazic acid are also seen to lead to a reduction in the selectivity of these peptides for the oxytocin receptor compared to the closely related kidney vasopressin (V_2) receptors. Thus, cyclo(Pro-D-2-Nal-Ile-D-Pip-Lys-N-Me-D-Phe) shows an IC_{50} of 13nM for the oxytocin receptor and 110 nM for the vasopressin (V_2) receptor[5]. It appears that the side chain of this residue might be spacially related to the important basic side chain in the C-terminal tail of the better known nonapeptide vasopressin antagonists. It is interesting that basic groups at this

L-365,209

same position, but differently constrained, result in peptides that also show bradykinin agonist activity[7].

The knowledge of structure-activity relationships in this series allowed the design of highly potent, water soluble analogs such as cyclo(Pro-D-2-Nal-Ile-D-Pip-Pip-D-His)[6]. In spite of being potent antagonists, none of the compounds of this class have shown significant oral availability, a property needed in any drug candidate planned for out-patient use.

The fact that most of the important receptor binding elements are focused in only 3 of the 6 residues of these cyclic peptides suggests that a non-peptide scaffold might be found to present them at the receptor. Unfortunately, approaches to the design of alternate scaffolds has not yet yielded general solutions. Through a process of receptor based screening of the Merck sample collection we have found a compound (I) that is of the limetic class[8,9]. Structure-activity studies have improved on the activity of this lead to produce compound II, which inhibits oxytocin binding to its receptor with a K_i

of 38 μM [8,9]. This compound shows good duration of action in both the rat and rhesus monkey. Antagonism of oxytocin is also seen after oral administration in the pregnant rhesus and intraduodenal administration in the rat[9].

It will be important for us to understand how this type of structure mimics the peptide so that design will become possible in the future. For this example we have a particularly constrained cyclic peptide and a limetic with only few degrees of rotational freedom. For the present, the nature of the structural relationship between this limetic and the peptide analogs must remain a matter of speculation that will require further experiments to interrelate the bioactive forms of these two molecules.

References

1. Veber, D.E., Peptides, In Smith, J.A. and Rivier, J.E. (Eds.) Peptides: Chemistry and Biology (Proceedings of the 12th American Peptide Symposium), ESCOM, Leiden, 1992, p. 3.
2. Pettibone, D.J., Clineschmidt, B.V., Anderson, P.S., Freidinger, R.M., Lundell, G.F., Koupal, L.R., Schwartz, C.D., Williams, J.M., Goetz, M.A., Hensens, O.D., Liesch, J.M. and Springer, J.P., Endocrinology, 225(1989)217.
3. Melin, P., Trojnar, J., Johansson, B., Vilhardt, H., Akerlund, M.J., Endocrinol., 111(1986)125.
4. Bock, M.G., DiPardo, R.M., Williams, P.D., Pettibone, D.J., Clineschmidt, B.V., Ball, R.G., Veber, D.F., Freidinger, R.M., J. Med. Chem., 33(1990)2321.
5. Freidinger, R.M., Williams, P.D., Tung, R.D., Bock, M.G., Pettibone, D.J., Clineschmidt, B.V., DiPardo, R.M., Erb, J.M., Garsky, V.M., Gould, N.P., Kaufman, M.J., Lindell, G.F., Perlow, D.S., Whitter, W.L. and Veber, D.F., J. Med. Chem., 33(1990)1843.
6. Williams, P.D., Bock, M.G. Tung, R.D., Garsky, V.M., Perlow, D.S., Erb, J.M., Lundell, G.F., Gould, N.P., Whitter, W.H., Kaufman, M.J., Clineschmidt, B.V., Pettibone, D.J., Freidinger, R.M. and Veber D.F., J. Med. Chem., 35(1992)3905.
7. Pettibone, D.J., Clineschmidt, B.V., Lis, E.V., Ransom, R.W., Totaro, J.A., Young, G.S., Bock, M.G., Freidinger, R.M., Veber, D.F. and Williams, P.D., Eur. J. Pharmacol., 196(1991)233.
8. Evans, B.E., Leighton, J.L., Rittle, K.E., Gibert, K.F., Lundell, G.F., Gould, N.P., Hobbs, D.W., DiPardo, R.M., Veber, D.F., Pettibone, D.J., Clineschmidt, B.V., Anderson, P.S. and Freidinger, R.M., J. Med. Chem., 35(1992)3919.
9. Pettibone, D.J., Clineschmidt, B.V., Kishel, M.T., Lis, E.V., Reiss, D.R., Woyden, C.J., Evans, B.E., Freidinger, R.M., Veber, D.F., Cook, M.J., Haluska, G.J., Novy, M.J. and Lowensohn, R.I., J. Pharmacol. Exp. Ther.(in press).

Mechanisms underlying the anti-opioid effect of cholesystokinin octapeptide (CCK-8) in central nervous system

Ji-Sheng Han[a], Xiao-jin Wang[a], Li-juan Zhang[a], Jun-feng Wang[a] and Ming-feng Ren[b]

[a]Department of Physiology, Beijing Medical University, Beijing 100083, China
[b]Dept of Pharmacology, Chinese Academy of Medical Sciences, Beijing 100005, China

Introduction

Cholecystokinin octapeptide (CCK-8) has been known to play a dual role in physiology, a hormone in the periphery as well as a neurotransmitter or neuromodulator in the central nervous system. It is widely and abundantly distributed in numerous brain regions and in the spinal cord. An array of physiological functions have been attributed to central CCK, among which the antagonistic effect on opioid analgesia has been clearly defined by behavioral and electrophysiological studies[1]. However, the molecular mechanisms whereby CCK-8 exerted its anti-opioid effect remain obscure. The aim of this study was to analyze the mechanisms underlying the anti-opioid effect of CCK-8 at different levels of the signal transduction pathways, including the membrane receptors, G proteins, cAMP, inositol phosphates, phosphokinase C(PKC), intracellular calcium,etc.

Results and Discussion

CCK-8 antagonizes opioid analgesia mediated by μ- and κ- but not δ-receptors in the spinal cord of the rat
Previous studies performed in our laboratory revealed that centrally administered CCK-8 antagonized the analgesic effect produced by morphine or by endogenously released opioids during the period of electroacupuncture(EA) stimulation[2,3]. Since opioid receptors were known to be divided into three different types, the μ-, δ- and κ-receptor, it would be interesting to characterize which of the 3 opioid receptors is most susceptible to CCK antagonism. Experiments were performed in the rat using tail flick latency (TFL) as the endpoint of nociception. The analgesia produced by intrathecal (ith) injection of PL017, a specific μ-opioid agonist, could be markedly antagonized by CCK-8 at a dose as small as 4 ng. Similar effect was observed when 66A-078, a specific κ-opioid agonist was used instead of the μ-agonist. In contrast, analgesia produced by intrathecal injection of the δ-opioid agonist DPDPE could not be blocked by CCK-8 even at a dose as high as 40 ng. Since the effect of CCK-8 could be totally reversed by the CCK receptor antagonist proglumide, this effect of CCK-8 is most likely mediated by CCK receptors rather than acting directly on opioid receptor.

Modification by CCK-8 of the binding of μ-, δ- and κ-opioid receptors

Previous study has shown that CCK-8 suppressed the binding of opioid receptor to the universal opioid agonist [^{3}H] etorphine (Wang et al. Life Sci 1989; 45:117). In the present study, highly selective tritium labelled agonists for μ- (DAGO), δ- (DPDPE) and κ-(U69,593) opioid receptor respectively were used to clarify which type(s) of opioid receptor in rat brain membrane is suppressed by CCK-8. In the competition experiments, CCK-8 suppressed the binding of [^{3}H] DAGO and [^{3}H]U69,593 but not that of [^{3}H]DPDPE to the respective opioid receptor. This effect was blocked by the CCK antagonist proglumide at 1 μM. In the saturation experiments, CCK-8 at concentrations of 0.1 to 1.0 nM decreased the Bmax of [^{3}H]DAGO binding sites without affecting the Kd; on the other hand, CCK-8 increased the Kd of [^{3}H]U69,593 binding without changing the Bmax. The results suggest that CCK-8 inhibits the binding of μ- and κ-, but not δ-opioid receptors via the activation of CCK receptors, which is in line with the findings that CCK-8 suppressed the analgesia induced by opioid agonists acting on μ- and κ-receptors, but not that on δ-receptor.

Evidence supporting a direct interaction between opioid receptor and CCK receptor

Two possibilities exist to explain the phenomena of CCK suppression of the binding of opioid receptor with its agonist: (a) receptor-receptor interaction, (b) interaction via post-receptor intracellular events. To rule out the second possibility, we used the opioid antagonist [^{3}H] naloxone which has no intrinsic activity, therefore would not induce post-receptor events. Radioreceptor assay with [^{3}H]naloxone in rat brain homogenate revealed two populations of [^{3}H]naloxone binding sites, a high affinity site and a low affinity site. CCK-8 at 10, 100 and 1000 nM dose-dependently increased the Kd and decreased the Bmax of high affinity site, with a concomitant slight decrease in Bmax of the low affinity site. The results suggest a direct receptor-receptor interaction, leading to a disabled opioid binding capability.

Uncoupling of opioid receptors from their relevant G proteins

The effect of opioids are known to be mediated by the G protein (Gi). Uncoupling of the opioid receptor with Gi would inevitably result in a decrease in post-receptor activities. Saturation experiments were performed with the universal opioid agonist [^{3}H]etorphine ([^{3}H]Et) in rat brain synaptic membranes. Scatchard analysis revealed that CCK-8 at 10 nM increased the Kd and decreased the Bmax of the high affinity Et binding sites. GTPrS, the hydrolysis-resistant GTP analogue capable of dissociating the G protein from its receptor, *per se* increased the Kd of opioid binding without affecting the Bmax. In the presence of both CCK-8 and GTPrS, the binding parameters of [^{3}H]Et were essentially the same as were in the condition of CCK-8 alone, suggesting that CCK-8 and GTPrS are acting through one and the same mechanism. GTPrS at 1-100 μM dose-dependently decreased the [^{3}H]Et binding. In the presence of CCK-8, however, the slope of the dose-response curve for GTPrS to suppress opioid agonist binding became flat. The two curves plateaued at the same level. The results suggesnt that CCK-8 may uncouple opioid receptors from their relevant G protein, resulting in a decreased capability of opioid binding and blockade of the post-receptor signal transduction.

The influence of CCK-8 and three opioid agonists on spinal cAMP content

It is well known that cAMP is involved in mediating opioid effects. It is therefore relevant to evaluate whether the effect of opioids on CNS cAMP content is affected by CCK-8. The spinal cAMP content was measured with ^{125}I-cAMP radioimmunoassay. No significant changes in spinal cAMP content were found 10 min after intrathecal injection of 5-40 ng CCK-8. In contrast, 25 ng PL017 (μ-agonist) and 20 μg DPDPE (δ-agonist) induced a remarkable decrease in spinal cAMP content. Injection of 300 ng of 66A-078, the κ-agonist, produced a slight decrease in cAMP copntent. The decrease in spinal cAMP content induced by the 3 opioid agonists could not be reversed by CCK-8 (at a dose as high as 20 ng) administered 10 min after the opioids. These results seem to indicate that central cAMP is not involved in the mechanisms of the anti-opioid effect of CCK-8.

Activation by CCK-8 of the phosphoinositide (PI) signalling system in CNS

Neonatal-rat brain (minus cerebella) cells were dispersed with trypsin. The intact cells were incubated with [^{3}H]inositol for 3 h, the amount of radio-labelling being 1.3%. The labelled cells were then stimulated with agonists in the presence of 10 mM LiCl. Under 1 mM carbachol stimulation, there was an increase in IP_3 in brain cells 10 min after stimulation, peaked at 30 min, and then decreased gradually, approaching the baseline at 45 min. A very similar time course was obtained for 10 nM CCK-8 in stimulating PI turnover. IP_3 content tended to increase from 5 min to 30 min after stimulation, peaked at 30 min, and decreased thereafter. The dose-response curve for incubated brain cells revealed that IP_3 formation increased when the concentration of CCK-8 was increased from 0.1 to 10 nM. A further increase of the CCK-8 concentration to 100-1000 nM, however, resulted in a gradual decrease in IP_3 formation. The results provide a direct evidence for CCK-8 to stimulate PI turnover, and to increase IP_3 content in dissociated rat brain cells.

Effect of opioid ligands and CCK-8 on the intracellular free calcium concentration in dissociated rat brain cells

In enzymatically dissociated brain cells prepared from neonatal rats, KCl at concentration of 25 and 50 mM produced a significant increase in intracellular free calcium concentration $[Ca^{2+}]i$, and this increase could be prevented by verapamil or nifedipine (10 μM) known to block voltage-sensitive calcium channel.

Opioid receptor agonist ohmefentanyl (OMF, μ-), DPDPE (δ-) and 66A-078 (κ-) at concentration of 10 nM to 1 μM produced a marked suppression of the Ca^{2+} influx induced by high K^+ depolarization without changing the $[Ca^{2+}]i$ level in resting cells.

Specific opioid receptor antagonist β-FNA (μ), ICI 174964 (δ) and nor-BNI(κ) exert no significant influence on the resting $[Ca^{2+}]i$. However, the suppressive effect of OMF, DPDPE and 66A-078 on high K^+ depolarization-induced increase in $[Ca^{2+}]i$ was markedly reversed by their respective antagonist β-FNA, ICI174864 and nor-BNI.

CCK-8 at concentrations of 0.3, 3, and 30 nM dose-dependently mobilized Ca^{2+} from intracellular stores. While CCK-8 30 nM did not affect significantly the increase of $[Ca^{2+}]i$ following high K^+, it did reverse the suppression of high K^+ induced increase $[Ca^{2+}]i$ caused by μ-agonist OMF and κ-agonist 66A-078, but not that by δ-agonist

DPDPE. This effect of CCK-8 can be observed in calcium-free medium. It is thus obvious that while opioid ligands suppress $[Ca^{2+}]i$ by blocking voltage-operated Ca^{2+} influx, the anti-opioid effect of CCK-8 seems to be operated via mobilization of Ca^{2+} from intracellular stores. That CCK-8 does not antagonize the $[Ca^{2+}]i$ lowering effect of δ-agonist seems to fit in with the *in vivo* finding that δ agonist induced analgesia was not antagonized by CCK-8, the underlying mechanisms of which deserve further investigation.

PKC antagonism of opioid analgesia

Hydrolysis of PI produces both IP_3 and diacyl glycerol (DAG). Since phorbol ester TPA has been known to mimic DAG in stimulating PKC, TPA was used to estimate the effect of an increase in DAG on opioid analgesia. TPA injected intrathecally in 5 cumulative doses (6.25-100 ng at 10 min intervals) produced no significant changes in the TFL. However, pretreatment wiht TPA markedly attenuated the analgesia elicited by 10 ng of PL017. Analogesia induced by intrathecal injection of 20 μg of DPDPE was also antagonized by TPA in a dose-dependent manner between 12.5 and 50 ng. Higher dose of TPA (50 and 100 ng) were needed to suppress the analgesia elicited by κ-agonsit 66A-078 (300 ng). The mechanisms whereby TPA suppresses opioid analgesia are not clear. It may be related to an activation of PKC, which increases the calcium conductance (resulting in a rise in $[Ca^{2+}]i$ level) and a decrease in potassium conductance (leading to a tendency of depolarization), both effects are against that of opioids.

Conclusions

1. CCK-8 exhibits a potent antagonistic effect on opioid analgesia, especially on the analgesia induced by μ- and κ-opioid agonists.

2. CCK-8 decreases the Bmax of μ-receptor and lowers the affinity of κ-opioid receptor, without affecting the δ-receptor.

3. The suppressive effect of CCK-8 on opioid binding may occur via interaction between opioid receptor and CCK receptor located on one and the same neuron.

4. CCK-8 may uncouple the opioid receptor with its relevant G protein, resulting in a decrease of the binding capability of the receptor and a blockade of the transmembrane signal transduction.

5. CCK-8 stimulates the phosphatidylinositide signal system, which raises intracellular level of IP_3 thereby releases free calcium from intracellular calcium storage and increases $[Ca^{2+}]i$ to counteract the effect of opioids which decreases the $[Ca^{2+}]i$.

6. Along with an increase in IP_3, CCK-8 may also cause an increase in PKC which works in the same direction of increasing intracellulalr free calcium level.

7. cAMP signal system seems not to be invloved in the anti-opioid effect of CCK-8.

Acknowledgements

This work was supported by the National Natural Science Foundation of China, No. 938900705 and 3880312, and a grant from NIDA,USA, DA03983.

References

1.   Han, J.S., In: Multiple Cholesystokinin Receptors in CNS, Dourish et al. Eds. Oxford, 1992, pp.480-502.
2.   Han, J.S., Ding, X.Z. and Fan, S.G., Neuropeptides 5(1985)339.
3.   Han, J.S., Ding, X.Z. and Fan, S.G., Pain, 27(1986)101.

B-50/GAP-43 mRNA expression in adult rat hippocampus

Tong Tang[a], Loes H. Schrama[b] and Willem Hendrik Gispen[b]

[a]Shanghai Institute of Biochemistry, Shanghai, 200031, China
[b]Rudolf Magnus Institute and Institute of Molecular Biology, Utrecht University,
3508 TB Utrecht, The Netherlands

Introduction

The neuronal growth-associated protein B-50/GAP-43 is a nervous system specific phosphoprotein[1]. It correlates with axonal outgrowth in nerve development and regeneration, and also plays important roles in phosphoinositide metabolism and neuronal plasticity[2]. Many efforts have been made to elucidate the localization of B-50/GAP-43. It has been shown that its expression is heterogeneous in rat brain. Furthermore, this expression is demonstrated to be cell-selective in some regions , for examle, in olfactory system[3].

In this study, we analyzed the subregional expression of B-50/GAP-43 mRNA in hippocampus by in situ hybridization histochemistry. Furthermore, the cell-selectivity of B-50/GAP-43 mRNA expression was investigated in this area.

In situ hybridization histochemistry was performed according to Verhaagen et al. [3] with some modifications. Briefly, 12 μm thick rat brain cryo-sections were post-fixed in 2% paraformaldehyde. After washed with triethanolamine/anhydride acetic acid and Tris/glycine, the sections were hybridized with ^{35}S labeled B-50/GAP-43 antisense probe at 62°C. The non-specific adsorption was washed out by 2X SSC/80% formamide and 0.1X SSC successively. The sections were then exposed to Hyperfilm β-max or NTB-2 nuclear emulsion, Cellular and quantitative analysis of B-50/GAP-43 mRNA expression in hippocampus were both performed under the microscope.

Results and Discussion

The expression of B-50/GAP-43 mRNA, like the expression of B-50/GAP-43 protein, was proved to be heterogeneous in adult rat brain(data not shown). The heterogeneous expression was also found in hippocampus. The signal density over CA4 subregion was the highest, while that over DG subregion being the lowest(Table 1). Due to the pre-synaptic localization of B-50/GAP-43 in mature neuron, we predicted an abundant B-50/GAP-43 in CAI and DG areas where the fibers efferented from CA3 and CA4 area terminate, respectively. Actually, similar results were reported previously[4]. Our study showed the B-50/GAP-43 expression on transcriptional level and the results are consistent with those previous reported.

Because of the pre-synaptic localization of B-50/GAP-43 in neuronal membrane, it is

Table 1 *Quantification of B-50/GAP-43 mRNA expression in hippocampus*

Subregions	Mean±SEM (n=12)
DG	31.9±2.3%
CA4	100.0±5.1%
CA3	85.4±5.6%
CA2	44.6±3.6%
CA1	61.1±4.5%

The densities of B-50/GAP-43 in situ hybridization signal over hippocampal subregions in adult rat were recorded and analyzed. Values are expressed as percentage of that over CA4 area.

very difficult to detect the cellselectivity of B-50/GAP-43 expression by immunohistochemistry, By in situ hybridization, a newly developed method to detect mRNA experssion on regional and cellular level, we could easily determine the cell-selectivity of B-50/GAP-43, We found that the expression of B-50/GAP-43 appeared to be restricted to pyramidal cells and big cells within the hilar cells(Fig.1A, B). The granule cells displayed no B-50/GAP-43 expression (Fig. 1C). The physiological roles of this differential expression are currently under study.

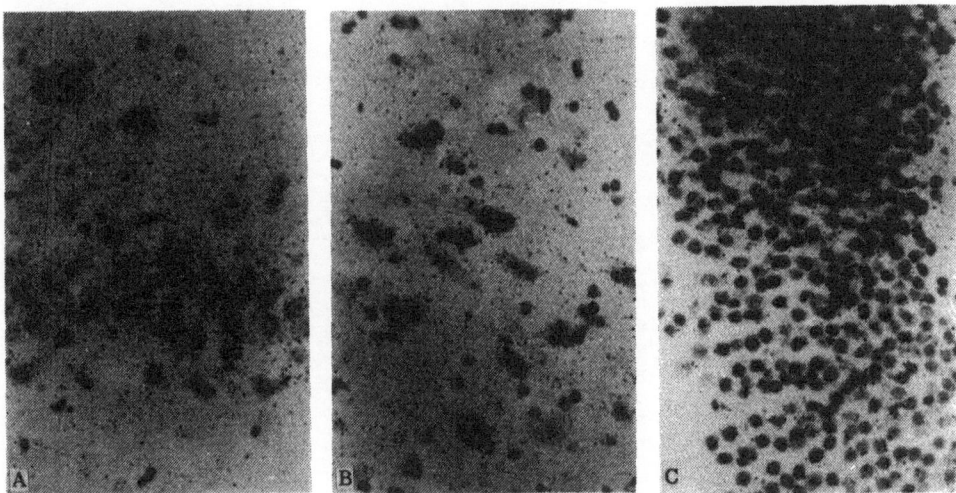

Fig. 1. Cell-selectivity of B-50/GAP-43 mRNA expression in adult rat hippocampus. Pyramidal cells(A) and bid cells(B) in hilus expressed large amount of B-50/GAP-43 mRNA, but granule cells expressed no significant B-50/GAP-43 mRNA(C).

In conclusion, B-50/GAP-43 mRNA expression in hippocampus was heterogeneous and also cell-selective. Pyramidal cells in CA3 and CA4 areas expressed the highest level of B-50/GAP-43 mRNA, whereas those in CA1 and CA2 areas espressed much less. The dentate gyrus granule cells expressed no B-50/GAP-43 mRNA. The big but not other hilar cells in hilus expressed B-50/GAP-43 mRNA.

References

1. Kristjansson, G.I., Zwiers, H., Oestreicher, A.B., and Gispen, W.H., J.Neurochem., 39(1982)371.
2. Skene, J.H.P., Annu. Rev. Neurosci., 12(1989)127.
3. Verhaagen, J., Oestreicher, A.B., Gispen, W.H., and Margolis, F.L., J. Neurosci., 9(1989)683.
4. Oestreicher, A.B., and Gispen, W.H., Brain Res, 375(1986) 267.

Interaction of Met-enkephalin-Arg^6-Phe^7 with phosphatidylserine

Maria D'Alagni^{a,b}, Maurizio Delfini^b, Maurizio Paci^c and L. Giorgio Roda^d

^{a}*Centro di Studio per la Chimica dei Recettori e delle Molecole Biologicamente Attive, C.N.R., Roma*
^{b}*Dipartimento di Chimica, Università di Roma "La Sapienza", Roma*
^{c}*Dipartimento di Scienze e Tecnologie Chimiche, Università di Roma "Tor Vergata"*
^{d}*Dipartimento di Medicina Sperimentale, Università di Roma "Tor Vergata", Roma*

Opioid peptides bind proteic receptors deeply embedded in the plasma membrane. In addition, they directly bind phospholipids, specifically phosphatidylserine (p) [1]. One of them, Met-enkephalin-Arg^6-Phe^7 (MenkAP), was studied with special regard to the conformation it assumed in water solution in the absence and in the presence of P. Circular dichroism (CD) spectra of MenkAP, with and without P, were recorded at different pH values (Fig.1). At pH 6.8, its CD profile is characterized by a positive dichroic band located at 215 nm. In the presence of P this band shifts to 219 nm, while molar ellipticity is significantly increased. At pH7.1, P increases the measured ellipticity, even though the ellipticity measured with and without P is considerably decreased compared with that measured at pH 6.8. A further decrease of molar ellipticities is found at pH 7.8. However, at this pH CD spectra of the system MenkAP-P are characterized by molar ellipticity values lower than those of MenkAP.

Nuclear magnetic resonance (NMR) spectra were recorded at pH 7.8. Spin-lattice relaxation times measured in the

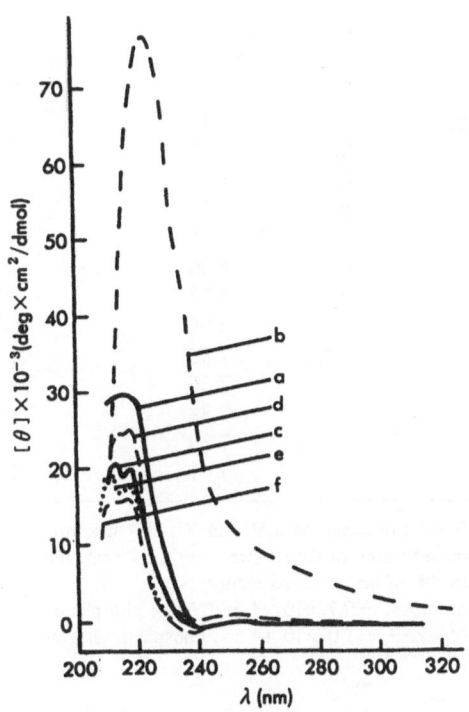

Fig. 1.   CD spectra of MenkAP (4.2×10^{-4}M) in the absence and in the presence of P. T=38.0°C.
Curve a: MenkAP in 2.0×10^{-3} M Tris·HCl, pH=6.8; curve b: MenkAP plus P, buffer as in a; curve c: MenkAP in 2.0×10^{-3} M Tris·HCl, pH=7.1; curve d: MenkAP plus P, buffer as in c; curve e: MenkAP in 1.0×10^{-3} M phosphate buffer pH=7.8; curve f: MenkAP plus P buffer as in e.

Table 1 *Proton chemical shifts(δ) and spin-lattice relaxation times (T₁)ᵃ*

Residue/Protons		MenkAPᵇ		MenkAP-Pᶜ	
		δ	T₁	δ	T₁
Tyr¹	α	3.88	1.539	3.90	0.411
	β₁	2.99	0.392	3.05	0.300
	β₂	2.98	0.356	"	"
	δ	7.25	1.338	7.25	0.632
	ε	6.96	2.285	6.97	0.834
Gly²,Gly³	α	3.97,3.96	0.466,0.416	3.96,3.95	0.378,0.398
Phe⁴	α	---	---	---	---
	β₁	3.16	0.250	---	---
	β₂	3.11	0.292	---	---
Met⁵	α	4.48	1.159	4.48	0.688
	β	2.03	---	2.03	---
	γ	2.52	0.424	2.52	0.378
	δ	2.17	2.087	2.17	1.187
Arg⁶	α	4.38	1.042	4.38	0.612
	β₁	1.88	0.311	1.88	0.125
	β₂	1.76	0.316	1.77	0.105
	γ	1.66	0.382	1.66	0.276
	δ	3.22	0.452	3.27	0.356
Phe⁷	α	4.56	1.228	4.59	---
	β₁	3.27	0.398	---	---
	β₂	2.96	0.364	---	---
Phe⁷,⁴	δ,ε,ζ	7.30	1.288	7.36	0.833
		7.36	1.297	7.38	0.861
		7.38	1.640	7.41	0.983
		7.39	1.530	7.46	0.960
		7.41	1.457	7.48	0.918

ᵃ, Results obtained on a Varian XL-300 spectrometer; δ in ppm referred to HDO placed at 4.75 ppm after standardization of the spectrometer to external tetramethylsilane; T, in s. The T₁ standard deviations are less than 5% of the observed values;
ᵇ, MenkAP, $c=9.72\times10^{-4}$M, in 0.01 M phosphate buffer pH=7.74 at 38.0±0.1°C;
ᶜ, MenkAP, $c=1.07\times10^{-3}$M and phosphatidylserine, $c=0.321\times10^{-3}$M in 0.01 M phosphate buffer pH=7.83, at 38.0±0.1°C.

presence of P with respect to those measured in its absence show a "multi-point" interaction. The presence of lipid induces a broadening of all proton resonances, more evident for all β protons and for the H-α of Phe⁷. As reported in Table 1, P also reduces relaxation times. This effect is uneven and particularly evident in the case of Tyr residue protons, Met and Arg H-α and of the δ,ε and ζ protons of both Phe residues.

Peptide conformation has been analyzed by NOESY. NOEs have been observed between the Tyr H-α and Arg β-protons and between the β-protons of Tyr and those of Phe⁷ (Fig.2). These NOEs could be attributed to a folded conformation, probably

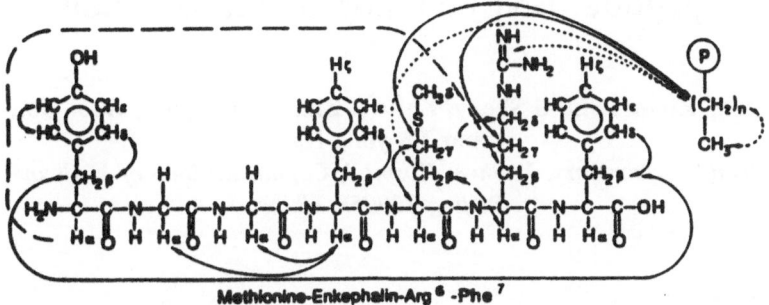

Methionine-Enkephalin-Arg[6]-Phe[7]

Fig. 2. Observed NOEs of MenkAP in the absence and in the presence of P.Dashed line indicates a weak NOE, continuous lines indicate medium NOEs and dotted lines indicate strong NOEs.

stabilized by hydrogen bonds. NOEs also indicate that the peptide-lipid interactions leave the conformation of the peptide chain substantially unchanged. In fact, the same NOEs of MenkAP without P have been recorded. Other NOEs (see in Fig. 2), correspond to the interactions of β and γ protons of Met and γ protons of Arg with those of the P hydrocarbon chains.

Reference

1.   Roy, S., Zhu, Y.X., Lee, N.M. and Loh, H.H., Biochem. and Biophys. Res. Com., 150(1988)237.

Effects of central and peripheral brain natriuretic peptide (BNP) on aldosterone secretion

Wan-jiang Zhang, Shu-li Sheng, Zhi-wei Zhao, Chun-yuan Guo, Hou-xi Ai and Jin-yong Han

Peptide Laboratory, Xuanwu Hospital, Capital Institute of Medicine, Beijing 100053, China

Introduction

BNP plays an important role in the regulation of aldostreone (Ald) secretion, but no one has reported that cerebral BNP and peripheral BNP have any difference in their regulation action. In order to answer this question, we have compared the effects of BNP administrated by icv or BNP given by iv alone or associated with angiotensin II(AT-II), adrenocorticotropin (ACTH) and arginine vasopressin (AVP) on the regulation of Ald secretion.

Results and Discussion

All hormone levels in blood were determined by RIA. Our results showed that:

1. Intravenous (iv) injection of BNP (5 µg/3 ml/h) depressed significantly the concentration of AT-II ($2.10\pm1.05\times10^3$ ng/L, n=9, in NS group; $0.98\pm0.58\times10^3$ ng/L, n=6, in BNP group, p<0.05) and Ald ($5.37\pm0.83\times10^{-6}$ mM, n=9, in NS group; $4.21\pm1.19\times10^{-6}$ mM, n=6, in BNP group, P<0.05). Intracerebroventricu-lar (icv) injection of BNP (2 µg in 10 µl NS) did not influence plasma level of AT-II in rats ($0.39\pm0.18\times10^3$ ng/L, n=9, in NS group; $0.33\pm0.25\times10^3$ ng/L, n=6, in BNP group, P>0.05), although plasma level of Ald was decreased ($4.27\pm1.08\times10^{-6}$ mM, n=9, in NS group; $2.08\pm0.91\times10^{-6}$ mM, n=6, in BNP group, P<0.01);

2. AT-II, ACTH and AVP given by iv (5 µg/3 ml/h) could increase the concentration of plasma Ald. BNP (5 µg/3 ml/h) inhibited the stimulation of ACTH and AVP on Ald secretion, although BNP effect for AT-II had not been found (Fig.1).

3. BNP (2µg in 10µl NS) administered by icv depressed the concentration of plasma Ald. Icv injection of AVP (2µg in 10µl NS) could rise plasma concentration of Ald in rats, but ACTH (2µg in 10µl NS) and AT-II (2µg in 10µl NS) individually with BNP (2µg in 10µl NS) could induce paradoxically stimulation on plasma Ald secretion (Fig.2).

4. Icv injection of BNP (2µg in 10µl NS) did not influence Ald secretion stimulated by iv infusion of AT-II, ACTH or AVP (5µg/3ml/h).

From above results, one can see that there are certain differences between central BNP and peripheral BNP on the regulation of Ald secretion. Cerebral ACTH and AT-II can induce the increase of plasma Ald only in cooperation with BNP.

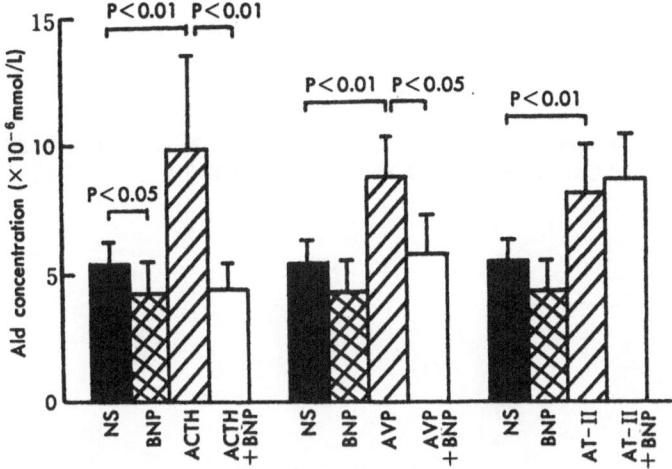

Fig. 1. *Effects of peripheral BNP on aldosterone secretion. Wistar rats were infused intravenously with NS(n=8), BNP(n=7), ACTH(n=7) or ACTH+BNP(n=7) respectively. Each bar represents the mean ±SD.*

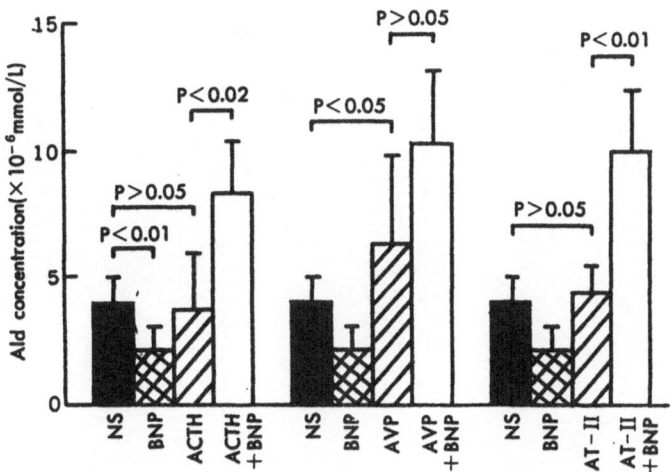

Fig. 2. *Effects of central BNP on aldosterone secretion. Wistar rats were injected intracerebroventricularly with NS(n=9), BNP(n=7), ACTH(n=7) or ACTH+BNP(n=7) respectively. Each bar represents the mean ±SD.*

Acknowledgement

This work was supported by National Natural Scientific Grants.

References

1. Anderson, J.V., Struthers, A.D., Payne, N.N., Slater, J.D.H. and Bloom, S.R., Clin. Sci., 70(1989)507.
2. Brands, M.W. and Freeman, R.H., Am. J. Physiol., 254(1988)R1011.

Synthesis of human growth hormone releasing factor (GHRF) analogue with high activity

Zhi-yong Tao, Jun Xin, Zhi-juan Ji, Yu-hong Guo, Guang-xing Wang
and Shu-li Sheng

*Department of Peptide and Endocrinology Research, Beijing Geriatric
Clinical & Research Center, Capital Institute of Medicine, Beijing 100053, China*

Introduction

A new human GHRF (1-29) analogue with high activity was reported by S.J. Hocart in 1991[1]. It had more than 650-fold activity than human GHRF (1-29) *in vitro* in rat. Its structure was human GHRF (1-29) with Tyr^1 replaced by His, L-Ala^2 by D-Ala, Asn^8, Ser^9 and Gly^{15} by Ala. In order to evaluated its use as a potential therapeutical drug in the clinic on the treatment of dwarfism, it was synthesized by manual stepwise solide-phase method with t-Boc strategy.

Human GHRF (1-29) analogue:

His-D-AlaAspAlaIlePheThrAlaAlaTyrArgLysValLeuAlaGlnLeuSerAlaArgLysLeuLeuGln
AspIleMetSerArg-NH_2

Results and Discussion

The synthesis was executed on BHA resin with styrene-1%-divinylbenzene. N^{α}-protection for all amino acids was by the t-Boc group and side chain blocking groups were as follows: Arg and His, Tos; Ser and Thr, Bzl; Asp, o-Bzl; Lys, Z; Tyr, 2-Br-Cbz. Deprotection of N-Boc at every step was carried out by 25% TFA-DCM (V/V) for 30 minutes at room temperature. t-Boc-His was coupled by carbodiimide method using NMP as solvent. All other residues were coupled by active ester method (preactiveated DDC-HOBT method). In all casas the 2.5 equivalents of the activated derivatives were used. The coupling procedures were monitored by Kaiser's ninhydrin test [2] and repeated 1-3 times to assure the completion of the reaction. The procedure per cycle of the synthesis was shown in Table 1.

After the completion of the synthesis, Boc group was first removed by TFA, the synthetic protected peptide-resin was treated with anhydrous hydrogen fluoride-anisole (10:1,V/V). The crude peptide was purified by reversed phase C8 HPLC [colume: AQUAPORE OCTYL 20 micron 10×100 mm; eluent: gradient from 20%-50% of slution B (0.1% TFA in acetonitrile) into solution A (0.1% TFA in water) in 50 min; flow rate: 4 ml/min; temp: 25°C; detective wavelength: 214 nm; retention time: 22.27 min]. The required fraction was collected and lyophiylized. The purified peptide was analysed by analytic HPLC. The amino acid analysis date of the purified peptide were as follow: Asp 2.20 (2), Thr 0.93 (1), Ser 1.74 (2), Glx 2.40 (2), Ala 6.33 (6), Val 1.16 (1), Met 0.75

Table 1 *The procedure per cycle of the synthesis*

No.	Reagents[a]	Times washed	Mixing time(min)
1	DCM	3	4.5
2	25% TFA- DCM-Indole[b]	1	1.5
3	25% TFA- DCM-Indole[b]	1	30
4	DCM	4-6	6-9
5	100% EtOH	1	1.5
6	5% DIEA-DCM (V/V)	2	3.0
7	DCM	3	4.5
8	Boc-amino acid active ester in DMF[c]	3-4 h	
9	DCM	2	3.0
10	DCM	2	3.0
11	monitor[2]		
12	Recouple if necessary by repeating steps	8-11	

a. The volume of each reagent used was 15 ml for 1 g BHA resin.
b. 1mg indole/ml TFA-DCM.
c. t-Boc-His(Tos) was coupled by DCC mothod using NMP as solvent.

(1), Ile 1.84 (2), Leu 3.97 (4), Tyr 0.82 (1), Phe 0.97 (1), Lys 1.66 (2), His 0.72 (1), Arg 2.81 (3).

The biological activity of this analogue on the secretion of GH in rat *in vitro* and *in vivo* is in progress.

References

1. Hocart, S.J. Murphy, W.A., and Coy, D.H., In Smith, J.A. and Rivier, J.E. (Eds.) Peptides: Chemistry and Biology (Proceedings of the 12th American Peptide Symposium), ESCOM, Leiden, 1992, p. 44.
2. Kaiser, E., Colescott, R.L., Bossinger, C.D., and Cook, P.I. Anal. Biochem. 34(1970)595.

Pyrogenic effect and muramyl peptides

Shi-yi Liu[a], Yi Zhang[a] Jie-cheng Xu[b], Min-zhu Zhang[b] and Ling-ling Cheng[b]

[a]Shanghai Institute of Physiology, Chinese Academy of Sciences,
Shanghai 200031, China
[b]Shanghai Institute of Organic Chemistry, Chinese Academy of Sciences,
Shanghai 200032, China

Introduction

Factor S isolated from sleep-deprived goats was first reported 25 years ago [1]. It has recently been known that these substances are muramyl peptides (MPs) (NAG-1, 6-anhydro-NAM-Ala-Glu-DAP-Ala) and muramyl dipeptide (MDP) (NAM-L-Ala-D-isogln) [2]. MPs are the monomeric building blocks of bacterial cell wall peptidoglycan and are powerful immunostimulants. MDP is known as the minimal chemical structure that can substitute for mycobacteria as the immunostimulatory component of Freund's complete adjuvant. In addition, both of them are thought to induce biological effects via the actions of interleukin-I (IL-1) [2]. Since MPs and MDP exert both somnogenic and immunomodulatory effects, whether immunomodulation is involved in normal sleep is an interesting question that merits wide attention. Nevertheless, they also exert significant pyrogenic effect and their somnogenic effect has been questioned recently [3], so whether MPs and MDP play a role in normal sleep still remains in debate.

Results and Discussion

Some new analogs of MDP and muramyl tripeptide (MTP) have been synthesized by us in order to explore whether some of them could exert similar somnogenic effect as MDP, but with negligible or no pyrogenic effect. For the synthesis of dipeptides or tripeptides, we use D-glutamic acid, γ-tert butyl ester as the starting material to avoid the preparation of isoglutamine by a tedious procedure in advance, as well as after the construction of dipeptides or tripeptides, the C-terminal moiety was converted to D-isoglutamine through amination. The bioassay of pyrogenic and somnogenic effects of these analogs were performed via mesodiencephalic intraventricular infusion in 32 rabbits. In order to avoid the influence of circadian rhythm etc., experiments were performed only during morning hours in a semi-soundproof chamber with 45 lux of illumination and a constant noise background of 72 db SPL. Measurements of middle back skin temperature were done by a specially-equipped multiple thermographic device model WMSY-01, which allows long-term automatic recordings. Results show that no pyrogenic effect was seen with MDP-III (N-Acetylmuramyl-γ-aminobutyryl-D-isoglutamine) when compared to MDP (increased by 1.4°C) and MTP-II (N-Acetylmuramyl-L-ananyl-D-isoglutaminyl-γ-oxalysine) (increased by 0.6°C) with the same dosage (0.3 µg, i.c.v./animal) in rabbits (Tab.1). The pyrogenicity of MDP usually

Table 1 *Effects of Muramyl Dipeptide (MDP), MDP-III and MTP-II (0.3μg/animal, i.c.v.) on middle back skin temperature in rabbits*

	MDP (M±SE)	MTP-II (M±SE)	MDP-III (M±SE)
Control	36.14±0.50 (N=8)	35.15±0.70 (N=8)	35.42±0.74 (N=8)
1-2 h	36.56±0.42** (N=8)	35.15±0.72 (N=8)	35.33±1.05 (N=8)
3-4 h	37.34±0.53** (N=8)	35.57±0.84* (N=8)	35.54±1.07 (N=8)

*P<0.05   **P<0.001

started after 50-60 min of infusion and then increased gradually to the maximun. MTP-II exhibited much less pyrogenicity than that of MDP and started after about 120 min of infusion. The sleep-inducing effect of MDPs and MTPs was less significant when compared to Asp⁵-α-DSIP or Phe⁵-DSIP [4], but MDP-III exhibited more somnogenic effect than MDP or MDP-II.

Acknowledgements

The study was supported by grants from the National Natural Science Foundation of China (9389007) and Chinese Academy of Sciences (875001).

References

1. Pappenheimer, J.R., Miller, T.B. and Goodrich, C.A., Proc. Natl. Acad. Sci. U.S.A., 58(1967)513.
2. Krueger, J.M. and Karnovsky, M.L., Ann. N.Y. Acad. Sci., 496(1987) 510.
3. Kovalzon, V., Obal, Jr.F., Sary, G., Kalikherich, V., Mikhaleva I. and Andronova,T., In "Sleep 86", Koella, W.F. et al. Eds., Gustav Fricher Verlag, Stuttgart, 1988, pp.171.
4. Liu, S.Y., Li, C.X. and Xu, J.Z., In Endogenous Sleep Substance and Sleep Regulation. Inoue,S. & Borbely, A.A. Eds., VNU, Utrecht. 1985, pp.41.

The interactions between human recombinant interleukin-2 (hrIL-2) and opioid receptors

Jian-ze Li[a], De-he Zhou[a], Xin-yuan Liu[b] and Zhi-qiang Chi[a]

[a]Shanghai Institute of Materia Medica, Chinese Academy of Sciences, Shanghai 200031, China

[b]Shanghai Institute of Biochemistry, Chinese Academy of Sciences, Shanghai 200031, China

Introduction

Interleukin-2(IL-2), also called T-cell growth factor, is a glycosylated polypeptide, which plays a critical role in regulation of the immune system. Recently, it had been reported that pretreatment of naloxone decreased or antagonized the behavioral and electrocorticogram (ECoG) sedative activity induced by injection of IL-2 and hrIL-2 into the locus coeruleus [1]. The discovery suggested that IL-2 and opioid receptors might be functionally coupled. It has also been reported that an opioid agonist 2-n-pentyloxy-2-phenyl-4-methyl-morphine (PM) inhibited the production of IL-2-like activity via opioid receptor mechanisms [2]. In our study, it was found that hrIL-2 (here used is [Ala^{125}]-IL-2, i, e, cysteine in 125 position of normal IL-2 was replaced by alanine) inhibited the specific binding of opioid radioligand to opioid receptor by competition assay in rat brain membranes. Using the bioassay, hrIL-2 antagonized the depressive effect of morphine and DAGO on the electrically evoked contraction of guinea pig ileum and the depressive effect of DPDPE on mouse vas deferens. In mouse hot plate test, hrIL-2 (icv) showed higher analgesic activity than morphine. Naloxone could partly reverse the analgesic effect of hrIL-2.

Results and Discussion

Competitive binding assay

The radioligands employed were specific mu agonist [^{3}H]ohmefentanyl, delta agonist [^{3}H]DPDPE and general opioid ligand [^{3}H]etrophine. Their concentrations selected in this assay were 0.5 nM, 3.0 nM and 0.5 nM respectively. HrIL-2 inhibited the specific binding of [^{3}H]ohmefentanyl, [^{3}H]DPDPE and [^{3}H]etrophine to opioid receptor in rat brain membranes. The inhibitory effects were dose dependent and the IC$_{50}$ values were 0.60 μM, 0.58 μM and 3.81 μM respectively. As their Ki values were compared, which were 0.19 nM, 0.36 nM and 0.99 nM respectively, the potency of inhibitory effects of hrIL-2 on the bindings of these radioligands to their receptors could be estimated: hrIL-2 inhibited the binding of [^{3}H] ohmefentanyl more potently than that of [^{3}H] DPDPE or [^{3}H] etrophine, in that order.

Bioassay

In guinea pig ileum (GPI) which is rich in mu receptor, hrIL-2 showed antagonism effect against the depressive effects of mu agonists morphine and DAGO, on the electrically evoked contraction of GPI. The Ke values were 0.10 μM and 0.09 μM respectively. The mu antagonist activity of hrIL-2 is less potent than that of naloxone, which had the Ke values of 1.8 nM and 1.9 nM respectively. In the mouse vas deferens (MVD) which is rich in delta receptor, hrIL-2 antagonized the depressive effect of DPDPE on electrically evoked contraction of MVD. The Ke value was 0.26 μM, being higher than that of naloxone (Ke=26 μM). The results showed that hrIL-2 had mu and delta antagonist activities.

Mouse hot plate test

HrIL-2 (0.08-2.0 nmol), after injection into the cerebral ventricle, induced analgesic activity with the ED_{50} value of 0.3 nmol/mouse, which is 3 times less than that of morphine. The analgesic effect appeared within 5 minutes and lasting between 45 to 120 minutes, depending on the dose. Treatment of naloxone (1 nmol, 7 minutes afterwards) completely reversed the analgesic effect of morphine and partly reversed the effect of hrIL-2. This indicated that the analgesic effect of hrIL-2 may be partly related to opioid receptor mechanisms.

Conclusion

HrIL-2 inhibited the specific bindings of mu agonist [^{3}H]ohmefentanyl, delta agonist [^{3}H]DPDPE and general opioid ligand [^{3}H]etrophine to rat brain membranes. In the bioassay, hrIL-2 showed mu and delta antagonist activity. Using mouse hot plate test, hrIL-2 had the analgesic effect.

Acknowledgement

This project is supported by the National Natural Science Foundation of China (No. 938900706).

References

1.   De Sarro, G.B., Masuda, Y., Ascioti, C.,   Audino, M.G. and   Nistico, G., Neuropharmacology, 29(1990)167.
2.   Hadjipetrou-Kourounnakis, L.,   Karagounis, E., Rekka, E. and Kourounakis, P.,   Scand. J. Immunol., 29(1989)449.

Session III
Immunopeptides

Chairs: James P. Tam
Vanderbilt University
Nashville, Tennessee, U.S.A.

and

Saburo Aimoto
Osaka University
Osaka, Japan

Design of HIV protease inhibitors based on the transition state concept

Yoshiaki Kiso

*Department of Medicinal Chemistry, Kyoto Pharmaceutical University, Yamashina-ku,
Kyoto 607, Japan*

Introduction

The human immunodeficiency virus type-1(HIV-1), the causative agent of acquired immunodeficiency syndrome (AIDS), codes for a virus-specific aspartic protease responsible for processing the gag and gag-pol polyproteins and for the proliferation of the retrovirus. The HIV-1 protease functions as a homodimer and can recognize Phe-Pro and Tyr-Pro sequences as the cleavage site, but mammalian aspartic proteases do not have such specificity. These features provided a basis for the rational design of selective HIV protease-targeted drugs for the treatment of AIDS and AIDS-related complex.

Results and Discussion

Design of Substrate-based Inhibitors

We focused on the Phe-Pro scissile site, which is a unique structure for HIV-1 protease in the design of substrate-based HIV protease inhibitors (Fig. 1 and 2).The hydroxy-methylcarbonyl (HMC) isostere was incorporated as a transition-state mimic at P1 site in a heptapeptide amide, Ser-Phe-Asn-Phe-Pro-Ile-Val-NH$_2$, similer to the TF/PR and p17/p24 sequences. In order to obtain smaller inhibitors, we deleted P$_4$ Ser and replaced P$_3$ Phe with the isosteric 3-phenylpropionic acid. Moreover, we replaced P$_3$ Phe with the

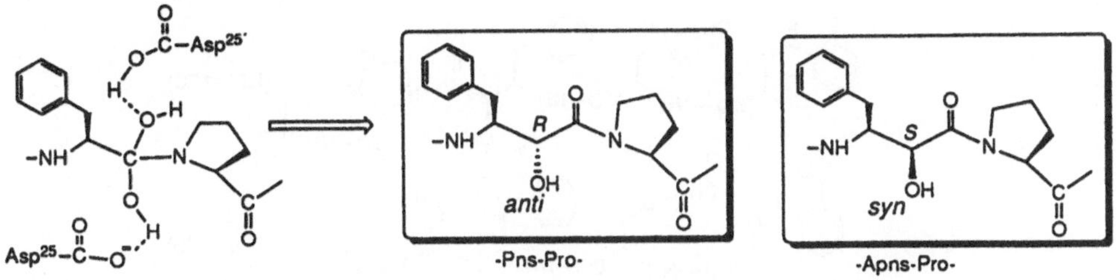

Fig. 1. The Phe-Pro transition state in HIV-1 protease and P$_I$-P$_I$' Pns-Pro and Apns-Pro with the hydroxymethylcarbonyl (HMC) isostere. Pns=phenylnorstatine=(2R,3S)-3-amino-2-hydroxy-4-phenylbutyric acid; Apns=allophenylnorstatine=(2S,3S)-3-amino-2-hydroxy-4-phenylbutyric acid.

87

isosteric benzyloxycarbonyl group, and deleted P$_3$' Val and replaced P$_2$' Ile with the isosteric t-butyl amine. We also designed symmetric-type inhibitors containing HMC structure at the symmetric axis based on the dimeric character of the HIV protease.

The protease inhibitory activities of HMC isostere-containing peptides (Table 1) were examined using the chemically synthesized[Ala^{67,95}]-HIV-1 protease [1] and the synthetic substrate, Ac-Arg-Ala-Ser-Gln-Asn-Tyr-Pro-Val-Val-NH$_2$.

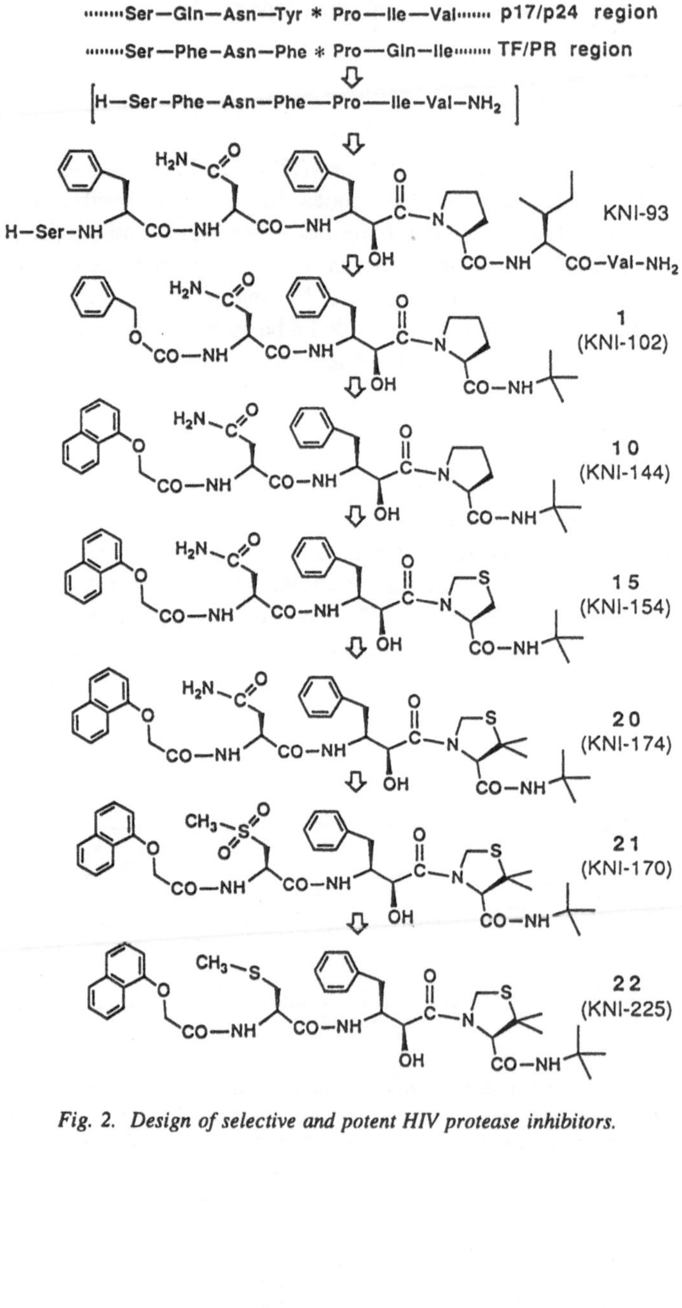

Fig. 2. Design of selective and potent HIV protease inhibitors.

Table 1 *Protease inhibitory activities of HMC compounds (IC$_{50}$, nM)*

No	P$_4$	P$_3$	P$_2$	P$_1$	P$_1$'	P$_2$'	P$_3$'	HIV Protease	Pepsin
1A(KNI-122)	Ser-	Phe-	Asn-	Pns-	Pro-	Ile-	Val-NH$_2$	100	>10,000
1S(KNI-93)	Ser-	Phe-	Asn-	Apns-	Pro-	Ile-	Val-NH$_2$	5	>10,000
2A		Pp-	Asn-	Pns-	Pro-	Ile-	Val-NH$_2$	3,000	>10,000
2S(KNI-81)		Pp-	Asn-	Apns-	Pro-	Ile-	Val-NH$_2$	468	>10,000
3A		Pp-	Ser-	Pns-	Pro-	Ile-	Val-NH$_2$	>10,000	N.D.
3S		Pp-	Ser-	Apns-	Pro-	Ile-	Val-NH$_2$	1,594	N.D.
4A		Pa-	Ser-	Pns-	Pro-	Ile-	Val-NH$_2$	>10,000	N.D.
4S		Pa-	Ser-	Apns-	Pro-	Ile-	Val-NH$_2$	5,041	N.D.
5A		Pp-	Asn-	Chns-	Pro-	Ile-	Val-NH$_2$	>10,000	N.D.
5S		Pp-	Asn-	Achns-	Pro-	Ile-	Val-NH$_2$	1,999	N.D.
6A		Z-	Asn-	Pns-	Pro-NHBu^t			>10,000	N.D.
6S(KNI-102)		Z-	Asn-	Apns-	Pro-NHBu^t			89	>100,000
7A		Val-	Val-	Pns-	Phe-	Val-	Val-NH$_2$	350	4,000
7S		Val-	Val-	Apns-	Phe-	Val-	Val-NH$_2$	2,600	>10,000

Pp= 3-phenylpropionyl; Pa= phenylacetyl; Achns= allocyclohexylnorstatine= (2S,3S)-3-amino-4-cyclohexyl-2-hydroxybutyric acid; Bu^t= t-butyl; N.D.=not determined.

Unexpectedly, the (2S)-hydroxymethylcarbonyl (HMC) inhibitor (KNI-93; **1S**: *syn* diastereomer) containing allophenylnorstatine (Apns) was more active against HIV-1 protease than the *anti* diastereomer (KNI-122, **1A**) containing phenylnorstatine (Pns), in contrast to renin inhibitors [2] which show a preference for the *anti* diastereomer.

Compound **2S** (KNI-81) containing Apns, in which P$_4$ Ser was deleted and P$_3$ Phe was replaced by the isosteric 3-phenylpropionic acid, exhibited substantial inhibitory activity, and was more active than the *anti* diastereomer (**2A**). Replacement of P$_2$ Asn with Ser (compound **3S**) decreased the inhibitory activity, but compound **3S** was also more active than the *anti* diastereomer (**3A**). Furthermore, incorporation of a phenylacetyl group at P$_3$ site (compounds **4A** and **4S**) decreased the potency, and also the *syn* configuration of the hydroxyl group was preferred. Replacement of P$_1$ Apns with allocyclohexylnorstatine (Achns) (compounds **5S**) reduced the inhibitory potency. The Achns-containing inhibitor also was more active than the Chns-containing inhibitor (**5A**), in contrast to renin inhibitors. The Apns-containing tripeptide **6S** (KNI-102), in which P$_3$ Phe was replaced by the isosteric benzyloxycarbonyl group, P$_3$' Val deleted and P$_2$' Ile replaced by the isosteric t-butylamine, exhibited higher activity compared to the pentapeptide **2S**. It was surprising that such a small compound as **6S** was more potent than the longer compound **2S**. On the other hand, the Pns-containing tripeptide **6A** exhibited little activity even at a concentration of 5µM.

As shown in Table 1, HMC-Pro inhibitors of HIV protease did not practically inhibit pepsin but the symmetric-type compound containing P_1' Phe (**7A**) inhibited pepsin, which showed that Apns-Pro inhibitors of HIV protease were selective. Especially, the tripeptide containing Apns-Pro (**6S**; KNI-102) was a potent and highly selective HIV protease inhibitor, causing no pepsin inhibition even at a concentration of 80 µM[3].

Lead Optimization

Having identified the tripeptide derivative KNI-102 as a lead compound, we studied lead optimization to find a highly selective and potent HIV protease inhibitor, KNI-174, with anti-HIV activity [4]. A further structure-activity relationship study considering penetration across cell membrane's and behaviour *in vivo* resulted in the generation of highly potent protease active sitetargeted anti-HIV agents. Here, we describe Apns-containing HIV protease inhibitors, kynostatin (KNI)-227 and kynostatin (KNI)-272, which exhibit extremely potent antiviral activity against a wide spectrum of HIV isolates[5].

Combinations of each preferred side chain led to highly selective and potent HIV protease inhibitors (Table 2), such as KNI-174 (IC_{50} = 2.8 nM), KNI-225 (IC_{50} = 2.7 nM) and KNI-170 (IC_{50} = 2.6 nM), with little inhibition of other aspartic proteases, porcine pepsin (IC_{50} > 10,000 nM for each inhibitor) and human plasma renin (IC_{50} > 100,000 nM for each inhibitor). These compounds exhibited potent antiviral activities against HIV-1 in CD4$^+$ATH8 cells.

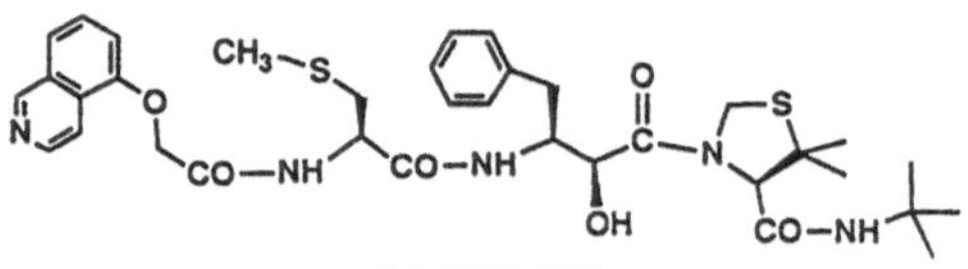

25 (KNI-227)

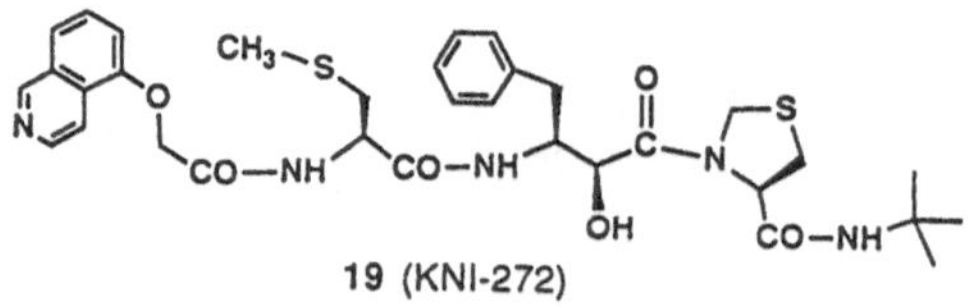

19 (KNI-272)

Fig. 3. KNI-227 and KNI-272.

The behaviors of compounds *in vivo*, such as penetration across the cell membrane and nonspecific adsorption in blood, are important factors for *in vivo* antiviral activity. Therefore, considering the subtle balance of lipophilicity-hydrophilicity and molecular size, we incorporated the 5-isoquinolinyloxyacetyl(iQoa) moiety at the P_3 position and combined it with each preferred side chain. Such modifications resulted in compounds, KNI-227 and KNI-272 (IC_{50}=0.01 µM for both compounds), highly active against clinical HIV-1 isolates in phytohemagglutinin-stimulated peripheral blood mononuclear cell assays (Table 3), which might correspond relatively well to *in vivo* antiviral activity. These antiviral activities against clinical HIV-1 isolates were more than 10-fold potent compared to a C_2 symmetric protease inhibitor, A-77003 [6]. As expected from the protease active site-targeted antiviral mechanism, these also exhibited potent antiviral activities (IC_{50}= 0.01 µM for both compounds) against AZT-insensitive clinical HIV-1 isolates [7].

Two compounds, KNI-227 and KNI-272 (Fig. 3), were highly potent HIV-1 protease inhibitors with little inhibition of other aspartic proteases such as human plasma renin (IC_{50} > 100µM) and porcine pepsin (IC_{50} > 10 µM), and showed potent antiviral activities against the infectivity and cytopathic effect of HIV strains, including HIV-1$_{LAI}$,

Table 2 *HIV-1 Protease inhibition and antiviral activities against HIV-1$_{LAI}$ in ATH8 cells*

	Structure					IC_{50}^a	IC_{50}^b	TC_{50}^c	TC_{50}/IC_{50}
	P3	P2	P1	P1'	P2'				
Compound	Phe	-Asn	-Phe	-Pro	-Ile	(nM)	(µM)	(µM)	
KNI-102	Z-	Asn-	Apns-	Pro-	NHBu^t	89	1.1	>20	>18
KNI-153	Qc-	Asn-	Apns-	Pro-	NHBu^t	20	1.0	>20	>20
KNI-144	Noa-	Asn-	Apns-	Pro-	NHBu^t	12	0.9	>20	>22
KNI-154	Noa-	Asn-	Apno-	Thz-	NHBu^t	8.8	0.5	>50	>100
KNI-174	Noa-	Asn-	Apns-	Dmt-	NHBu^t	2.8	0.4	30	75
KNI-225	Noa-	Mta-	Apns-	Dmt-	NHBu^t	2.7	0.2	7	35
KNI-170	Noa-	Msa-	Apns-	Dmt-	NHBu^t	2.6	0.1	16	160
KNI-227	iQoa-	Mta-	Apns-	Dmt-	NHBu^t	2.3	0.1	40	400
KNI-272	iQoa-	Mta-	Apns-	Thz-	NHBu^t	6.5	0.1	>50	>500

[a]HIV-1 protease inhibitory activity determined using a synthetic [Ala^{67,95}]-HIV-1 protease, as previously reported[1,2].
[b]50 percent inhibitory concentration against cytopathic effect of HIV-1.
[c]50 percent toxic concentration of the compound.
Abbreviations: Z=Benzyloxycarbonyl, Apns=(2S,3S)-3-amino-2-hydroxy-4-phenylbutyric acid, Bu^t= t-butyl, Thz=L-thiazolidine-4-carboxylic acid, Dmt=L-5, 5-Dimethylthiazolidine-4-carboxylic acid, Noa=1-naphthoxyacetyl, Qc=quinolin-2-ylcarbonyl, Msa=L-methane-sulfonylalanine, Mta=L-mtheylthioalanine, iQoa=5-isoquinolinyloxyacetyl.

Table 3 *Antiviral activity of KNI compounds against AZT-sensitive and -insensitive clinical HIV-1 isolates in target PHA-PBM*

Compound	TC_{50}(µM)[a]	AZT-sensitive strain IC_{50}(µM)[b]	AZT-insensitive strain IC_{50}(µM)[b]
KNI-154	77	0.14	0.16
KNI-174	32	0.13	0.17
KNI-225	8	0.03	0.03
KNI-170	19	0.03	0.02
KNI-227	49	0.01	0.01
KNI-272	>80	0.01	0.01

[a]TC_{50} values were assessed by [^{3}H]-thymidine incorporation assay.
[b]IC_{50} values were determined based on levels of p24 gag protein production.

HIV-1$_{RF}$, HIV-1$_{MN}$ and HIV-2$_{ROD}$, as tested in CD4$^+$ ATH8 cells. The 50% inhibitory concentrations (IC_{50}) of KNI-227 were 0.1, 0.02, 0.03 and 0.13 µM, while those of KNI-272 were 0.1, 0.02, 0.04, and 0.14 µM, respectively. From the viewpoint of the action mechanism, the active site-targeted HIV protease inhibitors have reason to exhibit activities against a wide spectrum of HIV strains, including HIV-2.

Interestingly, a relatively low-lipophilic and small-sized tripeptide derivative, KNI-272, combined with iQoa moiety and L-thiazolidine-4-carboxylic acid (Thz) residue, exhibited highly potent antiviral activities and low cytotoxicity ($TC_{50}>80$ mM). Ready availability due to the simple synthetic procedure (Fig. 4) of the tripeptide derivatives and the excellent antiviral properties indicate that KNI-227 and KNI-272 are promising candidates as selective anti-AIDS drugs.

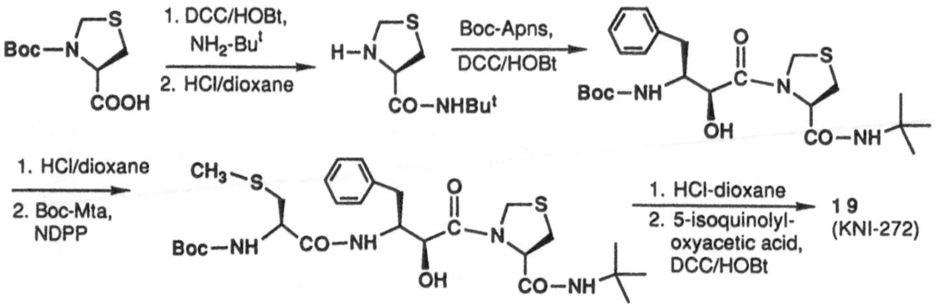

Fig. 4. Synthetic Scheme of KNI-272(19). Pure compound 19 was conveniently synthesized by the solution method in a stepwise manner and readily obtained. Abbreviations: Boc=t-butoxycarbonyl, DCC=dicyclohexylcarbodiimide, HOBt=N-hydroxybenzotriazole, NDPP=norborn-5-ene-2,3-dicarboximido diphenyl phosphate.

References

1. Mimoto, T., Imai, J., Tanaka, S., Hattori, N., Takahashi, O., Kisanuki, S., Nagano, Y., Shintani, M., Hayashi, H., Sakikawa, H., Akaji, K. and Kiso, Y., Chem. Pharm. Bull. 39(1991)2465.
2. Iizuka, K., Kamijo, T., Harada, H., Akahane, K., Kubota, T., Umeyama, H., Ishida, T. and Kiso, Y., J. Med. Chem., 33(1990)2707.
3. Mimoto, T., Imai, J., Tanaka, S., Hattori, N., Kisanuki, S., Akaji, K. and Kiso, Y., Chem. Pharm. Bull., 39(1991)3088.
4. Mimoto, T., Imai, J., Kisanuki, S., Tanaka, S., Hattori, N., Takahashi, O., Katoh, R., Yumisaki, T., Sakikawa, H., Akaji, K. and Kiso, Y., In Suzuki, A., (Ed.) Peptide Chemistry 1991, Protein Res. Found, Osaka, 1992, pp.395-400.
5. Mimoto, T., Imai, J., Kisanuki, S., Enomoto, H., Hattori, N., Akaji, K. and Kiso, Y., Chem. Pharm. Bull., 40(1992)2251.
6. Mitsuya, H., Yarchoan, R., Kageyama, S. and Broder, S., FASEB J., 5(1991)2369; Kageyama, S., Weinstein, J.N., Shirasaka, T., Kempf, D.J., Norbeck, D.W., Plattner, J.J., Erickson, J. and Mitsuya, H., Antimicrob. Agents Chemother., 36(1992)926.
7. Kageyama, S., Mimoto, T., Murakawa, Y., Nomizu, M., Ford, H., Jr., Shirasaka, T., Gulnik, S., Erickson, J., Takada, K., Hayashi, H., Broder, S., Kiso, Y. and Mitsuya, H., submitted.

Studies on the detection of antibodies of HIV-1 by synthetic peptides

Zhe-yu Cao[a], Shao-qing Chen[a], Xiao-jun Zhu[a], Yun Gu[a], Jie-cheng Xu[a], Qing-liang Liu[a], Mei-di Ye[b], Wei Zhu[b], Chang-long Wu[b]

[a] *Shanghai Institute of Organic Chemistry, Chinese Academy of Sciences, Shanghai 200032, China*
[b] *Shanghai Institute of Biological Products, Ministry of Public Health, Shanghai 200052, China*

Introduction

The short dodecapeptide I (Fig. 1) corresponding to amino acid sequence 598–609 of gp41, the transmembrane glycoprotein of HIV-1, have been shown consistently high specificity and sensitivity for detecting antibodies of HIV-1. There are three Gly and one Ser residues in this peptide, so this peptide is very flexible. The results of secondary structure and antigenic determinant predictions do not show that there are secondary structure or antigenic determinants in this peptide fragment. Chang-Yi Wang pointed out that the peptide region covering TrpGlyCysSer contributes to the antigenic configuration and John W.Gnann Jr. guessed that a disulfide bond perhaps formed between two Cys residues. It was reported by Lacroix on the Vth International Conference on AIDS that the reactivity was enhanced 1.2 to 43 times by air oxidizing some peptides which contain CysSerGlyLysLeuIleCys fragment. But the intermolecular link between Cys residues of the peptides would also enhance the reactivity.

Leu Gly Ile Trp Gly Cys Ser Gly Lys Leu Ile Cys	(I)
Leu Gly Ile Trp Gly Cys Ser Gly Lys Leu Ile Cys	(II)
Leu Gly Ile Trp Gly Ser Ser Gly Lys Leu Ile Ser	(III)
Leu Gly Ile Trp Gly Met Ser Gly Lys Leu Ile Mel	(IV)
Ile Trp Gly Cys Ser Gly Lys Leu Ile Cys	(V)
Ile Trp Gly Cys Ser Gly Lys Leu Ile Cys	(VI)
Ac-Trp Gly Cys Ser Gly Lys Leu Ile Cys	(VII)
Ac-Trp Gly Cys Ser Gly Lys Leu Ile Cys	(VIII)
Trp Gly Cys Ser Gly Lys Leu Ile Cys	(IX)
Trp Gly Cys Ser Gly Lys Leu Ile Cys	(X)

Fig. 1. Synthetic peptides I–X.

Table 1 *Reactivity of synthetic peptides with mixed standard serum samples from AIDS patients*

Peptides (Ag)		I	II	III	IV	V	V*	VI	VII	VIII	IX	X
serum dilution												
1:50	+	1.166	0.223	0.101	1.518	1.738	1.232	0.363	0.300	0.175	0.147	0.276
	–	0.443	0.520	0.089	0.445	0.584	0.312	0.464	0.130	0.285	0.155	0.365
1:200	+	0.341	0.066	0.092	0.540	1.028	0.409	0.150	0.148	0.019	0.119	0.026
	–	0.130	0.146	0.082	0.162	0.182	0.075	0.269	0.124	0.082	0.131	0.118

* This petide was obtained by reducing peptide VI with DTT.

Results and Discussion

The peptides I, III, IV, V, VII and IX have been synthesized by Merrifield solid phase peptide synthesis method with the acid-labile tert-butyloxycarbonyl (BOC) group for temporary amino terminal protection and acid-stable groups for the protection of side chains. Peptide-resins were cleaved by anhydrous hydrogen fluoride and extracted with 10% acetic acid. Peptides were purified by gel filtration on Sephadex G25 and Sephadex LH-20. The composition of the peptides was confirmed by amino analysis. The peptides I, VII and IX were oxidized by 5% to 20% dimethyl sulfoxide water solution and purified by gel filtration to obtain peptides II, VIII and X respectively. The peptide V was oxidized by air in water and purified to obtain peptide VI then the peptide VI was reduced by dithiothreitlo (DTT) and purified to regain peptide V.

The reactivity of all peptides obtained with standard serum samples from AIDS patients have been determined by ELISA method. The results are shown in Table 1.

As shown in the table, the peptide I and V manifest higher reactivity and the peptide VII shows less reactivity. The peptides II, VI, VIII and X obtained by oxidizing peptides I, V,VII and IX do not show any reactivity, but the peptide V*, regained by reducing peptide VI with DTT shows higher reactivity again. However, the peptide VI, the analogue of peptide I obtained by replacing Cys with Met which is unable to form cyclopeptide impurily, showed higher reactivity.

Acknowledgement

This work was supported by the National Natural Science Fundation of China.

Conformational studies on synthetic peptides from the principal neutralizing domain of HIV-1 Gp120 that bind to CD4 and enhance viral infectivity

**Carlo Di Bello, Monica Dettin, Silvia Tormene, Rossella Roncon, Andrea Bagno
and Silvio Bicciato**

Institute of Industrial Chemistry, University of Padua, Padua, Italy

Introduction

The key event in HIV infection is the binding between the C-terminal portion of the envelope viral glycoprotein gp120 and the V1 region of the cellular antigen CD4, which have been tentatively located within residues 402-428 and 35-53 respectively. On the other hand, a gp 120 domain required for HIV mediated syncytia formation and for eliciting neutralizing antibodies, which is contained in a disulfide-bridged loop spanning residues 303-338 in the third hypervariable region, represents the principal neutralizing determinant (PND) of gp120.

In previous studies [1,2] we have demonstrated that synthetic peptides, corresponding to PND sequences of different HIV-1 isolates, are specifically recognized by a site distinct from the high affinity gp120-binding site of CD4. Interestingly, a peptide designed from the HIV-1 MN strain (GE3) is able to enhance viral infection and a HTLV-IIIB derived peptide-analogue (GE1) is at least ten-fold less efficient, while no effect is shown by other tested peptides. This enhancing effect occurs in the early stage of infection and is not strain restricted.

A correlation between the structure and the interesting biological properties shown by these PND-derived synthetic peptides is presented in this paper. The experimental data, obtained by different physico-chemical techniques, are compared to theoretical structural predictions.

Results and Discussion

We have recently observed that peptides derived from the PND of different HIV-1 isolates (see Table 1) interact with a CD4 site adjacent to ,but distinct from, the gp120 binding site. Moreover, peptides patterned on MN and IIIB strains enhance viral infection with different efficiency while other similar peptides show no effect [1,2].

We have also deserved that minor structural modifications may result in dramatic changes in the biological activities of the peptides. In fact, substitution of the C-terminal free carboxyl function (peptide GE3) by an amide function (peptide GE3-NH$_2$) results in an 80% reduction of the biological activity and in the complete loss of CD4 binding capacity [2].

HIV-1 infection requires specific binding of its envelope glycoprotein gp120 to CD4,

Table 1

Peptide	Sequence
GE1	H-AsnAsnThrArgLysSerIleArgIleGlnArgGlyProGlyArgAlaPheValThrIleGlyLysIleGly-OH
GE2	H-AsnAsnThrArgLysSerIleThrLysGlyProGlySArgValIleTyrAlaThrGlyGlnIleIleGly-OH
GE2-NH$_2$	H-AsnAsnThrArgLysSerIleThrLysGlyProGlySArgValIleTyrAlaThrGlyGlnIleIleGly-NH$_2$
GE3	H-TyrAsnLysArgLysArgIleHisIleGlyProGlyArgAlaPheTyrThrThrLysAsnIleIleGly-OH
GE3-NH$_2$	H-TyrAsnLysArgLysArgIleHisIleGlyProGlyArgAlaPheTyrThrThrLysAsnIleIleGly-NH$_2$
GE1Y	H-TyrAsnThrArgLysSerIleArgIleGlnArgGlyProGlyArgAlaPheValThrIleGlyLysIleGly-ON
GE6	H-ArgIleGlnArgGlyProGlyArgAlaPheValThrIleGlyLys-OH
GE7	H-ProGlyArgAlaPheValThrIleGlyLys-OH

a surface antigen participating in the recognition of target cells by helper T-cells [3]. It is generally agreed that the amino acid sequences involved in the binding are localized in the V1 domain of CD4 and in the C-terminal part of gp120 [4-6]. However, other regions of gp120 have been suggested to act as conformation-inducing determinants or to play a direct role in the binding process [7,8]. In addition, the 303-338 region of gp120, representing the PND of gp120, is distinct from the CD4-binding site and is required for the production of syncytia and appears to determine the virus tropism [9]. It is obvious that the high variability in the viral envelope protein may play a critical role in virus infectivity and antigenicity. However, in spite of being situated in a hypervariable region, the central portion of PND is highly conserved in dfferent HIV-1 isolates [10-12]. The amino acids flanking this sequence are variable and antibodies elicited by peptides designed from PND of different HIV-1 strains are viral-variant specific and often neutralize only the homologous virus [9-11]. Moreover, monoclonal antibodies to PND *in vitro* block syncytia formation and cell-free infection, but not virus binding.

Recent studies by La Rose et al. [12] have suggested that there are constraints on PND variability that have been attributed, by a neural network approach, to the presence of a β-strand/type-II β-turn/β-strand/α-helix structural motif. Our theoretical prediction, obtained by the Chou and Fasman method[13], confirms two probable β-sheets around positions 313-316 and 322-326, and a β-turn for the 317-320 sequence. Interestingly, this latter region for GE3 corresponds to a HIGP sequence, whereas for the other 22mer peptides (e.g. GE1, GE2) it corresponds to QRGP or TRGP sequences. This could possibly lead to different β-turn types and consequently reflect the difference in binding and in biological activity observed among our synthetic peptides.

In order to further clarify at the molecular level, possible correlations between observed structural and biological effects, a physico-chemical investigation using CD, FT-IR, 1D and 2D NMR techniques has been performed. Each technique has complimented a piece of information as far as the structural features of these peptides are concerned. In particular, while no sign of a secondary structure could be detected in aqueous slution, a clear tendency to assume an ordered organization of an α-helix and /or β-turn type is evident in a hydrophobic environment. Preliminary NMR data indicate that one of the peptides (GE3-NH$_2$) can assume a β-turn type structure in DMSO. These findings confirm the prediction made by La Rose et al. [12] which locates the β-turn in the highly conserved GPGR sequence.

Further studies in progress will allow us to better define the type of secondary structure(s) adopted by these peptides and, possibly, to correlate the structure organization with their behavior with respect to interactions with CD4.

Acknowledgements

Work supported by grants from the Italian Ministry of Health, Progetto AIDS, Istituto Superiore di Sanita and by the Italian National Research Council (CNR), Progetto FATMA.

References

1. De Rossi, A., Pasti, M., Mammano, F., Panozzo, M., Dettin, M., Di Bello, C. and Chieco-Bianchi, L., Virology, 184(1991)187.
2. Autiero, M., Abrescia, P., Dettin, M., Di Bello, C. and Guardiola, J., Virology, 185(1991)820.
3. Sattentau, Q.J. and Weiss, R.A., Cell, 52(1988)631.
4. Jameson, B.A., Rao, P.E., Kong, L.I., Hahn, B.H., Shaw, G.M., Hood, L.E. and Kent, S.B.H., Science, 240(1988)1335.
5. Lamarre, D., Capon, D.J., Karp, D.R., Gregory, T., Long, E.O. and Secaly, R.P., EMBO J., 8(1989)3271.
6. Lasky, L.A., Nakamura, G., Smith, D.H., Fennie, C., Shimasaki, C., Patzer, E., Berman, P., Gregory, T. and Capon, D.J., Cell, 50(1987)975.
7. Dowbenko, D., Nakamura, G., Fennie, C., Shimasaki, C., Riddle, L., Harris, R., Gregory, T. and Lasky, L., J. Virol., 62(1988)4703.
8. Pert, C.B., Hill, J.M., Ruff, M.R., Berman, R.M., Robey, W.G., Arthur, L.O., Ruscetti, F.W. and Farrar, W.L., Proc. Natl. Acad. Sci., 83(1986)9254.
9. Javaherian, K., Langlois, A.J., McDanal, C., Ross, K.L., Eckler, L.I., Jellis, C.L., Profy, A.T., Rusche, J.R., Bolognesi, D.P., Putney, D.S. and Matthews, T.J., Proc. Natl. Acad. Sci., 86(1989)6768.
10. Goudsmit, J., Debouck, C., Meloen, R.H., Smit, L., Bakker, M., Asher, D.M., Wolff, A.V., Gibbs, C.J. and Gajdusek, D.C., Proc. Natl. Acad. Sci., 85(1988)4478.
11. Palker, T.J., Clark, M.E., Langlois, A.J., Matthews, T.J., Weinhold, K.J., Randall, R.R., Bolognesi, D.P. and Haynes, B.F., Proc. Natl. Acad. Sci., 85(1988)1932.
12. La Rosa, G.J., Davide, J.P., Weinhold, K., Waterbury, J.A., Profy, A.T., Lewis, J.A., Langlois, A.J., Dreesman, G.R., Boswell, R.N., Shadduck, P., Holley, L.H., Karplus, M., bolognesi, D.P., Matthews, T.J., Emini, E.A. and Putney, S.D., Science, 249(1990)932.
13. Chou, P.Y., Fasman, G.D. and Weiss, R.A., Ann. Rev. Biochem., 47(1978)251.

Synthetic gp41 peptides as sensitive and specific diagnostic reagents in HIV-1 infection

Li-li Guo[a], Ran Zhang[a], Hong Zhao[a], Wei Cui[b], Chun-hua Pan[a] and Guan-fu Zhu[a]

[a]Institute of Microbiology and Epidemiology, Beijing 100850, China
[b]Jilin University, Changchun 130023, China

Introduction

Synthetic peptides instead of HIV used as antigen to detect the antibodies against HIV have some advantages: safe to handle, easy to prepare, high sensitivity, specificity and stability. Peptides from the region of the transmembrane protein gp41, part of the env-gene products, appear to be useful as antigens for anti-HIV tests since an early antibody response against the env-gene products is induced and, if compared with antibodies against gag-gene products, persists in clinical AIDS[1]. An independent and strongly immunogenic region of the transmembrane protein was found around amino acid 600[2].

We examined the region from amino acid 578 to 607 with 4 overlapping peptides with regard to their possible use as diagnostic reagents in HIV-1 infection.

Results and Discussion

The following four peptides:

LGIWGCSGKLICTTA	(1)
YLKDEELLGIWGCSGKLICTTA	(2)
ARILAVERYLKDEELLGIWGCSGKLICTTA	(3)
ARILAVERYLKDEELLG	(4)

have been synthesized by 4170 Biolynx Automatic Peptide Synthesizer with Uitrosyn A (peptides 1,2,3) and by manual method with P-benzyloxy benzyl alcohol resin (peptide 4) according to Fmoc-solid phase peptide synthesis procedure. Through deprotecting with 1M TMSBr-Thioanisole/TFA and purifying with Sephadex G-25, crude peptides showing main peak on RP-HPLC were obtained (Fig.1).

The immunoactivities of the synthetic peptides were examined by indirect ELISA. 1µg of peptide was applied to each well of a 40-well flat-bottom polystyrene plate, which was subsequently blocked with 1% BSA in PBS. Anti-HIV positive serum diluted 1:50 was incubated in the wells. Bound human antibodies were detected with horseradish peroxidase conjugated monoclonal anti-human Fc-gamma, and TMB was used as substrate. When peptide competition was performed, peptide or HIV was added to the anti-HIV positive serum before adding the serum to the well.

98

Table 1 *Reactivity of synthetic peptides with sera*

sera\peptide	1	2	3	4
P1	0.53	0.53	1.25	0.49
P2	0.00	0.10	1.14	0.97
P3	0.01	0.46	0.54	0.08
N1	0.01	0.00	0.00	0.00
N2	0.02	0.01	0.01	0.03
N3	0.00	0.02	0.04	0.01

Each of the four peptides was used to test sera from 3 HIV-infected persons (P1,P2,P3), and 3 serum samples from healthy blood donors (N1,N2,N3,) (Table1). Only peptide 3 could reacted with all sera from 3 individuals. This may be due to incomplete epitopes of short peptide 1,2 and 4.

Furthermore we determined the serum reactivity to peptide 3 using another 52 anti-HIV antibody-positive sera, serum samples from anther 150 healthy blood donors. The peptide ELISA assay had sensitivity (number of positives diagnosed divided by number of confirmed positives) of 100% and specificity (number of negatives diagnosed divided by number of confirmed negatives) of 100%. Peptide 3 or HIV antigen inhibition assay also demonstrated that the peptide ELISA was specific.

Our results show that the 30-amino acid peptide (peptide 3) can be used as a highly sensitive and specific reagent for the diagnosis of HIV-1 infection.

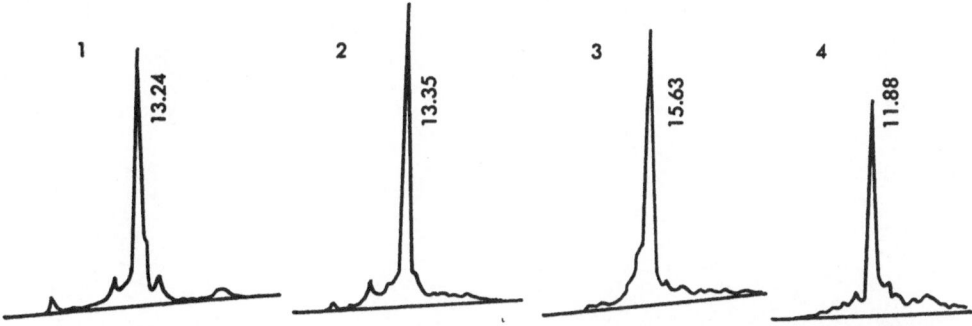

Fig. 1. *HPLC profile of synthetic peptides. Instrument: Perkin Eimer integral 4000; Column: PE-pack C18 5/120 4.6mm×15cm; Mobile phase: A: 0.05% TFA in water; B: 0.05% TFA, 100% CH₃CN; Gradient: 20-50% B in 30 min(1); 20-60% B in 30 min(2); 20-60% B in 30 min(3); 27-50% B in 50 min(4); Detection: 280 nm(1,4); 275 nm(2,3).*

Acknowledgement

The study was supported by grant from the National Natural Science Foundation of China(38900005).

L.L. Guo et al.

References

1.	Saah, A.J., Farzadegan, H., Fox, R., Nishanian, P., Charles, R., Rinaldo, J.R., Phair, J.P., Fahey, J.L., Lee, B.T.H. and Polk, B.F., J. Clin. Microbiol., 25(1987)1605.
2.	Wang, J.J.G., Steel, S., Wisniewolski, R. and Wang, C.Y., Proc. Natl. Acad. Sci. USA, 83(1987)6159.

The sporozoite surface antigen 2 of *Plasmodium yoelii* contains a T and B cell epitope cross reacting with the repetitive sequence of the *Plasmodium falciparum* circumsporozoite protein

Antonio S. Verdini[a][1], Sinella Le Moli[b], Assieh Mir[b], Chantal Tougne[c], Antonello Pessi[a], Paul Henri Lambert[c], Giuseppe Del Giudice[c] and Giampietro Corradin[b]

[a]*Eniricerche, 00015 Monterotondo, Italy*
[b]*Institute of Biochemistry, University of Lausanne, 1066 Epalinges, Switzerland*
[c]*WHO-Immunology Research and Training Centre, Department of Pathology, University of Geneva, 1211 Geneva 4, Switzerland*

Introduction

Sporozoites, the first developmental stage of malaria parasites, contain a species- and stage-specific major surface antigen, called circumsporozoite (CS) protein, which uniformly covers their surface, and which is the major target for antibody and cellmediated immunity after natural and/or experimental infections[2]. All CS proteins are known to consist of a central region of tandemly repeated amino acid sequences. The *Plasmodium falciparum* repeats $(NANP)_n$, represent the major B cell epitope in humans and animals. NANP repeats also contain T-cell epitopes[3]. Recently, a second sporozoite surface antigen(SSP2) has been defined in the murine malaria parasite, *Plasmodium yoelii*. It contains two repetitive sequences, $(NNP)_n$ and $(NPNEPS)_n$ with no homology with the CS protein of this parasite. SSP2 has been shown to induce specific CD8+ cytotoxic T cells in mice immunized with P815 cells expressing this antigen, and to confer protection against an infectious challenge with *P. yoelii* sporozoites to mice that were injected with the P815 mastocytoma cell line expressing SSP2 and CS protein [4]. Since *P. yoelii* SSP2 and *P. falciparum* CS antigen repeats share the linear NPN motif, we investigated whether or not a cross-reactivity between the two antigens existed at the T cell and antibody level.

Results and Discussion

The peptides used in this study were prepared by solid-phase synthesis on polyamide resins following the Fmoc-t-butyl strategy [$Y(NANP)_3$, $Y(NANP)_4$, $(NNP)_4$, $(NNP)_{11}$, $Y(NNP)_4NNG$, $(NP)_6$, $(NP)_{11}DGA$] and by polycondensation of NANP pentachlorophenylester in DMSO [$(NANP)_{40}$]. C57 BL/6 mice immunized twice with $(NNP)_4$ gave high titers of IgG antibodies and lymph node cells from mice immunized with $Y(NNP)_4NNG$ exhibited a strong proliferative pattern when restimulated in vitro either with the same peptide or $Y(NANP)_3$. The proliferation was specific since ovalbumin-primed cells were not stimulated *in vitro* by $Y(NNP)_4NNG$. Similarly, the

Table 1 *Proliferation of (NANP)-specific T cell clones in the presence of various antigens*[a]

| Clone | | µg/ml | Antigens CPM×10⁻³ | | |
			Y(NANP)$_3$	Y(NNP)$_4$	(NP)$_6$
11/3B	Exp.1	0.9	130.3	0.4	--
		2.7	120.1	0.8	--
		8.1	60.2	10.5	--
		27	32.7	27.5	--
		81	30.1	60.1	--
	Exp.2	0.08	152.4	--	0.3
		2	72.7	--	21.4
		50	59.0	--	165.2
11/3B	Exp.1	0.9	270.6	20.3	
		2.7	300.1	42.1	
		8.1	223.2	93.7	
		27	175.6	123.8	
		81	140.4	155.3	
	Exp.2	0.4	148.2		0.2
		2	116.4		0.2
		50	45.3		41.2

[a] 10^4 T cell clones were added to $0.5×10^6$ irradiated spleen cells at the indicated concentration of antigen. 3H-TdR (1µCi/well) was added on day 3 of proliferation and cells were harvested 18 to 24 h later. Control cultures (no antigen added) exhibited a proliferation ranging from 0.3 to $0.5×10^3$ cpm.

Y(NNP)$_4$NNG-specific T-cell clone B10 strongly responded to Y(NANP)$_3$ and (NANP)$_{40}$. Conversely, (NANP)$_n$-specific T-cell clones 11/3B and 11/6C also proliferated when Y(NNP)$_4$NNG was added to the culture. These results, taken together, demonstrated the existence of an extensive cross-reactivity between (NNP)$_n$ and (NANP)$_n$ at the T-cell level. On the other hand, (NNP)-primed lymph node cells as well as (NNP) specific T-cell clone B10 did not proliferate that in the presence of (NP)$_6$ and (NP)$_{11}$DGA, peptide congeners that also contain the linear NPN motif. Conversely, (NANP)-specific T-cell clones 11/3B and 11/6C proliferated in the presence of (NP)$_n$ peptide (Table 1). Antibody cross-reactivities were studied using serum samples from mice immunized with (NANP)$_{40}$, (NNP)$_{11}$, (NP)$_{11}$ (Table 2) and from individuals living in *P. falciparum*-endemic areas.

The lack of inhibition of (NANP)$_{40}$ antibodies by (NP)$_{11}$ and (NNP)$_{11}$ and, conversely, the inhibition of anti (NP)$_{11}$ antibodies by (NANP)$_{40}$ cannot be reconciled with the hypothesis that the NPN linear motifs shared by all polypeptides determine the observed partial cross-reactivity. It is rather likely that (NANP)$_{40}$, by virtue of its extreme flexibility, can recognize anti-(NP)$_{11}$ as well as anti (NNP)$_{11}$ antibodies through conformational rearrangements that eventually produce bound polypeptide structures close to that of (NNP)$_{11}$ and (NP)$_{11}$[5]. On the other hand, because (NP)$_{11}$ and (NNP)$_{11}$ are less flexible and conformationally different from (NANP)$_{40}$ they are not able to properly rearrange and bind anti-(NANP)$_{40}$ antibodies.

Table 2 *Immunizations with synthetic sequential polypeptides*[a]

Immunogen	Competitor concentration (µg/ml)	Competitor peptide		
		$(NANP)_{40}$	$(NNP)_{11}$	$(NP)_{11}$
$(NANP)_{40}$	0.2	16*	1	3
	2	30*	−4	−2
	20	86*	9	2
	200	99*	13	1
$(NNP)_{11}$	0.2	4	−2	ND
	2	34*	0	ND
	20	75*	2	ND
	200	100*	35*	ND
$(NP)_{11}$	0.2	−3	−2	−4
	2	25	4	8
	20	76*	4	58*
	200	100*	15	100*

[a] Groups of C57BL/6 mice were immunized at the base of the tail with 20µg of $(NANP)_{40}$ or 50 µg of $(NNP)_{11}$ or $(NP)_{11}$ on day 0 (Freund's complete adjuvant) and day 21 (Freund's incomplete adjuvant). Sera taken on day 28 were diluted 1:1000 [$(NANP)_{40}$-immunized mice] or 1:50 [(NNP)11-and (NP)11-immunized mice] and incubated for 1.5 h with different concentrations of competitor peptides. Then, the ELISA was carried out on plates coated with the $(NANP)_{40}$ peptide (1 µg/ml). Results are expressed as percentage of competition, as compared with those obtained in wells which did not receive any competitor peptide. Competitions equal or higher than 30% were considered as positive (*). ND: not done.

In conclusion, T-cell cross reactivity between *P. yoelii* SSP2 and *P. falciparum* CS antigen repeats seems to arise from the shared NPN linear motif. On the contrary, immunization experiments with synthetic polypeptide models suggest a partial cross-reactivity between the two antigen repeat domains which may reflect their different conformations and flexibilities.

References

1.   Present address: Italfarmaco S.P.A., Via dei Lavoratori, 54, Cinisello Balsamo, Milano, Italy.
2.   Nussenzweig, V. and Nussenzweig, R.S., Adv. Immunol., 45(1989)283.
3.   Togna, A.R., Del Giudice, G., Verdini, A.S. et al., J. Immunol., 137(1986)2956.
4.   Kushmith, S., Charoenvit, Y., Kumar, S. et al., Science, 252(1991)715.
5.   Verdini, A.S., Chiappinelli, L. and Zanobi, A., Biopolymers, 31(1991)587.

Comparison of the antigenic activity of five hendecapeptides on trichosanthin and α-momorcharin

Yu Wang, Jia-xiang Wu, Shan-wei Jin, Hui-qing Xie
and Yue-zhen Yao
*Shanghai Institute of Organic Chemistry, Chinese Academy of Sciences,
Shanghai 200032, China*

Introduction

In order to search for epitopes of trichosanthin, we had synthesized seventeen peptides based on the distribution of regional hydrophilicity along the peptide chain of trichosanthin. Studies using enzymelinked immunosorbent assay (ELISA) by reacting rabbit anti-trichosanthin serum with the above peptides showed that peptide (I): AAGKIRENIPL, sequence 117-127, has definite antigenic activity[1].

Recently, four analogs of it have been synthesized, their sequences are shown in Fig.1.

A A G K I R E N I P L	(I)
A G K I R E N I P L G	(II)
G K I R E N I P L G L	(III)
A A G G L R E N I P L	(IV)
A A G K P R E K I P I	(V)

Fig. 1. List of synthetic analogs of peptide.

Peptides (I), (II), (III) are corresponding to the amino acid residues in trichosanthin sequences 117-127,118-128,119-129, respectively; peptide(IV) is an analogs of peptide (I) with two residues exchanged, and peptide(V) is locating on α-momorcharin sequence 117-127, which is just on the corresponding position as peptide(I) on trichosanthin.

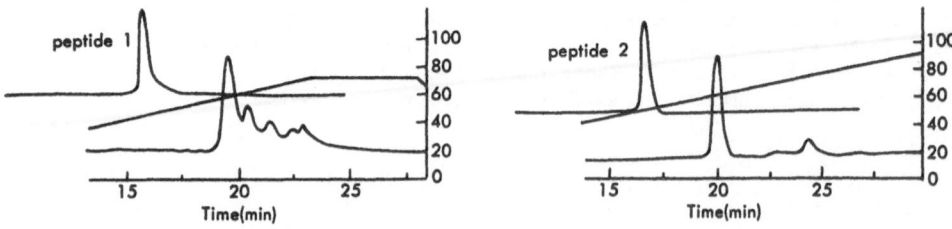

Fig. 2. RP-HPLC of peptide (I) and peptide (II) from Bondapak C18 Column: Bondapak C18(0.78×30 cm) solvent:A=0.1%TFA, B=CH3CN; Flow rate: 2.5ml/min.

Table 1 *Amino acids composition analysis of five hendecapeptides*

	N(1)	E(1)	G(1)	A(2)	I(2)	L(1)	P(1)	R(1)	K(1)
I	1.0	1.0	0.9	1.7	1.8	1.0	*	1.2	1.1

	N(1)	E(1)	P(1)	G(2)	A(1)	I(2)	L(1)	K(1)	R(1)
II	1.2	1.2	*	2.0	1.0	1.9	0.9	0.9	0.8

	G(2)	K(1)	I(2)	R(1)	E(1)	N(1)	P(1)	L(2)	
III	1.9	1.2	1.8	1.0	1.2	1.0	*	1.8	

	N(1)	E(1)	G(2)	A(2)	I(1)	L(2)	R(1)	P(1)	
IV	1.1	1.3	2.0	1.9	1.2	1.8	1.2	*	

	A(2)	G(1)	K(2)	P(2)	R(1)	E(1)	I(2)		
V	1.6	1.0	1.7	*	1.0	1.1	1.7		

Table 2 *Reactivity of synthetic peptides with rabbit anti-TCS serum (First test)*

Sample	TCS	I	II	III	IV	V	REF*
ΔOD 1:500	1.452	0.746	0.666	0.618	0.750	0.083	0.004
ΔOD 1:1000	1.555	0.412	0.426	0.369	0.452	0.066	0.007

(Repeat test)

Sample	TCS	I	II	III	IV	V	REF*
ΔOD 1:500	1.409	0.635	0.581	0.502	0.562	0.033	0.025
ΔOD 1:1000	1.404	0.375	0.351	0.313	0.303	0.041	0.013

*: REF is also a hendecapeptide fragment of TCS, but is not an antigenic one.

Results and Discussion

The analogs were synthesized by the solid phase method, anhydrous HF-anisole were used to treat the peptides-resin products, the crude peptides were then purified with HPLC and identified with amino acid analysis, some were also checked with sequence analysis.

Studies using enzyme-linked immunosorbent assay (ELISA) by reacting rabbit anti-trichosanthin serum with these peptides showed that all the peptides, except(V), showed definite antigenic activity.

The fact that peptide(V) has no detectable antigenic activity is interesting, because it is known that trichosanthin and α-momorcharin have great similarity in their primary structure and function like ribosome-inactivating effect, but they have obvious different antigenic quality.

The results that all peptides 117-127,118-128 and 119-129 are antigenic suggest that region 117-129 may construct one of the epitopes of trichosanthin.

Reference

1. Wang, Y., Wu, J.X., Jin, S.W., Qian, R.Q., Zhang, W.J. and Yao, Y.Z., Investigation on antigenic determinants of trichosanthin using synthetic peptides. Acta Biochemica et Biophysica Sinica (to be published).

Effects of methionine enkephalin on splenocytes proliferation and electrophoresis rate in liver impaired mice

Jie Yu, Gang Li and Jian-zheng Hou

Department of Biochemistry, Xi'an Medical University, Xi'an 710061, China

Introduction

The receptors of opioid peptides have been proved to be present on some immune cells[1]. One attractive trend has emerged in the past decade indicating that some neuropeptides may be secreted and released from the central nervous system and peripheral organs during stress and play an important role in the regulation of the immune system[2]. Diseases, as liver impairment, has been proved to produce stress in the body[3]. In this experiment, liver of mice was impaired by feeding CCl_4 for 4 weeks and the effect of M-Enk on proliferative responses and electrophoresis of splenocytes from liver impaired mice were determined *in vitro*. From these studies, The immunomodulatory mechanisms of M-Enk during stress or disease may be further clarified.

Results and Discussion

Splenocytes from either liver impaired or control mice were cultured with M-Enk and Concanavalin A(Con A) for 2 days, splenocytes proliferation stimulated by Con A were significantly suppressed by M-Enk in both groups. There were considerable variations between groups in the degree of suppression obtained from ^{3}H-thymidine (^{3}H-TdR) uptake in the presence of 5 µg/ml Con A and different concentrations of M-Enk (Fig. 1). The suppressive percentage was 17.05% in the liver impaired group compared to 53.35% in the control group. The suppressive effects of M-Enk could be reversed by 10 µM naloxone (Table 1). The results suggest that liver impairment may influence either the expression of M-Enk receptors or numbers of the receptors on splenocytes. It also indicated that suppressive effect of M-Enk on the Con

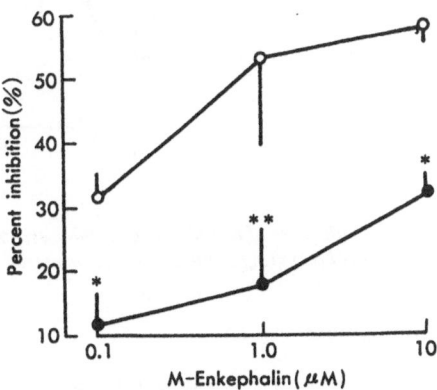

*Fig. 1. Percentages of suppression by M-Enk on Con A response. Splenocytes 2.5×10⁶/ml from impaired(•) and control(○) groups with Con A 5 µg/ml and different concentrations of M-Enk. n=3. *, P>0.05; **, P<0.05; ***, P<0.01 as compared to control group.*

Table 1 *Blackge of effects of M-Enk by naloxone*

M-Enke (μM)	Naloxone (μM)	10^{-2} × uptake of ^{3}H-TdR (cpm)
0	0	13±1.1
0	10	13±1.0*
1	0	11±0.9**
1	10	13±1.3*

*, P>0.05; **, P<0.05.

A response was mediated by opiate receptors.

The data on splenocytes electrophoresis demonstrated the correlation between the electrophoresis rate(ER) and the dosage of M-Enk (Fig. 2). In the absence of M-Enk, ER of splenocytes was significantly different in both groups (P<0.01). As M-Enk concentration was increased, ER rose in the impaired group, but fell in the control group. These results indicated the change in surface charge of cells from liver impaired group after adding M-Enk.

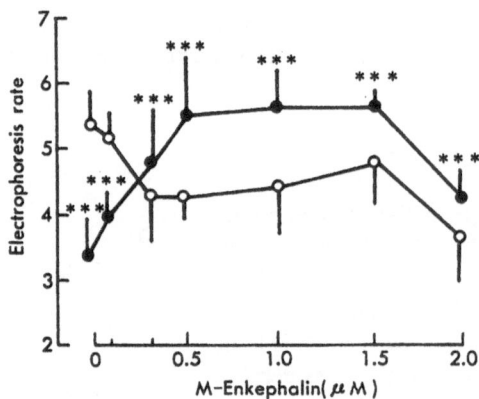

*Fig. 2. Influences of M-Enk on electrophoresis rate of splenocytes. Splenocytes 1×10⁷/ml from impaired (●) and control (○) groups with different concentrations of M-Enk. n=12. ***P<0.01 compared to control group.*

References

1. Mehrishi, J.N. and Mills, I.H., Clin. Immunol. and Immunopath., 27(1983)240.
2. Li, G. and Fraker, P.J., Acta Pharmacol. Sin., 10(1989)216.
3. Thornton, J.R. and Losowsky, M.S., BMJ., 297(1988)1241.

Evidence for *in vitro* effect of methionine enkephalin on macrophages from liver impaired mice

Gang Li, Jie Yu and Jian-zheng Huo

Department of Biochemistry, Xi'an Medical University, Xi'an 710061, China

Introduction

The plasma methionine enkephalin(M-Enk) concentration is increased in patient with liver diseases indicate an alteration of the immune function in these patients[1-2]. Since the effects of opioid peptide on modulation of immune system have been documented in human being and animals under normal condition, more interest has recently focused on stress. In this experiment, liver of mice were impaired by feeding CCl_4 for 4 weeks and the effect of M-Enk on migration of abdominal macrophages was determined *in vitro*.

Results and Discussion

Migration of macrophages in both liver impaired and control groups were suppressed by macrophages migration inhibitory factor(MMIF) produced from Con A-stimulated spleen lymphocytes, but the suppression might be reversed by adding 1 μM M-Enk in the reaction system($P<0.05$ in both groups) (Fig. 1). The effect of M-Enke was more obvious in the case of impaired group than that in control group, although the change did not significantly differ from each other.

It has been known that MMIF could inhibit the migration of macrophages *in vitro*, our

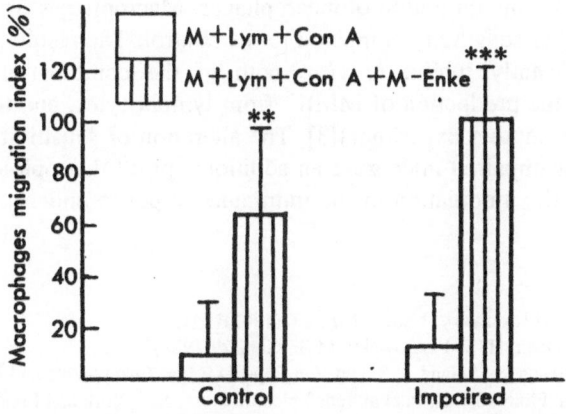

Fig. 1. Comparison of migration rates of macrophages incubated with or without M-Enk.
M: macrophages, Lym: lymphocytes. These data are the mean ± SD. n=3. or 4. *P>0.05, **P<0.05.
***P<0.01 compared to the data obtained from experiment with Con A.

109

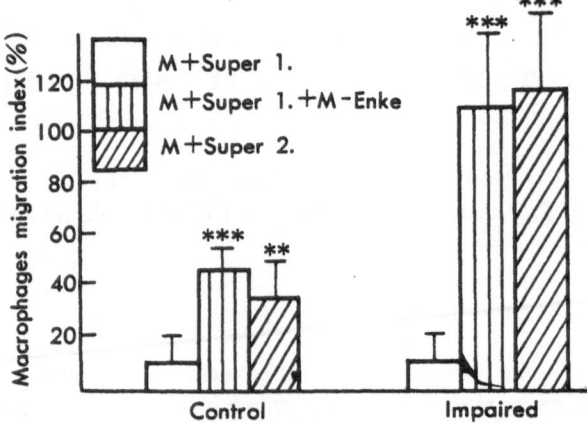

Fig. 2. Effect of M-Enk on production of MMIF and combination of MMIF with it's receptors. M: macrophages. Super: as supernatant obtained from either cultured only lymphocytes(Super 1.) or cultured lymphocytes preincubated with M-Enk (Super 2.). These data are the mean ± SD, n=3 or 4. **P<0.05, ***P<0.01 compared to the data obtaine from the experiment(M + super 1.).

results indicated that M-Enk might block the effect of MMIF and macrophages in liver impaired mice were more sensitive in response to the blockage.

In order to dertermine whether M-Enk affect the binding between MMIF and it's receptor on macrophages or interfere with the production of MMIF from lymphocytes, another experiment was performed. Macrophages were cultured in Con A-stimulated splenocytes supernatant and the distance of macrophages migration was measured in the presence of M-Enk. The data showed that M-Enk could inhibit the effect of MMIF on macrophages and macrophages from liver impaired mice were more sensitive than that from control(P<0.05) (Fig.2).

When lymphocytes were preincubated in 2.5 µg/ml Con A and M-Enk, the supernatant could no long inhibit the migration of macrophages. Macrophages from impaired group also showed a higher sensitivity compared to the control. The results suggested that M-Enk could significantly inhibit *in vitro* both of the combination of MMIF with macrophages and the production of MMIF from lymphocytes, and that was consistent with the results of another experiment[3]. The alteration of sensitivity of macrophages in the case of liver impaired mice gave an additional proof that opioid peptides take an important role on the modulation of the immmune response under stress.

References

1. Thornton, J.R. and Losowsky, M.S., BMJ., 297(1988)1241.
2. Thornton, J.R., Dean, H. and Losowsky, M.S., Gut., 29(1988)1167.
3. Brown, S.L., Tokada, S., Saland, L.S. and Van Epps, D.E., In: Enkephalins and Endorphins: Stress and Immune System. Plotnikoff, N. P. et al. (eds.) plenum Press, New York and London, 1986, pp.367-386.

Regulatory effects of substance p and somatostatin on natural killer activity and macrophage tumoricidal activity *in vitro*

Ye-ping Tian, Zheng-fang Zhou and Ling-li Zheng

Department of Immunology, Second Military Medical University,
Shanghai 200433, China

Introduction

It has long been recognized that natural killer (NK) cells and macrophages are closely involved in the antitumor immunity of individuals. The antitumor activities of NK cells and macrophages are regulated by interferon-γ (IFN-γ)[1] and some neuropeptides[2], such as β-endorphin [3]. In the present paper, we demonstrate that substance P (SP) and somatostatin (SOM) modulate the NK activity and the macrophage tumoricidal activity *in vitro*. The up-regulation of IFN-γ in the antitumor activities of NK cells and macrophages are suppressed by higher concentrations of SP and SOM. These data suggest that neuroendocrine transmitters may be important modulators in the antitumor immunity of human body.

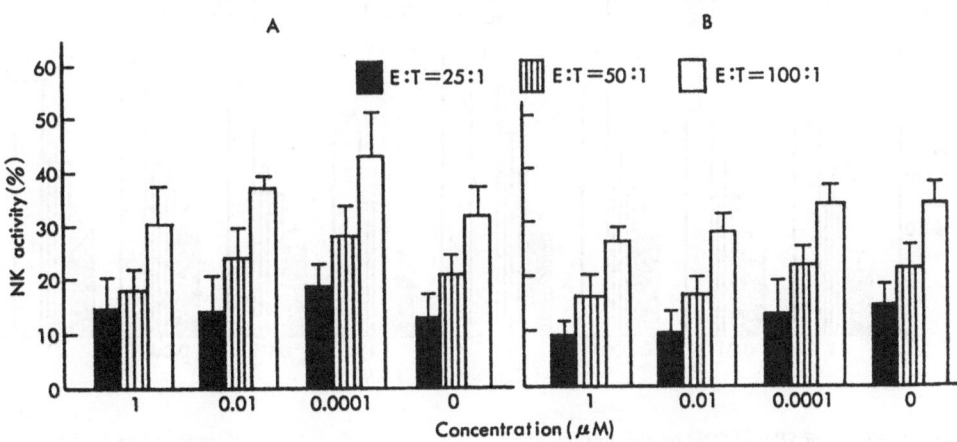

Fig. 1. Effects of SP and SOM on the NK activity of human spleen lymphocytes. # X±SD, n=6; (A) SP, (B)SOM.

Table 1 *Effects of SP on the tumoricidal activity of macrophages from human spleen*

Concentration of SP (μM)	Tumoricidal activity of macrophage[a] (%)	
	Treated without IFN-γ	Treated with IFN-γ
1	17.1±5.3	17.8±4.6**
0.01	20.3±2.9	21.0±2.7**
0.0001	22.1±2.8	25.1±3.5
0	18.6±3.4	28.3±3.6

[a] X±SD, n=4. ** P<0.02 as compared with control.

Results and Discussion

Lymphocytes and macrophages isolated from human spleen were suspended in RPMI 1640 medium with essential supplements and incubated with or without SP, SOM and recombinant human IFN-γ at 37°C in 5% CO_2 respectively. After 16-18h, the lymphocytes and macrophages were collected separately and used to examine the cytotoxicity against the erythroleukemia cell line, K562, by lactate dehydrogenase-release assay. Data are expressed as the mean of the cytotoxicity percentage ± SD.

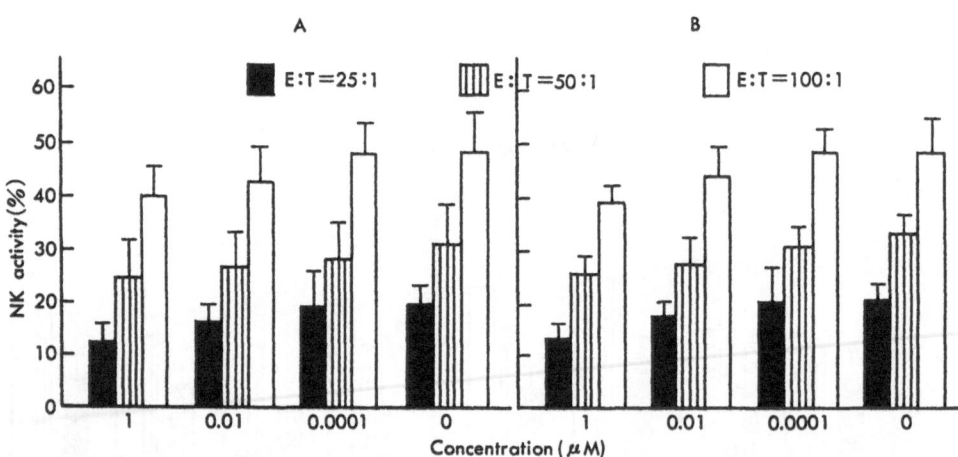

*Fig. 2. Effects of SP and SOM on the up-regulation of IFN-γ in the NK activity of human spleen lymphocytes.
X±SD, n=6. (A) SP, (B) SOM.*

Table 2 *Effects of SOM on the tumoricidal activity of macrophages form human spleen*

Concentration of SOM(μM)	Tumoricidal activity of macrophage[a] (%)	
	Treated without IFN-γ	Treated with IFN-γ
1	10.9±2.4*	17.5±6.4**
0.01	12.8±2.3*	22.1±3.5**
0.0001	17.2±3.6	28.2±4.2
0	18.1±3.5	29.2±3.2

[a] X±SD, n=4. * P<0.05, ** P<0.02 as compared with control.

Effects of SP and SOM on the NK activity

As shown in Fig. 1, the NK activity was significantly enhanced by 0.0001 μM SP at effector to target (E:T) ratios of 100:1, 50:1 and 25:1. Only at an 100:1 E:T ratio, 0.01 μM SP significantly enhanced the NK activity and no enhancement was observed at 1 μM SP in all E:T ratios. In contrast, 0.01 and 1 μM SOM significantly suppressed the NK activity at all E:T ratios and 0.0001 μM SOM appeared no suppressive action.

Effects of SP and SOM on the up-regulation of IFN-γ in The NK activity

The NK activity was significantly enhanced in the lymphocytes treated primely with 200 u/mol hrIFN-γ, but 0.01 or 1 μM SP (at all E:T ratios) resulted in a decrease in the NK activity of the lymphocytes treated primely with SP and 200 u/ml hrIFN-γ. The higher concentrations of SP exhibited a counteraction rather than cooperation or enhancement on the up-regulation of IFN-γ in the NK activity. After the pretreatment of lymphocytes with SOM and 200 u/ml hrIFN-γ, the up-regulation of IFN-γ in the NK activity was significantly counteracted by 1 μM SOM at all E:T ratios, but the counteraction of 0.01 μM SOM was not significant statistically (Fig.2).

Effects of SP and SOM on the macrophage tumoricidal activity

The tumoricidal capacity of macrophages was slightly increased by 0.0001 or 0.001 μM SP and no increase exhibited at 1 μM SP. 0.01 or 1 μM SOM significantly inhibited the tumoricidal capacity of macrophages. SP or SOM (0.01 and 1 μM) added simultaneously with 200 u/ml hrIFN-γ to the macrophage cultures could counteract the up-regulation of IFN-γ in the macrophage tumoricidal activity (Table 1-2).

It is demonstrated in these studies that the effects of SP and SOM on the NK activity and the macrophage tumoricidal activity are opposite. The lower concentrations of SP exhibit an augmentative action, whereas the suppressive action of SOM appears in the higher concentrations of SOM. This phenomenon may be beneficial to the maintenance of normal NK activity and macrophage tumoricidal activity in individuals.

It is noteworthy that the higher concentrations of SP and SOM are able to counteract the up-regulation of IFN-γ in the NK activity and the macrophage tumoricidal activity, and the further investigation should be taken in the effects of SP and SOM on the actions of other cytokines, such as interleukin-2 and tumor necrosis factor, which play an important role in antitumor immunity.

References

1. Pace, J.L., Russell, S.W., Torres, B.A., Johnson, H.M. and Gray, P.W.J., Immunol., 130(1983)2011.
2. Pawlikowski, M., Zelazowski, P., Dohler, K. and Stepien, H., Brain. Behav. Immun., 2(1988)50.
3. Mathews, P.M., Froelich, C.J., Sibbitt, W.L. Jr. and Brankhurst, A.D.J., Immunol., 130(1983)1658.

Knowledge-based computer modelling for the N-terminal domain of carcinoembryonic antigen(CEA)

Jing-chu Luo[b], Paul P. Bates[a] and Michael J.E. Sternberg[a]

*a Biomolecular Modelling Lab. Imperial Cancer Research Fund,
London WC2 3PX, UK*
bDepartment of Biology, Peking University, Beijeng 100871, China

Introduction

Computer modelling for biological macromolecules is an important tool to investigate structure-function relationship of enzymes, antibodies and other bioactive proteins and peptides. Knowledge-based computer modelling [1] has been successfully used to predict tertiary structure of proteins which have sequence similarity with their homologues whose three-dimensional coordinates have been solved by experimental studies, such as X-ray or NMR analysis. This method is becoming more practical as the three-dimensional coordinates in Protein Data Bank(PDB) grows rapidly recently. The fundamental step of this method is to establish sequence alignment between the unknown protein and its parent structure(s). It usually gives a satisfactory result if the sequence homology is above 40-50 percent. Difficulties may arise when the identical residues between two sequences are about 20-30 percent. In this case, conventional alignment programs can only give some initial indication, and knowledge of protein conformation and understanding of biological characteristics about the modelled system are needed to obtain a more accurate result.

This method was used in modelling the N-terminal domain of carcinoembryonic antigen (CEA). CEA is a highly glycosylated cell surface protein with molecular weight about 180,000 daltons and is regarded as tumor markers which are highly distributed in colon, breast and lung tumors. Monoclonal antibodies have been successfully used to locate CEA-marked tumor cells in clinical diagnostics. However, the exact location of antibodies at the surface of CEA is not clear, because there are no available three-dimensional information. Molecular modelling may provide some conformation features of this molecule to guide site-directed mutagenesis in studying the affinity of antibodies against this antigen.

Results and Discussion

The main steps of knowledge-based modelling are as follows:
    1) Collect information from experimental studies;
    2) Find template(s) by sequence analysis;
    3) Define main chains of secondary elements and loops;
    4) Complets side chains;

Table 1 *Sequence comparison of CEA-N and CD2-N, CD4 and Ig domains*

	CD2-N	CD4-N	CD4-2	REI-V	FAB-V	FAB-C
CEA-N	3.11	2.98	0.41	4.09	3.32	0.82

CD4-2 is the second domain of CD4; FAB-V is the variable domain of Fab; FAB-C is the constant domain of Fab.

5) Refine and evaluate the model;
6) Suggest structural information for biological experiments.

Various experimental research on CEA have been carried out in many laboratories, which yields a conclusion that this molecule belongs to immunoglobulin superfamily. Sequence analysis reveals that the extra-cellular region of CEA which consists of 643 amino acids can be divided into seven domains, each of which has sequence similarity with immunoglobulin(Ig) domains. The all β-sheet pattern which is a striking feature in Ig fold was proved by circular dichroism analysis. Modification of Trp residues of CEA gave evidence about the packing role of the β-sheets.

Based on these investigation, an all β-pattern of Ig-like fold for CEA domains can be proposed, though secondary structure prediction programs did not give explicit results. Nevertheless, difficulties arise in modelling the CEA-N domain, because the typical disulfide bond which connects the two sheets in Ig fold does not exist. Fortunately, this question was answered by the newly-solved NMR structure of the N- terminal domain of rat T lymphocyte CD2 antigen(CD2-N) which also belongs to Ig fold and has function and sequence similarities with CEA-N. Structure analysis of CD2-N shows that the two cysteines are replaced by Ile and Val which are tightly packed to each other to form the core of the molecule with other hydrophobic side chains of conserved residues such as the tryptophan at the center of the molecule.

Automatic sequence alignment was then performed between CEA-N and CD2-N, two domains of CD4 which is also a cell surface protein, and some Immunoglobulin variable and constant domains. Apparently, CEA-N should be categorized to Ig variable domain class based on the alignment scores (Table 1). In addition to CD2-N, the N-terminal domain of CD4 (CD4-N) and the variable domain of Bence-Jones immunoglobulin(REI-V) were chosen as templates to build up the framework of CEA-N model.

Although the global fold pattern remains the same, differences such as the lengths of β-strands can be found among the parent structures. In order to define the secondary structure component, structure alignment among the templates was carried interactively using the molecular modelling software from BIOSYM. Multiple sequence alignment was then followed by manually intervention to obtain the best result (Fig. 1). Several points were considered in this step: (1) Maintain the β-strands, i.e. all insertion and deletion should be in the loop regions; (2) Keep the conserved hydrophobic core; (3) Keep the key residues such as Trp and salt bridge link.

The main chains of β-sheet framework were defined using the average lengths of the strands by superimposing the templates, and the coordinates were assigned to CEA-N. The β-turn between A and B strands of CEA-N was taken directly from CD2-N, not

Table 2 *The templates and loop fragments used in modelling the CEA-N domain*

1-4	5-8	9-20	21-23	24-35	36-42	43-48	49-52	53-58
REI-V	2HLA	CD2-N	3MCG	CD2-N	4FAB	CD4-N	2MCP	CD4-N

59-63	64-68	69-70	71-75	76-92	93-98	99-106	107-109
REI-V	CD4-N	CD2-N	CD4-N	CD2-N	2HFL	CD2-N	3FAB

The numbers denote segments of residues of the CEA-N model. The PDB codes are used except for those that are defined in the text.

only because of the same length of these two turns, but also because of the same position of the Gly which is a good β-turn former. This conformation principle was also considered for the E-F loop. On the other hand, variable regions of other loops such as B-C, C-C' and F-G were fitted by searching a structure library which was compiled from a set of Ig fragments. Table 2 shows the corresponding templates and loop framents for each segment of the model.

Upon completing the main chains, side chains were placed using the conventional criteria. For identical or conserved residues, the orientation of parent side chains were taken. For non-conserved substitutions, the standard conformation was first considered and local energy minimizaiton was applied to remove sterical clashes.

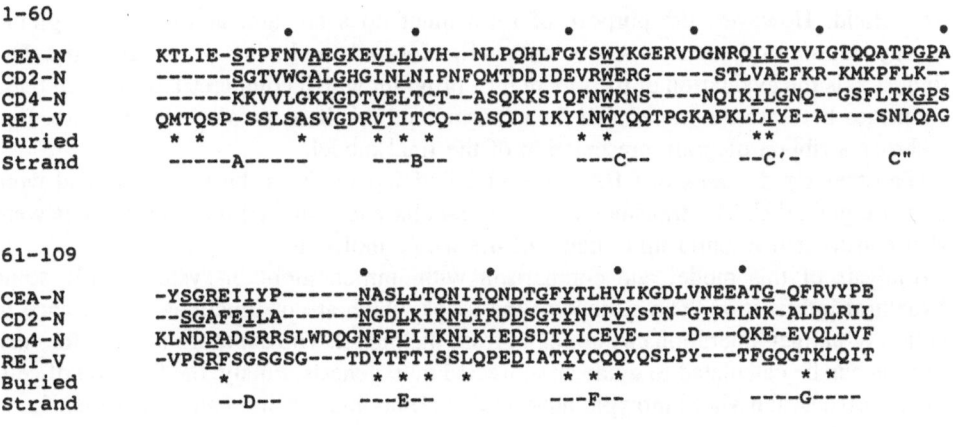

Fig. 1. Sequence Alignment of CEA N-terminal domain with the templates used for modelling[2]. Every tenth residue of CEA-N is marked by a dot. Identical residues are underlined, gaps are denoted by dashes. The asterisks indicate buried residues. The averaged secondary structure elements are labeled A to G.

117

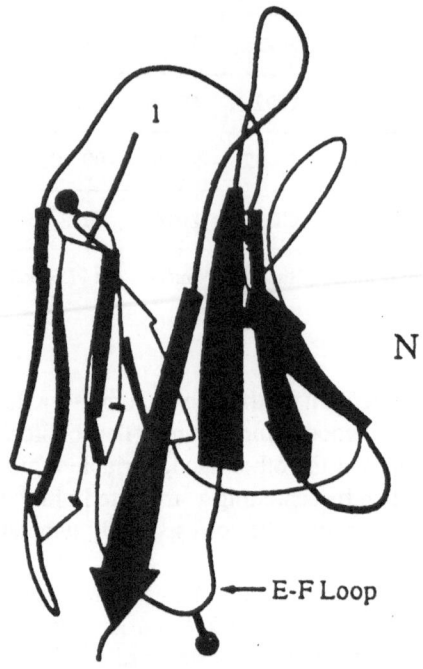

Fig. 2. The ribbon representation of CEA-N model.

The model of CEA-N was refined by energy minimization using molecular mechanics force field. However, the purpose of refinement does not aim at finding the global energy minimum. Instead, it serves as an efficient tool to obtain a good geometry. The final step is the evaluation of the model by a program which compares key characteristic quantities between the model and the experimental results from stucture database. Fig. 2 shows a ribbon diagram presentation of the final model.

The other six domains of CEA were modelled separately in the same way and were linked together[2]. The transmembrane alpha-helix and N-linked oligosaccharides were also constructed to build up a model of the whole molecule.

Analysis of this model and comparison with immunoglobulin system imply some structural information. The surface of the loops which are of functional importance in antibody-antigen interation can be found from the model. The exposure degrees for each residue can be calculated to guide site-directed mutagenesis. Finaly, the function of cell-to-cell association via homotypic adhesion or homodimerization can be postulated.

Knowledge of conformation information of proteins is crucial to sturctural and functional study in molecular biology, especially in systematic engineering of peptides and proteins. Computer-aided molecular modelling can be used to compensate the shortage of availability of experimental data. Knowledge-based modelling is one of the efficient approaches for many biological systems. The key point of this method as discussed in modelling the CEA-N domain is the understanding of the biological characteristics of the modelled system and the command of protein conformation principles. Computer softwares such as sequence alignment and secondary structure prediction programs as well as interactive molecular graphics and energy minimization packages can only be used as tools in various steps of model building. It is unreliable to build up a model merely based on the automatic computer procedures.

Acknowledgements

We thank Drs. P. Driscoll, J. Cyster, I. Campbell and A. Williams for the CD2 coordinates, and Sir Walter Bodmer, H. Durbin and Dr. F. Weba for helpful discussion. Luo J. thanks the support for his visit at ICRF from the China National Center of Biotechnology and Development, and the National Laboratory of Protein Engineering and Plant Genetic Engineering at Peking University.

References

1.   Blundell, T.L., Sibana, M.J., Sternberg, M.J.E. and Thornton, J.M., Nature, 326(1987)347.
2.   Bates, P.P., Luo, J. and Sternberg, M.J.E., FEBS Lett., 301(1992)207.

Poly(ethylene glycol) hybrids of RGD and YIGSR and their inhibitory effect on experimental metastasis

Koichi Kawasaki, Yuko Yamasiro, Machiko Namikawa, Tomohiro Murakami, Toyohiko Mizuta, Takao Hama and Tadanori Mayumi[a]

Faculty of Pharmaceutical Sciences, Kobe-Gakuin University, Ikawadanicho, Nishi-ku, Kobe 651-21, Japan
[a]Faculty of Pharmaceutical Sciences, Osaka University, Yamadaoka 1-6, Suita-shi, Osaka 565, Japan

Introduction

Laminin and fibronectin are glycoproteins, so-called cell adhesion proteins. Laminin contains a Tyr-Ile-Gly-Ser-Arg(YIGSR) sequence and fibronectin, Arg-Gly-Asp(RGD). Both YIGSR[1] and RGD[2] peptides were reported to inhibit experimental metastasis of B16 melanoma BL6 in mice. We synthesized analogs of these peptides to examine their metastasis-inhibiting effect.

Peptides analogs were prepared by the solid-phase method using N^{α}-Boc strategy. p-Methylbenzhydrylamine resin was purchased from the Peptide Institute, Osaka, Japan. The following groups were used for side-chain protection of amino acids: tosyl and nitro groups for Arg; benzyl group for Ser, Tyr and Glu; cyclohexyl group, for Asp. The final deprotection was performed by HF treatment and the products were purified by reverse-phase HPLC. Highly metastatic B16-BL6 cells were used to examine the inhibitory

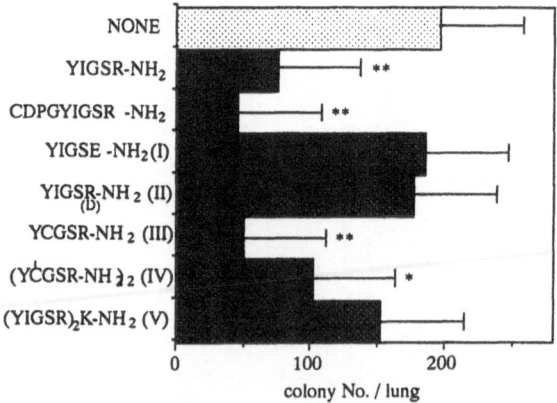

Fig. 1. Effect of various peptides on lung tumor colonizion. B16-BL6 cells ($1 \times 10^5/0.2ml$) were injected i.v. with or without admixing with 1mg of peptides into five mice per group. Lung tumor colonies were examined 21 days later. values were the mean±SD. *, P<0.05; **, P<0.01 compared with untreated control(MEM) by student's t-test.

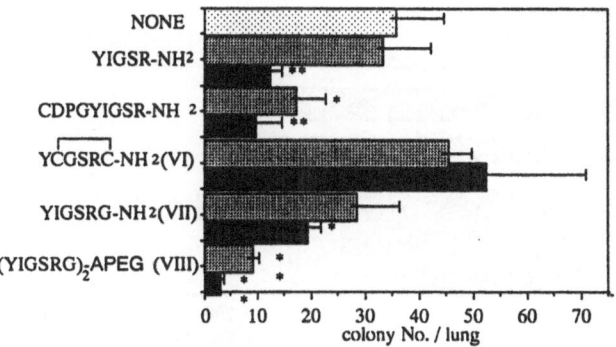

Fig. 2. *Inhibitory effect of synthetic peptides on the formation of lung metastasis. B16-BL6 cells($1 \times 10^5/0.2ml$) were injected i.v. with or without admixing with 0.3mg, 1.0mg or 2.0mg of peptides into five mice per group. Lung tumor colonies were examined 21 days later. Values were the mean±SD. *, $p<0.05$; **, $p<0.01$ compared with untreated control (MEM) by student's t-test.*

effect of synthetic peptides on lung metastasis in mice. In brief, B16-BL6 cells were injected into the tail vein of male mice receiving synthetic peptides. Three weeks later, the mice were killed and melanoma colonies in the lung were counted macroscopically.

Results and Discussion

The following YIGSR analogs were synthesized; YIGSE (I), YIGSR (R=D-from, II), YCGSR (III), (YCGSR)$_2$ (dimer through a disulfide bond, IV) and (YIGSR)$_2$K (dimer through Lys residue, V). Their inhibitory effect is shown in Fig.1. I, II and V were ineffective but III and IV were effective. The inhibitory effects of III and IV were as potent as that of CDPGYIGSR and YIGSR, respectively.

On the basis of recent reports that cyclic YIGSR[3] and poly(YIGSR)[4] are potent inhibitors of experimental metastasis, cyclic and polymer-bound YIGSR analogs were prepared and examined for inhibitory effects on experimental metastasis. YCGSRC, a cyclic peptide having an intramolecular disulfide bond, was prepared and its inhibitory

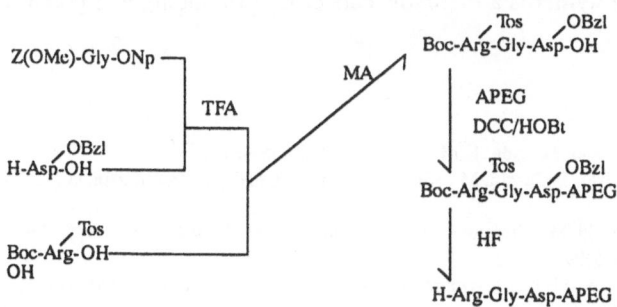

Fig. 3. *Synthetic scheme for RGD-APEG hybrid. MA, mixed anhydride method.*

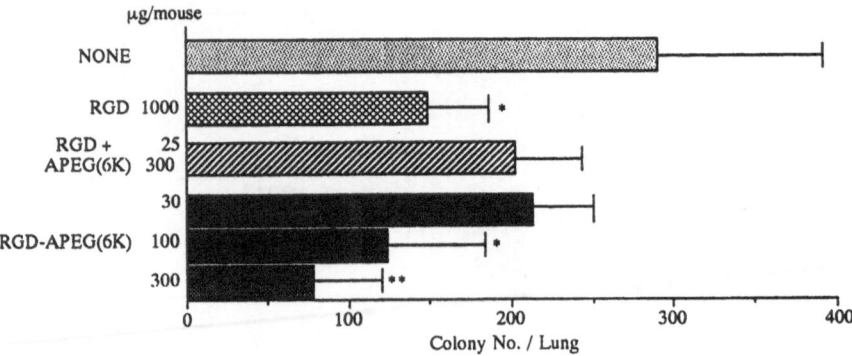

Fig. 4. Inhibition of lung colonization by RGD-APEG. The tumor colonization assay was carried out as described in Fig.2. Values were the mean ± SD. *, P<0.05; **, P<0.005 compared with untreated control by Student's t-test.

effect was examined. The cyclic analog was inactive as shown in Fig.2. Polymer-bound YIGSR analog was then prepared. Poly(ethylene glycol) (PEG) was employed as the polymer, because PEG has non-toxic, highly soluble, stable and non-immunogenic features. The hydroxyl group of PEG was converted into the amino group according to the procedure reported by Mutter[5], and couple with Fmoc-YIGSRG-OH by the diphenylphosphoryl azide(DPPA) method. The Fmoc group on the product was removed by piperidine treatment.

The inhibitory effect of the hybrid was more potent than that of YIGSRG. Next, RGD-PEG hybrid was prepared as shown in Fig.3; its inhibitory effect is shown in Fig.4.

For preparation of RGD hybrids, two different types of PEG were used: one was PEG #4000 (4K; MW 3000-3700) and the other, PEG #6000 (6K; MW 7300-9000). These PEGs were converted into amino PEG(APEG). Since RGD-APEG(6K) exhibited almost identical effects at 200 μg/mouse and 600 μg/mouse, it was further examined at more diluted concentrations. As shown in Fig.4, the inhibitory effect of 0.1 mg of hybrid 6K nearly equaled that of 1 mg of RGD. Thus it can be said that the inhibitory effect of hybrid 6K is 10 times and 230 times as potent as that of RGD in terms of weight and molar ratios, respectively. Polymer is a favorable carrier for prolonging and potentiating peptide activity.

References

1. Humphries, M.J., Olden, K. and Yamada, K.M., Science, 233(1986)467.
2. Iwamoto, Y., Robey, F.A., Graf, J., Sasaki, M., Kleinman, H.K., Yamada, Y. and Martin, G.R., Science, 238(1987)1132.
3. Kleinman, H.K., Graf, J., Sasaki, M., Kleinman, H.K., Yamada, Y., Martin, R.G. and Robey, F.A., Arch. Biochem. Biophys., 272(1989)39.
4. Saiki, I., Murata, J., Iida, J., Nishi, N., Sugimura, K. and Azuma, I., Br. J. Cancer, 59(1989)194.
5. Pillai, V.N.R. and Mutter, M., J. Org. Chem., 45(1980)5364.

Session IV
Receptor and recognition

Chairs: Yu-cang Du
Shanghai Institute of Biochemistry, CAS
Shanghai, China

and

Lawrence J. Slieker
Lilly Coporate Laboratories
Indianapolis, Indiana, U.S.A.

Oligomerization *in vitro* of the extracellular domain of the insulin receptor by non-covalent interactions is accompanied by increased binding affinity, decreased binding capacity and curvilinear Scatchard plots

Jan Markussen, Lauge Schäffer and Mogens Christensen
Novo Research Institute, Novo Allé, DK-2880 Bagsvard, Denmark

Introduction

The association behavior of the insulin receptor has been studied using free receptors in solution either as insulin receptor halves ($\alpha\beta$, the heterodimer) or as holoreceptors ($\alpha_2\beta_2$, the heterotetramer) as well as receptors fixed in cells membranes. The successful association and reoxidation of receptor halves prepared by reduction of the holoreceptor requires the presence of insulin[1]. Cross-linking experiments using the bivalent reagent disuccinimidyl suberate showed that the insulin-induced conformational changes occur in the α subunits, since α_2 species were formed by cross-linking[2]. The insulin induced association into $\alpha_2\beta_2$ holoreceptor could be inhibited by iodoacetamide, whereas Mn^{2+} or Mg^{2+} plus ATP induced association was unaffected by the sulfhydryl reagent[3]. Ligand mediated covalent aggregation of soluble holoreceptors was demonstrated by SDS-polyacrylamide gel electrophoresis. Addition of the sulfhydryl reagent N-ethylmaleimide inhibited receptor oligomerization, indicating that disulfide bonds may be involved[4]. Ligand induced oligomerization of solubilized insulin receptors by non-covalent interactions was demonstrated using non-denaturing polyacrylamide gel electrophoresis, i.e. electrophoresis in the absence of SDS (native PAGE). Insulin binding properties and autophosphorylation properties did not vary in a parallel fashion, e.g. the high molecular species, possibly a trimer, showed the lowest insulin binding capacity and the highest degree of phosphorylation[5]. Similarly, a high tyrosine kinase activity and a low binding capacity of higher oligomers was demonstrated using radiation inactivation[6]. Ligand induced oligomerization is a general phenomenon among growth factor receptors. Thus, epidermal growth factor induces oligomerization of the soluble, extracellular domain of its receptor[7]. Using fluorescent insulin derivatives and fibroblastic cells, an insulin induced aggregation and internalization of the hormone receptor complex could be demonstrated at 23°C and 37°C[8,9]. The insulin induced decrease in receptor number, downregulation, was shown to be accompanied by an increase in binding affinity in rat hepatoma cells[10].

We wish to report a ligand-independent, non-covalent oligomerization of purified sIR[11], which is the soluble insulin receptor truncated in the β-subunit at the first amino acid of the transmembrane region, i.e. after residue 929 using the numbering system of Ebina (Ebina et al., 1985). By using native gel electrophoresis (without SDS) we observed oligomerization in highly concentrated samples of pure sIR stored in the frozed state and during deglycosylation in 3 M urea. As crystallization from a population of various aggregates is unlikely to occur, formation of aggregates should be avoided. Treatments that would reverse oligomers into the monomeric form were sought. We report the separation of oligomers from the monomeric form by gel filtration and their binding properties.

Results and Discussion

Figure 1 shows he chromatogram of a gel filtration of sIR after storage at pH 7.8 in the frozen state at −18°C for 9 months, concentration 25 μM (8.5 mg/ml). Despite the pronounced appearance of two peaks the separation is incomplete, as the first pool still contains the monomeric form and the last pool still contains traces of the oligomers, as demonstrated by native PAGE. However, a partial separation has been achieved. The insulin binding was characterized for the proteins in pools I and IV by Scatchard plots (see Fig. 2A). Both the monomeric form in pool IV and the oligomeric forms in pool I show curvilinear plots, in contrast to the linear plot obtained using freshly prepared sIR. The initial slopes of the fractions of stored sIR are steeper than that of fresh sIR, indicating an increase in binding affinity. Concomitantly, binding capacity decreases from the 2 molecules of insulin per molecule of sIR found in fresh preparations[11]. Treatment of freshly prepared sIR with neuraminidase in an acetate buffer at pH 5.0 caused very little oligomer formation, and the slopes of the Scatchard plots of neuraminidase-treated and untreated sIR are both linear and not significantly different (Fig. 2B). Freezing and storage at 18°C for only 2 weeks followed by neuraminidase treatment also resulted in heavy oligomerization. The Scatchard plots of the monomeric and oligomeric fractions from the gel filtration of this combined treatment are shown in Fig. 2C.

The curves are very similar to those obtained from storage alone (Fig. 2A). Besides storage, treatment of freshly prepared sIR with combinations of glycohydrolases causes fomation of oligomers (Fig. 2D). The treatment combined neuraminidase at pH 5.0 followed by glycopeptidase F plus O-glycosidase in 3 M urea at pH 7.5. Obviously, both storage in the frozen state and partial deglycosylation in 3M urea lead to oligomerization

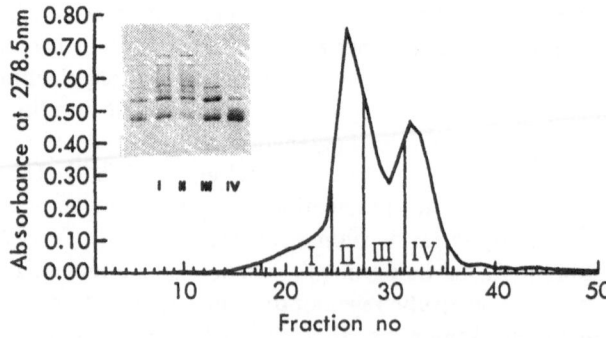

Fig. 1. Gelfiltration of sIR after storage, using a column of Sephacryl S-300 and Tris/HCl buffer at pH 7.8. The insert shows native, gradient PAGE (4-15%) of stored sIR (left lane), and of pools I-IV.

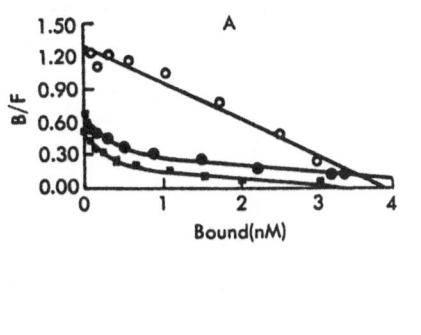
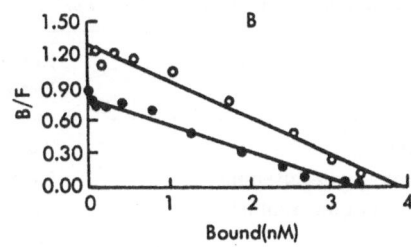
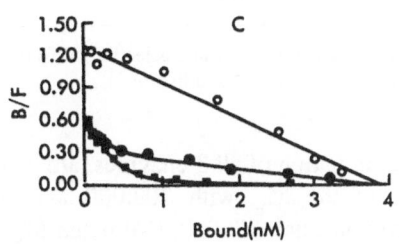
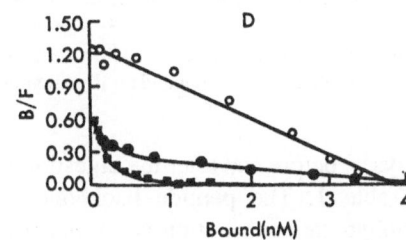

Fig. 2. Scatchard plots of monomeric and oligomeric sIR fractions obtained by gel filtrations after storage (2A), after neuraminidase digestion (2B), after storage and neuraminidase (2C), and after digestion with glycosidases in 3 M urea (2D).

and to an increase in binding affinity, most markedly in the oligomeric fractions but also in the remaining monomeric form.

The neuraminidase treatment efficiently removes sialic acid residues, as demonstrated by decreased mobility using native gel electrophoresis, and by a change in isoelectric pH from a broad smear around pH 4.5 to a narrow band at pH 5.8 as determined by isoelectric focusing (Fig.3). Two other glycohydrolases, glycopeptidase F and O-glycosidase which cleave at the root of the carbohydrate side chains, were used. Digestion was performed at pH 7.8 and in 3 M urea in order to make the cleavage sites as accessible as possible without denaturing the insulin binding capability. Nevertheless, the cleavage was incomplete, probably due to steric hindrance. The results of the carbohydrate

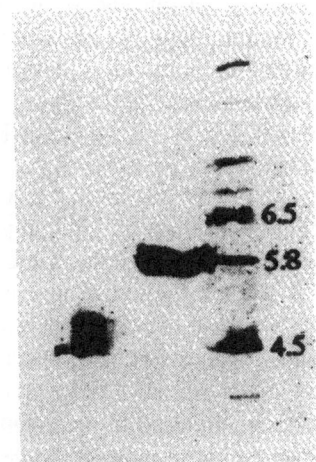

Fig. 3. Isoelectric focusing before (left lane) and after (right lane) neuraminidase digestion.

127

Table 1 *Analysis of carbohydrate after acid hydrolysis of sIR before and after treatment with a combination of 3 glycohydrolases, i.e. neuraminidase, glycopeptidase F and O-glycosidase*

	Residues per molecule of sIR($\alpha_2\beta'_2$)	
	Native sIR	Hydrolyzed sIR
Fucose	7.4	1.8
N-acetyl galactosamine	6.6	5.4
N-acetyl glucosamine	110	87
Galactose	28.4	14.8
Glucose	5.6	1.6
Mannose	78.6	51.2
M_w of carbohydrate excl. sialic acid	42000	30000
Sialic acid (estimated)[a]	80	0
M_w of carbohydrate incl. sialic acid	65000	30000

a. The estimate of 80 residues of sialic acid/molecule sIR is based upon the shift in isoelectric pH from 4.5 to 5.8 as a result of neuraminidase digestion. Sialic acid is destroyed during acid hydrolysis.

analyses, before and after the digestion with the combination of all 3 enzymes, are given in Table 1. The peptide backbone contributes to the M_w with 208,000 and the carbohydrate contribution is estimated to be 65,000, in total 273,000. Estimated M_w of sIR by SDS electrophoresis yields 125,000 and 45,000 for the α- and β-subunits, respectively, or a total of 340,000.

Conclusions

The soluble, extracellular domain of the insulin receptor (sIR, -exon 11) binds 2 molecules of insulin and shows linear Scatchard plots with a K_d about 4 nM[11]. It oligomerizes during storage at $-18°C$ in concentrations of 10-25 mg/ml in Tris/HCl at pH 7.8 and by partial deglycosylation in 3 M urea using a mixture of glycopeptidase F and O-glycosidase, After oligomerization both the monomeric and oligomeric forms of sIR bind to insulin more like the human insulin holoreceptor (hIR), showing curvilinear Scatchard plots and lower K_d as estimated from the steep slope. The oligomers were separated by gelfiltration and visualized by native gel electrophoresis in the absence of SDS. In the presence of SDS only a single band, corresponding to the monomeric form, was observed, showing the non-covalent nature of the association. Attempts to convert the higher molecular oligomers to the monomeric forms with milder reagents than SDS, e.g. 3 M urea at pH 5 and pH 8.8, failed. Treatment of sIR with neuraminidase alone efficiently desialylates the molecules, since isoelectric focusing results in changes from a broad smear around pH 4.5 to a narrow band about pH5.8. Desialylation alone using fresh sIR results neither in oligomerization nor in curvilinearity of the Scatchard plot.

The formation of heterogeneous mixtures of sIR monomers and oligomers is likely to hamper attempts to crystallize sIR, whereas the conversion to desialylated molecules, more homogeneous with respect to isoelectric pH, should be beneficial for crystallization.

Acknowledgements

We are grateful to L. Drube, H.Kronstrom, and V.S. Pedersen for excellent technical assistance.

References

1. Böni-Schnetzler, M., Kaligiaan, A., DelVecchio, R. and Pilch, P.F., J.Biol. Chem., 263(1988)6822.
2. Waugh, S.M. and Pilch, P.F., Biochemistry, 28(1989)2722.
3. Wilden, P.A., Morrison, B.D. and Pessin, J.E., Biochemistry, 28(1989)785.
4. Chen, J.-J., Kosower, N.S., Petryshyn, R. and London, I.M., J. Biol. Chem., 261(1986)902.
5. Kubar, J. and Van Obberghen, E., Biochemistry 28(1989)1086.
6. Fujita-Yamaguchi, Y., Harmon, J.T. and Kathuria, S., Biochemistry, 28(1989)4556.
7. Lax, I., Mitra, A.K., Ravera, C., Hurwitz, D.R., Rubinstein, M., Ullrich, A., Stroud, R.M. and Schlessinger, J., J. Biol. Chem., 266(1991)13828.
8. Schlessinger, J., Schechter, Y., Willingham, M.C. and Pastan, I., Proc. Natl. Acad. Sci. U.S.A., 75(1978)2659.
9. Schlessinger, J., Van Obberghen, E. and Kahn, C.R., Nature, 286(1980)729.
10. Crettaz, M., Jiala, I., Kasuga, M. and Kahn, C.R., J. Biol. Chem., 259(1984)11543.
11. Markussen, J., Halstrom, J., Wiberg, F.C. and Schäffer, L., J. Biol. Chem., 266(1991)18814.

Synthesis and insulin-receptor binding activity of two insulin-related peptides

Cheng-fu Wang[a], Xin-tang Zhang[b] and Ching-I Niu[b]

[a] *Shantou University Medical College, Shantou 515031, China*
[b] *Shanghai Institute of Biochemistry, Chinese Academy of Sciences,*
Shanghai 200031,China

Introduction

On the basis of the hypothesis about the binding of insulin to its receptor, H-Phe-Phe-Val-Leu-Tyr-Gly-OH and Cyclo (Phe-Phe-Val-Leu-Tyr-Gly) were synthesized by the Boc solid-phase synthesis strategy. The synthetic peptides were detached by M TMSOTf-thioanisole/TFA and purified on silica gel 60 or Sephadex G10 and RP-HPLC. The linear peptide was cyclized by the azide method. Insulin receptor binding assay showed that the cyclohexapeptide had a definite capacity of binding to insulin-receptor, while the linear hexapeptide had no or very week such ability at high concentrations.

Results and Discussion

Boc-Phe-Phe-Val-Leu-Tyr(O-Bzl)-Gly-resin was synthsized with N^{α}-BOC protected amino acids. After cleavage from the resin by M TMSOTf-thioanisole/TFA[1] and purification on Sephadex G10 and RP-HPLC (Fig.1), H-Phe-Phe-Val-Leu-Tyr-Gly-OH was obtained.

Boc-Phe-Phe-Val-Leu-Tyr(O-Bzl)-Gly-resin was cleaved by hydrazine hydrate to produce Boc-Phe-phe-Val-Leu-Tyr(O-Bzl)-Gly-NHNH$_2$. After Boc was removed by TFA, the ensuing product was cyclized by the azide method with pyridine as the solvent. The cyclo-(Phe-Phe-Val-Leu-Tyr(O-Bzl)-Gly) was obtained after precipitating with ether. The Bzl was removed by M TMSOTf-thioanisole/TFA. After purification on silica gel 60 and RP-HPLC (Fig.2), cyclo(Phe-Phe-Val-Leu-Tyr-Gly) was obtained. No spot was visualized with ninhydrin.

The linear hexapeptide 176μg was dissolved in 30μl of CH$_3$CN and 350 ul of 150 mM PH7.5 Tris-HCl buffer. The cyclohexapeptide 160μg was dissolved in 40μl of CH$_3$CN, 420 μl of 150 mM, pH 7.5, Tris-HCl buffer. The insulin-receptor binding assay was perfermed according to the method discribed by Feng [2].

It is proposed that insulin has a hydrophobic domain related to the binding of insulin to its receptor. The insulin-receptor binding assay showed that the cyclohexapeptide has a definite capacity of binding to insulin-receptor, while the linear hexapeptide has no or very week such ability. The result suggests that the cyclohexapeptide is to some extent analogous to the hydrophobic domain of insulin, with regard to hydrophobicity and conformation.

Table 1 *Percent displacement of ^{125}I-Insulin by the linear hexapeptide and the cyclohexapeptide*

Peptides concerned	Concentration(μM)	Percent displacement(%)
Insulin		100
The cyclohexapeptide	48.0	26.6
The linear hexapeptide	62.2	9.0

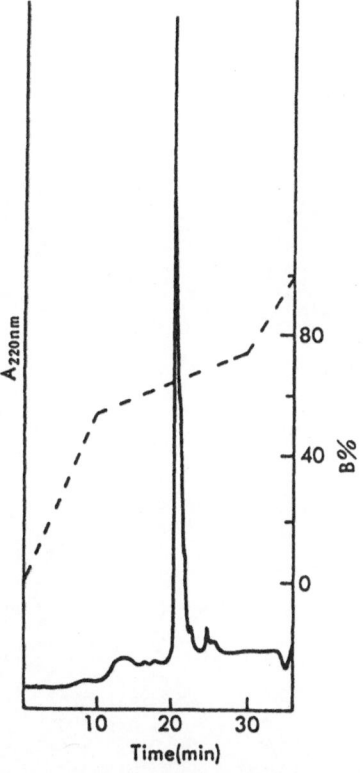

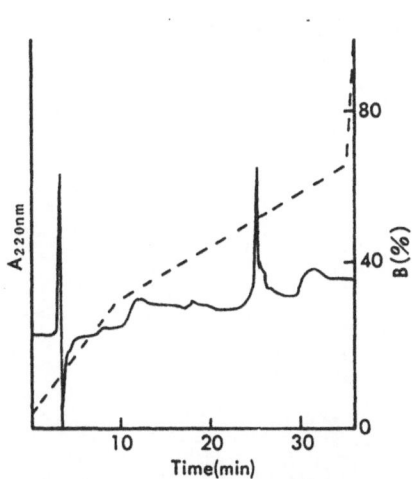

Fig. 1. *RP-HPLC analysis of H-Phe-Phe-Val-Leu-Tyr-Gly-OH. Column: Ultra-Sphere Octyl C18 (Beckman) (4.6×250mm); Solvent system: A=0.1% TFA; B=0.1% TFA-80% CH₃CN; Flow rate: 1 ml/min.*

Fig. 2. *RP-HPLC analysis of cyclo-(Phe-Phe-Val-Leu-Tyr-Gly). Column: Ultra-sphere Octyl C18 (Beckman) (4.6×250mm); Solvent system: A=0.1% TFA; B=0.1% TFA-80% CH₃CN; Flow rate: 1 ml/min.*

A number of peptides have been found to antagonize or enhance the functions of insulin[3]. This information suggests that the functions of insulin can be influenced or regulated by other peptides. Therefore, it is of significance to study the insulin-related peptides.

131

C.F. Wang, X.T. Zhang and C.I. Niu

Acknowledgement

This work is supported by the National Nature Science Foundation(3880193)

References

1. Fuji, N., J. Chem. Soc. Chem. Commun., (1987)174.
2. Feng, Y.M., Acta Biochem. Biophys. Sin., 14(1982)137.
3. Ma, J.N., Diabetes, 40(1991)1218.

The studies on synthesis, bioactivity and paramagnetism application of spin labelled angiotensin (AngII•R)

Xiao-yu Hu[a], Xiao-xu Li[a,*], Guo-lin Yang[a], Zhu-yin Wang[a], Chang-lin Li[b] and Jin-fen Lu[b]

[a]Department of Biology, Lanzhou University, Lanzhou 730000, China
[b]National Research Laboratories of Natural and Biomimetic Drugs, Beijing Medical University, Beijing 100083, China

Introduction

Spin labilling is an effective method that greatly deepened the knowledge in structure, movement and interaction of biological molecule. For example it had been used in studying the interaction of antigen and antibody successfully[1]. The spin labelled TP-5 and its bioactivity was reported recently. Results suggested that the stable free radical compounds are not only informational but also functional[2].

Ang II plays an important role both in maintenance of normal blood pressure and in occurrence of hypertension. We modified Ang II with a stable free radical for two purpose. One is about the possibility of labelled Ang II for a candidate of antagonist, the other is to study Ang II•R's ESR spectra before and after its binding to Ang II receptor and to detect interaction information of Ang II and its receptor because the different spectra of spin label on Ang II might be obtained by ESP determination when Ang II•R is in different state of restrain, finally to find an effective, simple method in studying the molecular environment of peptide and receptor interaction.

Results and Discussion

Synthesis of Ang II and Spin Labelled Ang II

Ang II was synthesized by SPPS method and Boc strategy. Spin labelled Ang II had been made. A stable free radical (2,2,5,5,-tetramethyl-3-ene-4-carboxyl-1-oxyl-pyrrolidin) was assembled to the first amino acid residue (Asp) of Ang by formation of amide bond. Purifications of Ang II and Ang II•R were carried out with gel filtration (Sephadex G-25). These peptides were shown to be 95% pure in HPLC. Calculated value for M.W. of Ang II were in agreement with that obtained by FAB MS [protonated molecular ion peak $(M+1)^+$ was 1047]. Amino acid composition were consistent with expected values (spectra of data are not shown). In ESR spectrum Ang II•R gave typical nitroxide free radical signal characterized by isotropic three peeks and suggested a successful labelling of Ang II. (Fig. 1c)

---

* Present address: Northwest Nationalities College, Lanzhou 730030, China

Table 1 *Bioactivity results of Ang II•R and Ang II*

		Ang II•R	Ang II	P value
Receptor Binding Assay	$Ic_{50}(M)$	1.5×10^{-8}	1.4×10^{-8}	>0.05
Action on blood vessel	$Ec_{50}(M)$	6.7×10^{-8}	2.0×10^{-8}	<0.05
Action on cholescyst	$Ec_{50}(M)$	4.3×10^{-7}	2.7×10^{-8}	<0.05
Action on blood pressure	Ed_{50} (μg/kg)	7.26	1.84	<0.01

Bioactivity of Ang II•R

I. Receptor binding assay

Ang II•R and Ang II of various concentrations were incubated with freshly prepared rat adrenal microsomes in Tris buffer containing ^{125}I-Ang II (25000 cpm/45ul) for 60mins at 37°C. Then ice-cold Tris buffer was added, the bound and free radioactivities were separated by filtration through a glass fiber filter. The trapped radioactivity was determined in a γ counter (Beckman 5500), together with data of unspecific binding and total binding, Ang II•R inhibited the binding of ^{125}I-Ang II concentration dependently. Value of Ic_{50} is listed in Table 1.

II. Action of blood vessel and cholescyst muscle

Respectively, the rat aorta helical strips and guinea pig cholescyst strips were prepared and bathed in Kreb's solution bubbled continuously with 5% CO_2 in O_2. The tissues were treated with Ang II•R of different concentrations and the cumulative contractive responses were recorded with a force-displacement transducer (NIHON KOHDEN) connected to a polygraph. Response was expressed as a percent of the maximal Ang II response by the same tissue. Ec_{50} of Ang II•R on both vessel and cholescyst were obtained from the effect-concentration curves (Table 1)

III. Action on rat blood pressure

Ang II•R of different dosages was injected (0.1 ml) through glossal vein of anesthetized rats whose carotid artery cathers were connected to a pressure transducer (NIHON KOHDEN) coupled to a polygraph and the changes in MAP were recorded. The effect on pressure was expressed as a percent of the maximal Ang II response and Ed_{50} of Ang II•R was obtained (Table 1).

IV. Observation on antagonism of Ang II•R

In rat aorta contraction model and pressure measurement model, Ang II•R was added or injected prior to Ang II, and its counteraction against Ang II was observed within the non agonist concentration of dosage and it was expressed as follows:

$$\left(1 - \frac{\text{observed contraction}}{\text{maximal contraction}} \right) \times 100\%$$

in contraction model and

$$\left(1 - \frac{\text{observed net increase of MAP}}{\text{maximal net increase of MAP}} \right) \times 100\%$$

in pressure model. The inhibition percent may reach 35% and 50% respectively.

ESR spectra measurement of Ang II•R

ESR spectrum of Ang II•R (Fig. 1c) was obtained on ESR spectrometer (Brucker ESP300) and it showed stable nitroxide free radical three peaks with isotropy, calculated τ_c value is $(5.6\pm0.5)\times10^{-11}$ sec (n=6). After binding reaction of Ang II•R (10^{-2}M) with Ang II receptor preparation (concentration of membrane protein: 200 mg/ml), the Ang II•R-receptor complex was gathered by centrifugation or by fast filtration and gave signals as Fig. 1b (centrifugation sample) and Fig. 1a (filtration sample) via ESR measurements. In spectrum b and a, the existence of anisotropy suggested the restrain of Ang II•R and immobilization peaks (indicated by arrows) revealed Ang II•R's tighten binding to receptor.

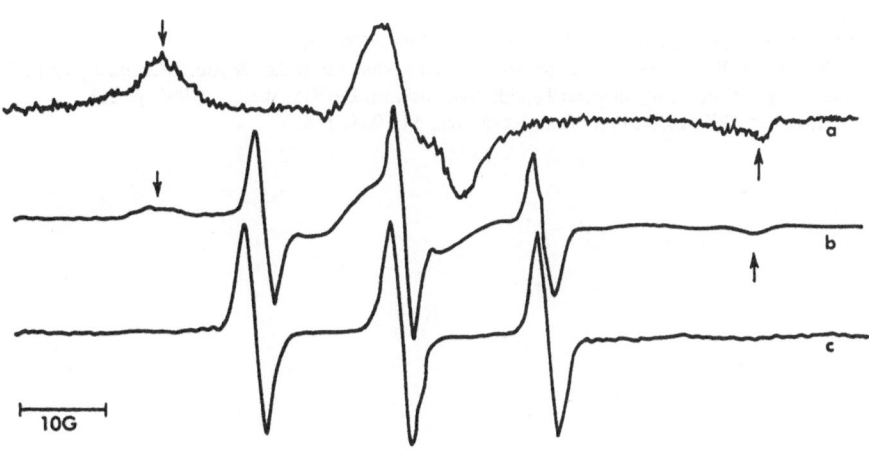

Fig. 1. ESR spectra of Ang II•R.

135

Conclusion

In blood vessel contraction model and blood pressure model, the decrease in activity of Ang II•R is owing to modification of N-terminal of Ang II. We suggest that the contribution of N-terminal to activity is larger than that to binding ability, and the site of binding and activity is not identical, above results also indicate Ang II•R's partial antagonism. Further study is needed to verify its possibility as antihypertension drug.

The high affinity of Ang II•R to receptor indicates that the labelling to Ang II almost does not interfere the Ang II molecular system, it is a suitable paramagnetical label. Immobilization signals in spectra a and b suggest that R• is restrained in Ang II•R-receptor complex, the N-terminal of Ang might be in the deep position of complex, so does the receptor activity site. The τ_c value of spectrum a is estimated to be 10^{-8} sec or so, the intensively prolonged τ_c value also reflects immobilization of R• group[3]. We also note that the magnetic field range in which strong immobilization peaks appeared in spectrum a is similar to that in spectrum b, while nonimmobilization peaks in spectrum b appeared in same magnetic field range as those in spectrum c. Spectrum b is overlapped one from spectrum a and c. In these spectra, immobilization peaks come from bound Ang II•R and the nonimmobilization peaks from unbound or absorbed Ang II•R. We also conclude that filtration method is more effective than centrifugation method for separating free Ang II•R from bound one.

Acknowledgement

This work was partly supported by National Laboratories of Natural and Biomimetic Drugs. Analysis Center of Lanzhou University helped us with FAB MS and HPLC.

References

1. Hsia, J. and Piette, L., Arch. Biochem. Biophys., 129(1969)296.
2. Hu, X., Young, K. and Wen, Y., In Smith, J.A. and Rivier, J.E. (Eds.) Peptides: Chemistry and Biology (Proceedings of the 12th American Peptide Symposium), ESCOM, Leiden, 1992, p. 235.
3. Stryer, L. and Griffith, O., Proc. Natl. Acad. Sci., 54(1965) 1785.

Insulin receptor binding, insulin receptor kinase activity, and stimulation of insulin-like growth factor II binding to isolated rat adipocytes

Xin-zhi Tang, De-min Liu, Jia-yi Wang, Wen-yan Niu and Fu-yun Dong
*Department of Biochemistry and Analysis Centre, Tianjin Medical College,
Tianjin 300070, China*

Introduction

Insulin-like growth factor II (IGF-II) is one of a family of growth factors closely related to insulin in its amino acid sequence and structure. It has been demonstrated that the receptor for IGF-II is distinct from the insulin and insulin-like growth factor I receptors. IGF-II receptor does not bind insulin. However. insulin promotes increased binding of IGF-II to its receptor on adipocytes. Receptor regulation by homologous hormone was demonstrated for IGF-II.

Preparation of fat cells

Fat cells were isolated from the epididymal fat pads of 150-200g male Sprague-Dawly rats by the method described by Rodbell. The fat pads were excised, chopped with scissors and put into a freshly made Krebs-Henseleit buffer, pH7.4 containing 3% bovine serum albumin and 1 mg/ml collagenase (type II, Sigma). Isolated fat cells were then filtered through nylon mesh and washed twice with the above buffer. The number of fat cells was estimated as described by Hirsch et al.

Binding of ^{125}I-IGF-II to IGF-II receptor

Purified IGF-II was iodinated (100-150 ci/g) by lactoperoxidase method as described by Oppenheimer. Isolated fat cells were incubated with or without 10 nmol porcine insulin for 15 min at 37°C . The cells were then incubated with 0.4 nmol ^{125}I-IGF-II for 40 min at 24°C in a total volume of 400µl. The amount of ^{125}I-IGF-II bound was determined by the oil flotation method.

Solubilization and partial purification of insulin receptor

After incubating isolated adipocytes with or without insulin (10ng/ml) for 15 min at 37°C, the reactions were terminated by adding solubilization buffer and immediately freezing the cells (for 1 h at −80°C). Cells were then thawed, homogenized, and the cell extract was further centrifuged 100,000g for 1 h at 4°C to collect the supernatant. The supernatant was applied to a column (0.5 × 3.0cm) of wheat germ agglutinin-agarose;

and after washing with Hepes buffer, pH 7.6, and 0.1% Triton X-100, the bound material was eluted with the above buffer containing 0.3 mol N-acetyl-D-glucosamine.

Determination of tyrosine kinase activity

Aliquots of partially purified insulin receptor were incubated with or without 10 nmol insulin for 30 min at 22°C in 50 nmol Hepes buffer, pH 7.6 the phosphorylation reaction was initiated by the addition of poly (Glu, Tyr) (4:1), and 5 μmol $[\gamma\text{-}^{32}P]$ ATP as described by Zick et al. The reaction was terminated by spotting 80μl of reaction solution onto Whatman No.3M paper. The radioactivity of the paper was determined in β-counter.

Results and Discussion

Insulin stimulation binding of ^{125}I-IGF-II to adipocytes

After cells were preincubated with insulin (10 nmol/ml) at 37°C for 15 min, ^{125}I-IGF-II was then added at 24°C for 40 min and binding analyzed. We found that at 1 nmol ^{125}I-IGF-II, binding in the presence of insulin is about increased by 4 fold (not shown). The stimulation of IGF-II binding is dose-dependent on insulin, as shown in Fig.1, Cells were preincubated with insulin at concentration ranging from 0 to 10 nmol for 15 min at 37°C before analysis of ^{125}I-IGF-II binding. The half-maximal effect is about 0.08 nmol, with saturation achieved at 1 nmol insulin. Scatchard the analysis of IGF-II binding to intact adipocytes indicates that this effect is due to an increase in receptor affinity, from K_a=0.047 $nmol^{-1}$ in the absence of insulin to K_a =0.16 $nmol^{-1}$ in the presence of 10 nmol insulin, without a change in the number of cell surface binding sites (Fig.2).

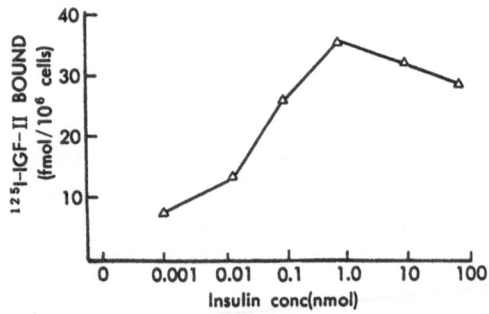

Fig. 1. Insulin dose-response curve for stimulation of ^{125}I- IGF-II binding to adipocytes.

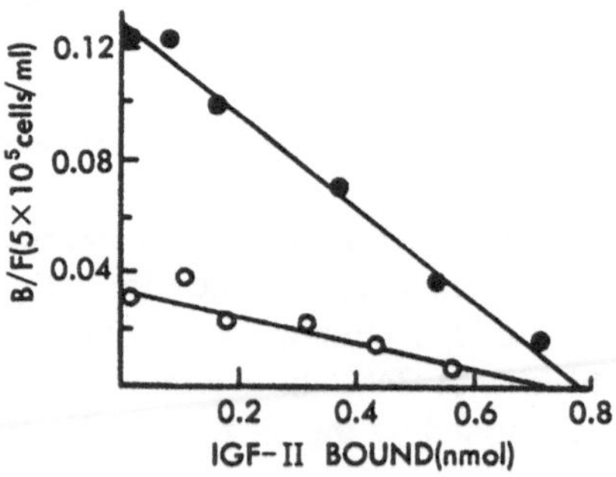

Fig. 2. Scatchard analysis of ^{125}I-IGF-II binding to intact adipocytes.

The effect of KCN on IGF-II binding to intact adipocytes

Isolated cells were incubated at 37°C for 15 min in the presence of 0 or 10 nmol insulin plus 5 additional min in the presence of KCN added to a final concentration of

1.0 mmol. ^{125}I-IGF-II binding was then measured at 24°C for 40 min and analyzed as described above. We found that insulin treatment appears to result in a 5-fold increase in apparent receptor number without a change in apparent receptor affinity (Fig.3). The result suggested that the presence of KCN during the ^{125}I-IGF-II binding blocked IGF-II receptor recycling and internalization, and IGF-II-receptor complexes remain at the surface of the cell.

Activation of the insulin receptor tyrosine kinase by insulin at its correlation with IGF-II binding

Adipocytes insulin receptor purified by wheat germ agglutinin chromatogarphy were assayed for tyrosine kinase activity using the synthetic substrate poly (Glu, Tyr 4:1) as described above. We found that insulin activated the insulin receptor tyrosine kinase, when intact adipocytes were incubated to insulin (10 ng/ml) at 37°C for 15min. By plotting the extent of activation of the tyrosine kinase versus the extent of augmentation of ^{125}I-IGF-II binding, a very significant positive correlation between tyrosine kinase activity and ^{125}I-IGF-II binding was observed (r=0.919, P<0.001), (n=6) (Fig.4). It was demonstrated that the stimulation of IGF-II receptor binding by insulin is mediated via activation of the insulin receptor tyrosine kinase.

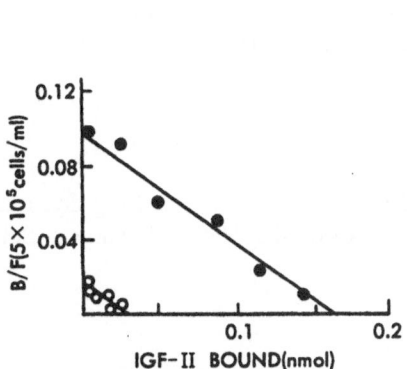

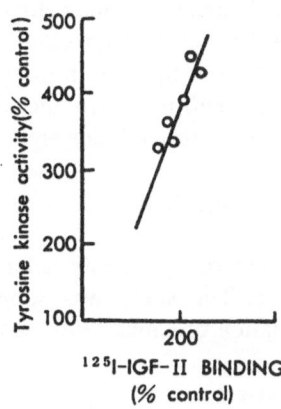

Fig. 4. Correlation between tyrosine kinase activity and ^{125}I-IGF-II binding to adipocytes.

Fig. 3. Scatchard analysis of KCN effect on specific ^{125}I-IGF-II binding to intact adipocytes.

The effect of dithiothreitol on insulin binding to adipocytes

Adipocytes were treated for 10 min at 37°C with 1 mmol dithiothreitol (DTT), and cells then were washed 3 times with Krebs-Henseleit buffer, pH 7.4 and resuspended for use in the insulin binding assay. Fig. 5 showed that a high concentration (1 mmol) of DTT caused about 90% inhibition of binding but there was no significant inhibition of insulin stimulation of receptor kinase activity (not shown). The data is consistent with the space receptor hypothesis.

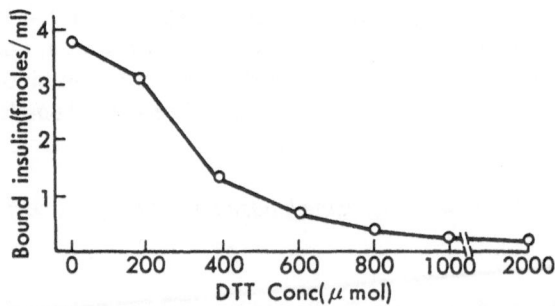

Fig. 5. The effect of dithiothreitol on insulin binding to adipocytes (5×10⁵ cells/ml) at 15°C for 1 h.

Summary

Insulin can activate the tyrosine kinase activity of the insulin receptor and up-regulate IGF-II receptor in intact rat adipocyte system during short-term incubating. Our study is complementary in indicating a significant positive correlation between the activation of insulin receptor-tyrosine kinase and IGF-II binding. Although the molecular mechanism of the insulin receptor kinase for inducing up-regulation of IGF-II receptor binding is still unknown, these observation further support the role of insulin receptor kinase in transmembrane signalling.

Acknowledgements

The authors gratefully acknowledge the assistance of Miss Gao Ling for making pictures. The study was supported by a grant from the National Natural Science Foundation of China (39070243).

References

1.   Werther, G.A., Hintz, R. L. and Rosenfeld, R. G., Horm. Metabol. Res., 21(1989)109.
2.   Douen, A.G. and Jones, M.N., Receptor Res., 10(1990)45.

Structure-function relationships in the thrombin receptor agonist peptides

Dong-Mei Feng[a], Daniel F. Veber[a], Tom Connolly[b] and Ruth F. Nutt[a]
[a]Departments of Medicinal Chemistry, and
[b]Biological Chemistry, Merck Research Laboratories, West Point, PA 19486, U.S.A.

Introduction

Thrombin interacts with its receptor on platelets to cause platelet aggregation. According to a novel mechanism of receptor activation [1], thrombin cleaves its receptor to create a new amino terminus, which can then act as a tethered ligand to activate the receptor. A 14-amino acid peptide derived from the sequence of the new amino terminus is able to fully activate human platelets and cause aggregation in the absence of thrombin. Structure-function studies of this agonist peptide are described.

Peptide analogs of Ser-Phe-Leu-Leu-Arg-Asn-Pro-Asn-Asp-Lys-Tyr-Glu-Pro-Phe-OH (TRP 42-55) were synthesized incorporating modification which would help elucidate structural features critical for thrombin receptor activation. Chemical syntheses were carried out by the solid phase method using an ABI 430 A instrument, HF deprotection,

Table 1 *Structure-function studies of thrombin receptor agonist peptide*

TRP	Amino Acid Sequence	Platelet Aggregation EC_{50} (mM)
42-55	Ser-Phe-Leu-Leu-Arg-Asn-Pro-Asn-Asp-Lys-Tyr-Glu-Pro-Phe-OH	10
	Phe-Ser-Leu-Leu-Arg-Asn-Pro-Asn-Asp-Lys-Tyr-Glu-Pro-Phe-OH	>800
42-51	Ser-Phe-Leu-Leu-Arg-Asn-Pro-Asn-Asp-Lys-OH	8
42-51	Ser-Phe-Leu-Leu-Arg-Asn-Pro-Asn-Asp-Lys-NH_2	2
42-48	Ser-Phe-Leu-Leu-Arg-Asn-Pro-NH_2	0.5
42-47	Ser-Phe-Leu-Leu-Arg-Asn-NH_2	0.2
42-46	Ser-Phe-Leu-Leu-Arg-NH_2	1
42-45	Ser-Phe-Leu-Leu-NH_2	185
	Residue substitutions with alanine in receptor peptide (42-46)	
Ala^{42}	Ala-Phe-Leu-Leu-Arg-NH_2	0.8
Ala^{43}	Ser-Ala-Leu-Leu-Arg-NH_2	>800
Ala^{44}	Ser-Phe-Ala-Leu-Arg-NH_2	1
Ala^{45}	Ser-Phe-Leu-Ala-Arg-NH_2	15
Ala^{46}	Ser-Phe-Leu-Leu-Ala-NH_2	70

and purification by HPLC. Products were characterized by HPLC (>99% purity), amino acid analysis and FABMS. Biological potencies of analogs were measured as peptide effects on the rate and extent of human blood platelet aggregation.

Results and Discussion

As shown in Table 1, the N-terminal dipeptide of TRP 42-55 is critical for receptor activation. Sequence inversion of these two residues results in the total loss of activation potency. C-terminal shortening of the 14-peptide to five residues retains receptor activating potency (TRP 42-46). C-terminal amides are more potent than acids. Ser-42 and Leu-44 can be replaced by Ala with complete retention of potency. The hydrophobic sidechain of Phe-43 is critical for receptor activation.

References

1.   Vu, T.-K.H., Hung, D.T., Wheaton, V.I. and Coughlin, S.R., Cell, 64(1991)1057.

Melanotropic peptides and melanoma cell receptors

Jin-wen Jiang[a], Shelley Nakamura[a], Shubh D. Sharma[b], Victor J. Hruby[b] and Mac E. Hadley[a]

[a]Departments of Anatomy and [b]Chemistry, College of Medicine, University of Arizona, Tucson, AZ 85724, U.S.A.

Introduction

α-Melanocyte stimulating hormone (α-MSH) stimulates pigment cells and is a well known regulator of melanocyte activity in the skin of many vertebrates [1]. α-MSH, Ac-Ser-Tyr-Ser-Met-Glu-His-Phe-Arg-Trp-Gly-Lys-Pro-Val-NH$_2$, increases the production of melanin in melanocytes and turns skin dark. Several synthetic congeners of α-MSH, e.g. [Nle^4,D-Phe^7]α-MSH, exhibited superpotent and prolonged activity compared to the native α-MSH in the tyrosinase assay [2,3].

Several groups have used receptor binding techniques to identify melanotropin receptors in mice and in several human melanoma cell lines [4,5]. However, the binding assay did not explain the lack of receptor expression in some of the human melanoma cell lines. Neither did this technique give information about the possible homogeneity or heterogeneity within individual cell lines, for it determines only the overall binding of the radioactive probe to all cells. We have synthesized macromolecular conjugates in which multiple copies of the hormone analog and multiple copies of a fluorophore were attached to a macromolecule [6]. We used a conjugate in which a MSH analog has been attached to polyvinylalcohol (PVA) through a disulfide (-S-S-) linkage to visually demonstrate the presence of melanotropin receptors in various human melanoma cells lines.

Results and Discussion

To determine whether melanotropin receptors exist in human melanoma cells, we used a number of human melanoma cell lines , both melanotic and amelanotic. We found: 1) The MSH-conjugate bound to all mouse and human melanoma cells (Table 1). 2) No binding to MCF-7 nor to normal mouse spleen and liver cells used as negative controls was observed. 3) Capping phenomena were found in most cells of both mouse and human melanoma, suggesting that the conjugate might subsequently become internalized. 4) DTT pretreated MSH-conjugate, which removes MSHs from the macromolecule, failed to bind to melanoma cells. 5) In competition experiments, incubation of MSH-conjugate with melanoma cells which were pre-exposed to unbound hormone, did not result in any fluorescence labeling. 6) All cells of every mouse and every human melanoma cell bound the fluorescent melanotropin-conjugate. These results suggest that there is little or no cellular heterogeneity between or within any one cell line, and that melanotropin receptors are specific membrane markers for cells of melanotic (melanocyte) origin. These rceptors may allow the site-specific delivery of diagnostic

Table 1 *Binding specificity of melanotropin receptors on various melanoma types by fluorescent MSH-macromolecular conjugate*

Cell Type	MSH-Conjugate[b]	DTT[c]-Pretreated Conjugate	Competition
Mouse B16 Melanoma	+	−	−
Human Melanoma:			
LH[a]	+	−	−
WC[a]	+	−	−
LR-1714[a]	+	−	−
LR-1649	+	−	−
LR-1650	+	−	−
Human Breast Cancer			
MCF-7	−		
Normal Mouse Cell Types			
Spleen	−		
Liver	−		

+ indicates fluorescence labeling; − indicates no fluorescence labeling; [a]amelanotic cell lines; [b]MSH-S-S-PVA-Fluorescein; [c]DTT=Dithiothreitol.

and chemotherapeutic drugs either by melanotropin-drug conjugates or appropriate antibodies.

Acknowledgement

This work supported in part by a research grant from the U.S. Public Health Service (DK-17420 to VJH).

References

1. Hadley, M.E., (1992) Endocrinology, 3rd Ed. Prentice-Hall, Inc., Englewood Cliffs, N.J.
2. Marwan, M.M., Abdel-Malek, Z.Z., Kreutzfeld, K.L., Hadley, M.E., Wilkes, B.C., Hruby, V.J. and Castrucci, A.M.L., Mol. Cell Endocrinol., 41(1985)171.
3. Abdel Malek, Z.A., Kreutzfeld, K.L., Marwan, M.M., Hadley, M.E., Hruby, V.J. and Wilkes, B.C., Cancer Res., 45(1985)4735.
4. Siegrist, W., Oestreicher, M., Stutz, S., Girard, J. and Eberle, A., J. Receptor Res., 8(1988)323.
5. Tatro, J.B., Atkins, M., Mier, J.W., Hardarson, S., Wolfe, H., Smith, T., Entwistle, M.L. and Reichlin, S., J. Clin. Invest., 85(1990)1825.
6. Sharma, S.D., Hruby, V.J., Hadley, M.E., Granberry, M.E. and Leong, S.P.L., In Smith, J.A. and Rivier, J.E. (Eds.) Peptides: Chemistry and Biology (Proceedings of the 12th American Peptide Symposium), ESCOM, Leiden, 1992, p. 599.

Isolation and purification of μ-opioid binding protein from rat brain

De-he Zhou[a], Xue-jun Xu[a], Hong-ping Zhang[a], Chong-hu Ni[a], Zhong-kwi Zeng[b] and Zhi-qiang Chi[a]

[a] *Shanghai Institute of Materia Medica, Chinese Academy of Sciences, Shanghai 200031, China*
[b] *Sichuan University, Chendu 610064, China*

Introduction

Isolation and purification of active subtype opioid receptor is an important and rather difficult problem in the neurosciences. The opioid receptors are extremely low of content in the brain and fragile on exposure to many of the relatively gentle detergents that have been successfully used to solubilize other receptors. In addition, opioid receptors are heterogeneous, at least three different classes are present in the brain or in peripheral tissues, μ-, δ- and κ- which differ from each other in their ligand selectivity. Most opioid ligands are not selective for a single receptor type, so it is difficult to isolate a specific subtype receptor. In studying on selective ligands to opioid receptor subtype, we found that ohmefentanyl(OMF) is a novel highly selective agonist for μ-opioid receptor. We thought that OMF could be used in purifying μ-opioid receptor. In the present study , we report that the isolation and purification of active μ-opioid binding protein by using OMF affinity chromatography column and Vicia Bungei Ohwi lectin (VBL) affinity chromatography column, and purified μ-opioid receptor was characterized.

Results and Discussion

Solubilization of opioid receptor from rat brain and binding assay with ^{3}H-OMF
By the method previously used in our laboratory, a crude rat brain membrane fraction(P_2) was prepared. Digitonin(1%) and Mg_2SO_4 (5mM) were added to P_2 fraction. After stirring for 45 min at 4°C and centrifugation at 100,000 × g for 60 min, the 1300-100 adsorbent resins were added to the supernatant at a ratio of 0.5g/ml. The mixture was stirred at 0°C for 60min to adsorb the excess detergent. The protein concentration were determined by a modified coomassie blue assay. Binding assay for ^{3}H-OMF to opioid receptor was performed by PEG precipitation and gel filtration method. The results showed some characteristics of the digitonin-solubilized opioid receptor were similar to those of P_2 fraction.

Two steps of affinity chromatography
The succinyl-DMF was covalently coupled to $NH_2(CH_2)_6NH$-agarose in the presence of carbodiimide in 50% aqueous ethylene glycol , pH5.5, at 20°C, stirring for 16 hours.

145

VBL is a lectin whose suppressive sugar was mannose. We have first found VBL to have the sugar affinity to opioid receptor. VBL was coupled to Sepharose by CNBr method. The digitonin-solubilized opioid receptor was applied to OMF affinity chromatography column and specifically eluted by OMF(10μM) in Tris buffer containing 0.1% digitonin. The elution was directly applied to VBL-Sepharose column and specifically eluted by 0.1 M α-methyl-D-mannoside in the same buffer. OMF and mannoside were removed from the elution component by 1300-100 resin. Thus the μ-opioid binding protein purified by two steps of affinity chromatography was enriched about 1400 fold as determined by ^{3}H-OMF binding assay.

Characterization of purified μ-opioid binding protein
In radio-ligand receptor assay, the binding of ^{3}H-OMF to purified opioid binding protein was specific and saturable. Scatchard analysis showed a linear plot with Kd value of 7.9 nM and Bmax=12.9 pmol/mg protein. The binding of ^{3}H-OMF to purified μ-opioid binding protein could be inhibited by morphine, naloxone and OMF with the corresponding IC_{50} (μM): 0.42,0.35 and 0.0023. ^{3}H-U69593, a selective for κ-opioid receptor and ^{3}H-DPDPE, a selective for δ-opioid receptor binding to purified protein were tested, but both bound capacities were extremely low.

By using the ELISA method, the purified protein could cross-react with the anti-idiotypic anti-OMF antibody which could specifically recognize and bind to μ-opioid receptor in rat brain membrane fraction (P_2). The results showed that purified μ-opiod binding protein should be μ-opiod receptor which had a molecular weight of about 45K as determined by SDS-PAGE.

Conclusion

The μ-opioid binding protein purified from rat membrane fraction(P_2) by using two steps of affinity chromatography was characterized as a μ-opioid receptor with molecular weight of 45K. The purified μ-opioid receptor was enriched about 1400 fold.

Acknowledgement

The project is supported by the National Natural Science Foundation of China (No. 938900706).

References

1. Wang, F. and Chi, Z.Q., Acta Pharmacol. Sin., 8(1987)490.
2. Xu, H., Chen, J. and Chi, Z.Q., Sci. Sin. [B], 28(1985)504.
3. Zhou, D.H., Zhang, H.P., Xu, X.J., Ni, C.H., Chen, H.W. and Chi, Z.Q., In Du, Y.C., Tam, J.P. and Zhang, Y.S. (Eds.) Peptides: Biology and Chemistry, ESCOM, Leiden, 1993, pp. 147–148.

Recognition and binding of anti-idiotypic anti-ohmefentanyl antibody to μ-opioid receptor

De-he Zhou, Hong-ping Zhang, Xue-jun Xu, Chong-hu Ni,
Hong-wei Chen and Zhi-qiang Chi
Shanghai Institute of Materia Medica, Chinese Academy of Sciences,
Shanghai 200031, China

Introduction

Antibodies that specifically recognize and bind membrane receptors were developed by the anti-ligand anti-idiotypic (anti-Id) method. The rational of anti-Id antibody formation is based on the ability of an antibody generated against a specific ligand to display ligand-binding characteristics similar to the natural receptor of the ligand. Ohmefentanyl(OMF) is a highly selective agonist for μ-opioid receptor, and has been used as an affinity chromatography ligand for purifying μ-opioid receptor from rat brain in our laboratory. In this study, we show that anti-Id technique is applicable to opiod receptor system. An anti-Id anti-OMF antibody was formed against anti-OMF antibody and their *in vitro* interaction with μ-opioid receptor was characterized.

Results and Discussion

Production and characterization of anti-OMF antisera

The antigen, OMF-Suc-BSA was synthesized by coupling succinyl-OMF to bovine serum albumin in the presence of carbodiimide in saline, pH 5.5, at 20°C for 5 h. The conjugate contained 16-18 mol of OMF per mol of OMF-Suc-BSA as determined by spectrophotometer. Three male white rabbits were immunized with 6 mg of OMF-Suc-BSA emulsified in complete Freund adjuvant injected in the hind foot pad. Subsequent booster injections were given in lymph node in hind leg biweekly in incomplete Freund adjuvant. At ten weeks after initial immunization, animals were bled. The anti-OMF antiserum analyzed by radioimmunoassay showed high-affinity specific binding to ^{3}H-OMF with a Kd of 2.3×10^{-7} M, it could also cross-react with succinyl-OMF and Fentanyl, but not with a variety of opiate ligands such as morphine, naloxone, levorphanol, p-7521 and U-50488H.

Production of anti-Id anti-OMF antibody

Immunogen, anti-OMF IgG was purified from rabbit anti-OMF antisera by Membrane Affinity Separation System (NYGENE Corp.). The characteristics of the purified anti-OMF IgG were quite similar to those of antisera. The anti-OMF IgG was emulsified with complete Freund adjuvant, and injected subcutaneously into six Guinea pigs. Five booster injection were done. The activity of the anti-Id anti-OMF antisera were analyzed

147

by immunoprecipitation on the Agarose gel plate.

Anti-Id anti-OMF antibody binding to opioid receptor
In order to study the interaction between anti-Id anti-OMF antibody and opioid receptor, the radio-ligand receptor binding assay were performed. The anti-Id anti-OMF antisera showed dose-dependent inhibition in binding of ^{3}H-OMF to rat brain membrane fraction(P_2) and digitonin-solubilized opioid receptor. At a 1:12 dilution of anti-Id antisera, the maximal inhibition of ^{3}H-OMF binding to P_2 fraction and digitonin-solubilized receptor were 68% and 64% respectively. In the same experimental conditions, the anti-Id antisera could not inhibit the ^{3}H-U$_3$ 69593 (a selective agonist for κ-opioid receptor) and ^{3}H-DPDPE(a selective agonist for σ-opioid receptor) binding to P_2 fraction. The results showed that anti-Id anti-OMF antibody could bind specifically with μ-opioid receptor, but could not bind with κ- or σ-opioid receptor.

Western blotting analysis of μ-opioid receptor
In order to purify μ-opioid receptor, the immunoaffinity chromatography gel were prepared. The anti-Id anti-OMF IgG was isolated using Protein-A Agarose column and coupled to CNBr-activated Sepharose. The eluent of digitonin-solubilized P_2 fraction adsorbed by the anti-Id anti-OMF IgG-linked affinity column were concentrated and analyzed by SDS-PAGE under reducing conditions. In order to identify the purified μ-opioid receptor, western blot was performed. The separated proteins were transferred from gel to nitrocellulose membrane by Semi-Dry Electrophoresis Transfer Cell(Bio-Rad) at 90 mA for 45 min. The nitrocellulose strip was blocked with 3% chicken egg albumin, followed treatment with anti-Id anti-OMF antisera, anti-OMF antisera, horseradish peroxidase linked-anti rabbit IgG goat IgG, A single band with an apparent molecular weight of about 50K was shown.

Conclusion
Anti-Id anti-OMF antibody could recognize and bind specifically with μ-opioid receptor, and it is a powerfully tool for studying and purifying opioid receptor.

Acknowledgement

The Project is supported by the National Natural Science Foundation of China (No. 938900706).

References

1. Satoh, T., Mori, M., Murakami, M., Iriuchijima, T., Yamada, M., Kobayashi, I. and Kobayashi, S., Neuropeptides, 18(1991)121.
2. Xu, H., Chen, J. and Chi, Z.Q., Sci. Sin.[B], 28(1985)504.

148

A nonlysosomal endocytic pathway of high density lipoprotein(HDL) by rat sinusoidal liver cell

Man-ping Wu, Pei-fang Chen, Qi-shan Chen, Mei-zhen Mei, Zu-de Xu and Da-mei Rong

Department of Biochemistry, School of Pharmacy, Shanghai Medical University, Shanghai 200032, China

Introduction

HDL takes part in the cholesterol reverse transport and has anti-atherosclerosis effect. Recently the study on its mechanism has been progressing rapidly. First, the HDL receptors on the membrane of liver and some peripheral cells have been isolated and purified[1], and live cells were found to intake selectively cholesterol ester (CE) in HDL. Second, hepatic lipase were found to takes part in this selective in take[2]. How can liver cells intake selectively CE in HDL and what happens after the binding of HDL to its receptor on liver cell? Searching for the answer to these questions are the purpose of our study.

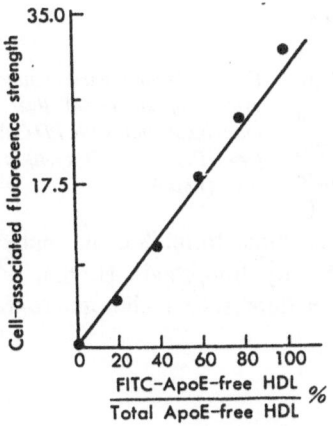

Fig. 1. Binding of FITC-ApoE-free HDL on rat sinusoidal liver cells in competition with unlabelled ApoE-free HDL.

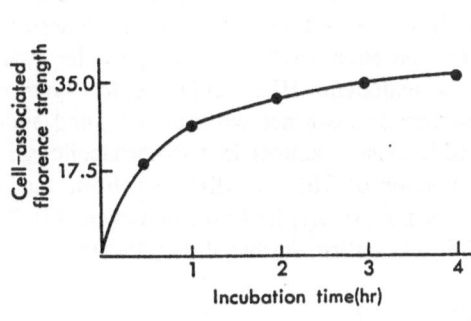

Fig. 2. Rat sinusoidal liver cells were incubated with FITC-ApoE-free HDL at 37°C in different time.

149

Table 1 *Cells-bound flnorescence strongth (FS) after incubation with FITC-HDL*

	incubation at 37°C	incubation at 0°C
(1) total cell-associated FS	32.84±0.78[a]	12.67±0.67
(2) cell-associated TCA soluble FS	0.075±0.078	
(3) (2)/(1)%	0.23	

a, X±SE, n=2.

Results and Discussion

We got HDL (d 1.063-1.211 g/ml) by ultra-centrifugation. In order to eliminate intervention of apoprotein E(ApoE) in HDL, we prepared ApoE-free HDL by heparinsepharose CL-4B affinity columm. We used fluorescein isothiocyanate (FITC) to label ApoE-free HDL. The maximum excitation and emission wavelenth of FITC-ApoE-free HDL is the same as that of FITC (494 and 519 nm respectively). FITC label doesn't affect the electrophoretic and biologic properties of natural HDL (Fig. 1-2). Therefore, we could use FITC-ApoE-free HDL as a tracer to study the intracellular events after receptor binding of HDL. Tab. 1 shows the cell incubation results. These results suggest that the endocytosis of HDL particles really happened, and endocytic HDL didn't enter lysosome, because endocytosis can not work at 0°C, and endocytic FICT-HDL existed almost in trichloroacetic acid (TCA)-insoluble form. So, the endocytic behavior of HDL is different from that of low density lipoprotein (LDL), which undergoes truely the lysosomal events[3]. The results of fluorescence photomicrographs (Fig.3) confirm further the endocytosis of HDL.

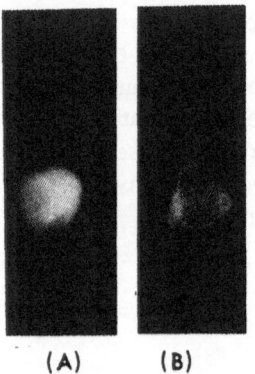

(A)   (B)

Fig. 3. Fluorescence photomicrographs of sinusoidal liver cells after incubation with FITC-ApoE-free HDL. A. at 37°C (×400); B. at 0°C (×400).

References

1. Tozuka, M. and Fidge, N., Biochem. J., 261(1989)239.
2. Wu, M.P., Jin, Y.P., Xu, T., Chen, B.F., Zhang, Q.Q. and Mei, M.Z., Acta Biochem. Biophys. Sinica, 22(1990)163.
3. Goldstein, J.L. and Brown, M.S., Annu. Rev. Biochem., 46(1977)897.

150

Characterization of insulin binding to rat spinal cord cell membrane

Xin-jian Jiang[a], Shu-qiong Lin[a], Shang-quan Zhu[b], Xin-tang Zhang[b], Ming-hua Xu[b] and Ying Ye[b]

[a]Department of Rehabilitation Medicine, Shanghai First People's Hospital, Shanghai 200085, China
[b]Shanghai Institute of Biochemistry, Chinese Academy of Sciences, Shanghai 200031, China

Introduction

Insulin receptors are widely distributed in the central nervous system (CNS)[1]. Kinetic evidences showed that characteristics of insulin receptor in the brain were similar to those in the peripheral. The present study was conducted to characterize insulin receptors in rat spinal cord cell membrane.

Results and Discussion

Binding of ^{125}I-insulin to cell membrane obtained by mechanically dissociated rat spinal cord was specific (Fig.1), time dependent, pH dependent and saturable, with optimum pH, pH7.8 (Fig.2) and reached equilibrium within 24 h at 4°C (Fig.3). The specific binding of ^{125}I-insulin was 2.79% at 4°C, within the range from 0.3% in pituitary to 4.6% in olfactory bulb[1]. At 37°C the nonspecific binding was rather high and the total amount of bound insulin was lower than that at 4°C. The addition of Triton-100 resulted in a decrease in bound insulin. The $t_{1/2}$ of dissociation was 5 min. At 37°C, addition of

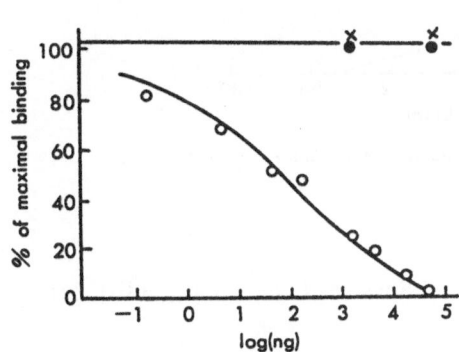

Fig. 1. Specificity of insulin binding in rat spinal cord cell membrane. O--O insulin; ●--● A-chain; ×--× Bovine growth hormone.

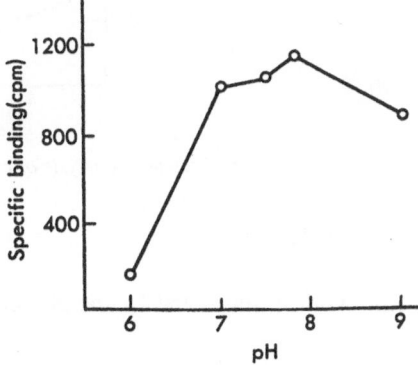

Fig. 2. Effect of pH on ^{125}I-insulin binding to the membrane.

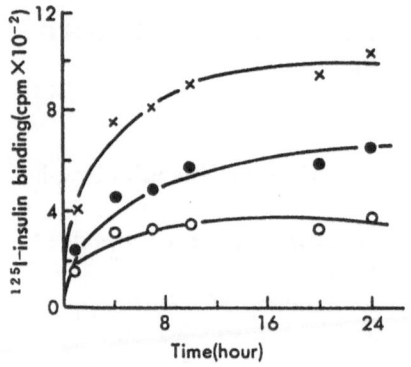

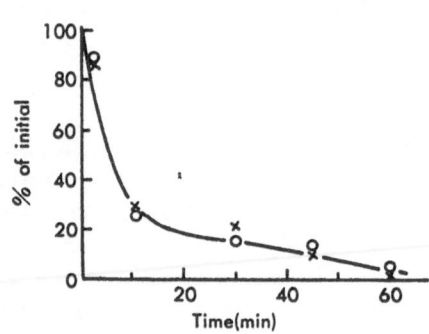

Fig. 3. Time course of insulin binding at 4°C. ×--× total bound; ●--● specific bound; O--O non-specific bound.

Fig. 4. Time course of dissociation of ^{125}I-insulin from membrane. O--O with 0.1 mg/ml insulin; ×--× withou insulin.

excessive unlabeled insulin did not increase the dissociation rate (Fig.4), suggesting that the receptors are not influenced by negative cooperative interaction. Scatchard analysis (Fig.5) revealed two classes of binding sites with different affinity. The dissociation constants (Kd) of high and low affinity sites are about 62.6 nM and 7000 nM respectively, and the binding capacities are 21 pmol/mg protein and 72.3 pmol/mg protein.

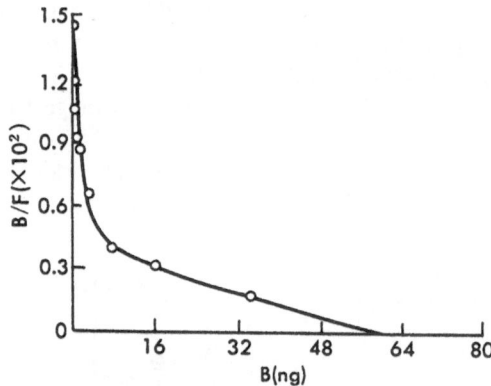

Fig. 5. Scatchard analysis of receptor binding in rat spinal cord cell membrane.

Acknowledgement

This work is supported by the National Natural Sciences Foundation (39070224).

Reference

1.   Havrankova, J. and Roth, J., Nature, 272(1978)827.

Characterization of bacterial membrane transport peptides using toxophoric agent, N-OH-Ala

Nam-joo Hong

Yeungnam University, Gyungsan City, Gyungbuk 712-749, Korea

Introduction

The factor that determines the recognition of peptides by microbial peptide transport systems has been the subject of numerous investigations[1]. Early indications that this process could be exploited therapeutically were demonstrated in E.coli. Their normally impermanent amino acid transport systems, were shown to enter these cells by a peptide carrier mechanism when incorporated into the backbone of a peptide by α-linkage on glycine[2]. Intracellular hydrolytic release of the glycine amino group allows the exiting group to be expelled and express its antibacterial properties. Many natural and synthetic examples involving peptides that contain growth inhibiting amino acids have been reported and considerable interest has been expressed in utilizing this approach as a means of developing novel chemotherapeutic agents[3-5]. In this paper, we describe the use of N-hydroxyalanine (N-OH-Ala) as the warhead component of peptides and the *in vitro* activity of these peptides against *E.coli*.

Results and Discussion

Peptides required for evaluation of membrane transport concept through permease were prepared by sloution method. In order to demonstrate the feasibility of such peptide carriers, DL-N-OH-Ala was selected under the assumption that N-hydroxyamino acids would be potent irreversible inhibitors of pyridoxal-9'-dependent enzyme[6], since DL-N-OH-Ala itself is slightly acitive (see Fig. 1) due to its poor transport peptide synthones containing D-amino acids are not, or are at best poor substrates for peptide permeases. In the dipeptide conjugate, L-Ala-Gly(α-DL-N-OH-Ala), whose DL-N-OH-Ala was linked through the primary amino group to the α-carbon of a glycine residue, the carboxyl terminal of the conjugate proved to be 26 times more active than the corresponding underivatized amino acid. In an attempt to establish the relative ratio between LL and LD-isomers, hydrolysis of N-hydroxyalanyl di- and tripeptides with aminopeptidase resulted in the formation of glyoxylate containing N-hydroxyalanine. The amount of glyoxylate released was measured with lactic acid dehydrogenase (LDH) and NADH. The result showed that a value of 43% of LL was obtained from the dipeptide and 29% of LLL was obtained from the tripeptide. This result coincides with NMR determinations from the peak intensities of the group at 2.0 ppm.

It is of interest to note that the tripeptide conjugate, L-Ala-L-Ala-Gly(α-DL-N-OH-Ala), proved to exhibit greatly reduced activity. This result is consistent with the

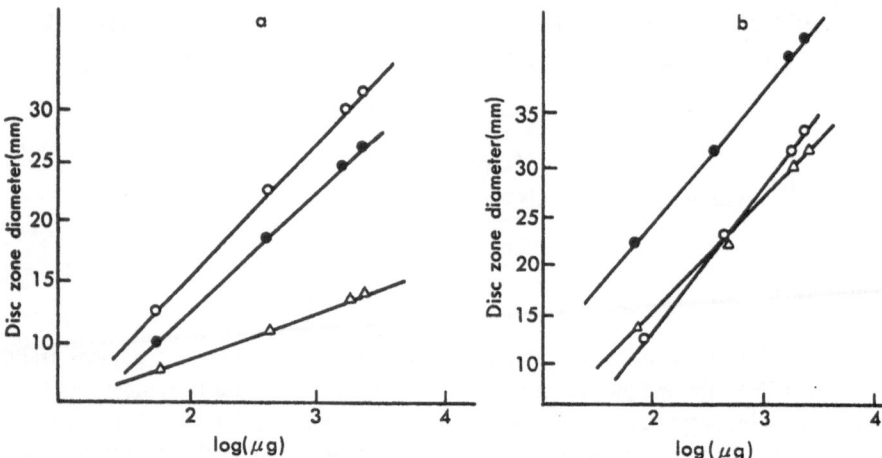

Fig. 1. *Activity of peptide derivatives. Solution of the compounds were added to filter paper discs (6mm diameter) and the discs were transfered to the plates. After incubation at 37°C for 8 h the diameter of the zones of inhibition were measured. (a) (O) DL-N-OH-Ala; (●) L-N-OH-Ala; (Δ) D-N-OH-Ala, (b) (O) DL-N-OH-Ala; (●) Ala-Gly (α-DL-N-OH-Ala); (Δ) Ala-Ala-Gly(α-DL-N-OH-Ala).*

possibility, that neither alanine recemase nor transaminase are inhibited by a mechanism based on an inactivation reaction in the cytosol of E. coli due to the law of mass action. The potent antagonism of the inhibitory tripeptide by the oligopeptide synthone could be speculated to be a result of an increased cellular concentration of alanine that could compete with DL-N-OH-Ala, presumably at a cytoplasmic enzyme site, and not a result of competition for the oligopeptide transport permeases.

Figure 1a shows that L-N-OH-Ala is more potent than D-N-OH-Ala. It could be explained that the L-N-OH-Ala peptide proved to be a more effective irreversible inhibitor against alanine recemase due to stability of the pyridoxal-P/inhibitor adduct. Another finding that D-amino acid oxidase utilized D-N-hydroxyalanine as a substrate is unexpected (Km: 2.17×10^{-1} M, γ_{max}: $1.0 \times^{-2}$ μM·min), since N-methylamino acids are neither substrates nor effective inhibitors of this enzyme[7].

This finding supports the peptide transport mediated entry of the inhibitory peptides, followed by release of N-OH-Ala.

Acknowledgement

This work was supported in part by a grant from the Korea Science and Engineering foundation.

References

1. Ringrose, P., in "Microorganisms and Nitrogen Sources", ed. by J. Payne, J. Wiley, Chichester and New York, N.Y., 1980, pp.641.

154

2.   Kingsbury, W., Boehn, J. and Gilvarg, C., Proc. Natl. Acad. Sci., U.S.A., 81(1984)4573.
3.   Payne, J.W., Barrett, K.J. and Shallow, D.A., FEMS Microbial Lett., 79(1991)15.
4.   Burston, D. and Mathews, D., Clinical Sciences, 79(1990)267.
5.   Sokol, P.P., J. Pharma. Exp. Ther., 255(1990)436.
6.   Cooper, L. and Griffith, W., J. Biol. Chem., 254(1979)2748.
7.   Meister, A. and Wellner, D., in "The enzymes" 2nd ed., Vol. 7, Academic press, New York (1964), pp.609-648.

Session V
New bioactive peptides

Chairs: Cheng-wu Qi
Shanghai Institute of Biochemistry, CAS
Shanghai, China

and

Jan Markussen
Novo Research Institute
Novo Alle, Bagsvaerd, Denmark

Isolation and sequencing of a new active peptide from the skin of Chinese frog *Rana Kuangwuensis*

Sheng-hai Tian[a], Shi-xiang Wu[a], Hua-qing Mu[a], Jia-cheng Hua[a], Guan-fu Wu[a] and Er-mi Zhao[b]

[a]*Shanghai Institute of Materia Medica, Chinese Academy of Sciences, Shanghai 200031, China*
[b]*Chengdu institute of Biology, Chinese Academy of Sciences, Chengdu 610015, China*

Introduction

It is believed to be an important approach to search for new neuropeptides from amphibian skin peptides due to the brain-gut-skin triangle, and the concentration of the peptides in amphibian skin is usually much higher than those in mammalian tissues. During our systematic study on active peptides from Chinese frog skin, recently a new bradykinin-like peptide was isolated from the skin of *Rana Kuangwuensis*. In this report we describe the isolation and sequencing of the peptide.

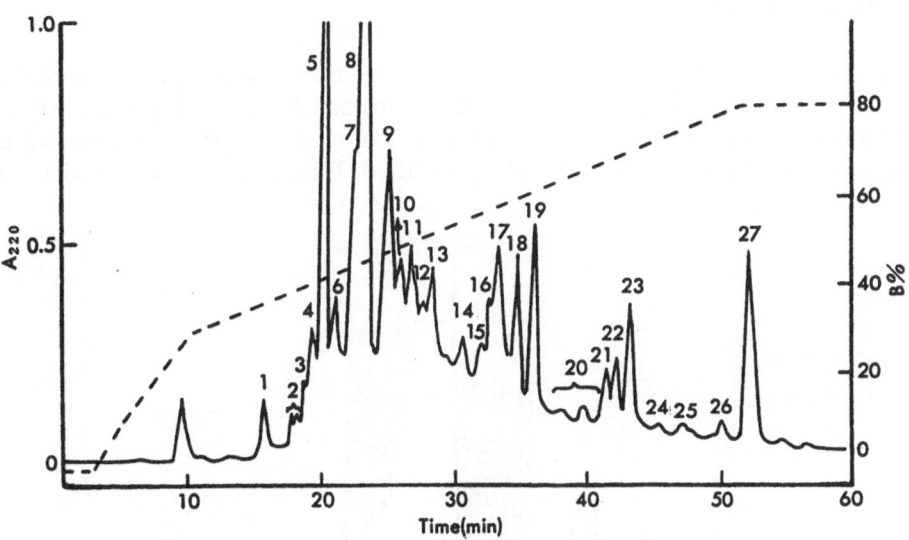

Fig. 1. RP-HPLC of 35% acetonitrile eluate from sep-pak C_{18} cartridges. Column: μ Bondapak C_{18} (0.78×30 cm); Solvent system: A=0.05%TFA; B=60%CH_3CN in 0.05%TFA; Flow rate: 1.2ml/min.

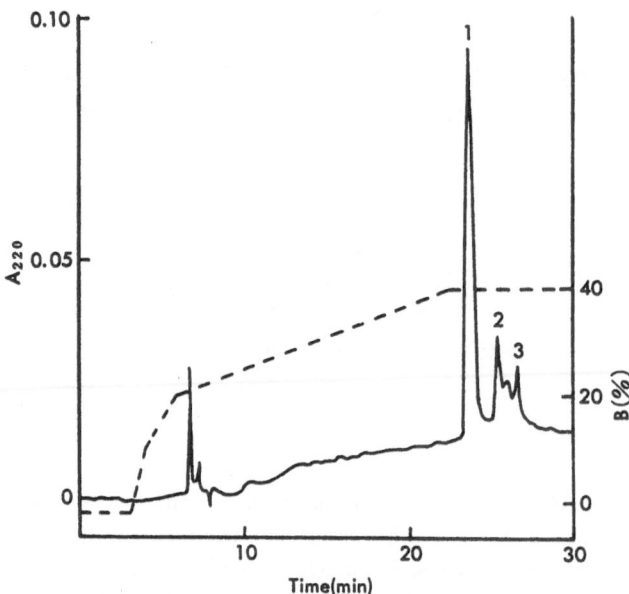

Fig. 2. RP-HPLC of the active peak(5) in Fig. 1. Column: μ Bondapeak C_{18} (0.39×30cm); Solvent system: A=0.05%TFA; B=60%CH_3CN in 0.05%TFA; Flow rate: 0.8ml/min.

Results and Discussion

The skin of frog, *Rana Kuangwuensis* was collected from Nan Jiang county of Sichuan province. Fresh skins removed from the frogs immediately after killing were extracted with methanol and 80% methanol respectively. The methanol extract was filtered and evaporated rotatively under vacuum, and lyophilized. The residue was dissolved in 0.1%

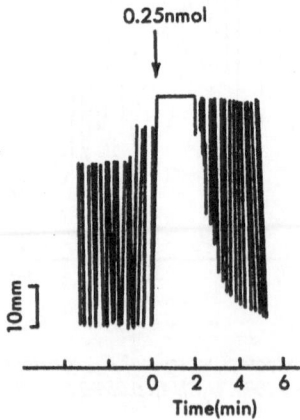

Fig. 3. Effect of peptide PK-1 on guinea pig ileum.

160

F₃CCOOH (TFA) at 4°C then centrifuged. The supernatant was passed through sep-pak C_{18} cartridge. The cartridges was washed with 0.1% TFA and eluated stepwisely with increasing CH_3CN concentration (15-80%). The eluates were lyophilized. The fractions were then subjected to bioassay for contractile activity on the guinea pig ileum (GPI). The fraction of 35% CH_3CN eluate was remarkably active. This fraction was purified with reverse phase high performance liquid chromatography (RP-HPLC) on a semi-preparative C_{18} column using TFA-CH_3CN system (Fig.1). Active peak 5 was further submitted to RP-HPLC on analytical C_{18} column using the same elution system (Fig.2). The active peak RK-1 was repurified.

The peptide RK-1 consists of the following amino acids: Ser(1), Gly(2), Pro(3), Phe(2), Arg(1). Gas phase sequencing revealed the N-terminal sequence of peptide RK-1 to be Gly-Pro-Pro-Gly-Phe-Ser-Pro-Phe-Arg. Studies of C-terminal of the new peptide are now in progress.

The preliminary result showed that this nonapeptide is similar to Bradykinin.

Acknowledgement

The work was supported by the National Natural Science Foundation of China.

References

1.   Erspamer, V. and Melchiorri, P., TIPS 1980. 391.
2.   Tang, Y.Q., Tian, S.H., Wu, S.X., Hua, J.C., Wu, G.F., Zhao, E.M. and Tsou, K., Science in China (Ser.B), 32(1989)570.

The structure modifications and functions of P6A

**Ming Zhao, Shi-qi Peng, Meng-shen Cai, Yu-feng Xu, Chao-shu Tang
and Li Zhang**
*National Laboratory of Natural and Biomimetic Drugs, Beijing Medical University,
Beijing 100083, China*

Introduction

The sequence for P6A derived from fibrinogen β-chain is Ala-Arg-Pro-Ala-Lys-OH (compound 7). Mehta and coworkers[1] reported on the dog coronary thrombosis model, i.v. drip administration of P6A increased coronary perfusion flow (CPF) significantly, but the mechanism is unknown. In order to elucidate the action mechanism of P6A, we synthesized P6A by solid phase and solution methods[2]. Based on the bioassay it was proved that P6A increased CPF dose-dependently on isolated perfused rat heart. Perfusion with 5×10^{-5}M of P6A increases the CPF comparable with the effect of 1×10^{-6}M of PGI$_2$ (49.7±1.4% and 60.7±3.3% increase respectively), with 1×10^{-4}M of P6A the CPF was increased by 70.9±70%. Preadministration with indomethacin significantly inhibited the effect of P6A, on the heart(75% inhibition), but can not completely abolish the coronary dilatation action of P6A, suggesting that the promoting of endogenous PGI$_2$ synthesis and release is one of the mechanisms of P6A-induced coronary dilatation. On the rat venous thrombosis model, i.v. drip or local administration of P6A significantly inhibited thrombosis, the weight of thrombus lowered by 36.9% and 53.9% respedtively in comparision with the control (in both cases P<0.01)[3]. The results imply that P6A may be a useful thrombolytic agent and excellent lead compound.

For studying the relationships between the conformations and activities of fibrinogen-related peptide a series of P6A derivatives were designed based on the hydrophobicity and stereo-effect of the side chain of amino acid. The synthetic compounds were evaluated with the relaxing effect on rat aortic strips treated with noradrenaline(NE). In this study the results of compounds 1–8 were reported.

Lys-Arg-Pro-Ala-Lys-OH	(1)
Arg-Arg-Pro-Ala-Lys-OH	(2)
Asp-Arg-Pro-Ala-Lys-OH	(3)
Asn-Arg-Pro-Ala-Lys-OH	(4)
Glu-Arg-Pro-Ala-Lys-OH	(5)
Glu-Arg-Pro-Ala-Lys-OH	(6)
Ala-Arg-Pro-Ala-Lys-OH	(7)
Gly-Arg-Pro-Ala-Lys-OH	(8)

The protective intermediates 9–16 corresponding to compounds 1–8 were prepared by using solution method (Fig.1). The protective peptides 9–16 were deprotected with 1 N sodium hydroxide in methanol/water at 0°C for 2 h, and with liquid hydrogen fluoride at 0°C for 1 h, then purified with Sephadex G 10 to provide corresponding compounds 1–8 respectively. The data were listed in Table 1.

Boc-Lys(ClZ)-Arg(Tos)-Pro-Ala-Lys(ClZ)OMe (9)
Boc-Arg(Tos)-Arg(Tos)-Pro-Ala-Lys(ClZ)OMe (10)
Boc-Asp(OcHex)-Arg(Tos)-Pro-Ala-Lys(ClZ)OMe (11)
Boc-Asn-Arg(Tos)-Pro-Ala-Lys(ClZ)OMe (12)
Boc-Glu(OcHex)-Arg(Tos)-Pro-Ala-Lys(ClZ)OMe (13)
Boc-Gln-Arg(Tos)-Pro-Ala-Lys(ClZ)OMe (14)
Boc-Ala-Arg(Tos)-Pro-Ala-Lys(ClZ)OMe (15)
Boc-Gly-Arg(Tos)-Pro-Ala-Lys(ClZ)OMe (16)

The sequences for compounds 1–16 were confirmed with FAB-MS, and based on both of the FAB-MS and amino acid analysis their structures were determined. Usually, under FAB-MS there will be three fragmentation means for peptides and they can be represented as Fig. 2.

Based on the molecular ions and/or the fragments, the sequences of peptides can be

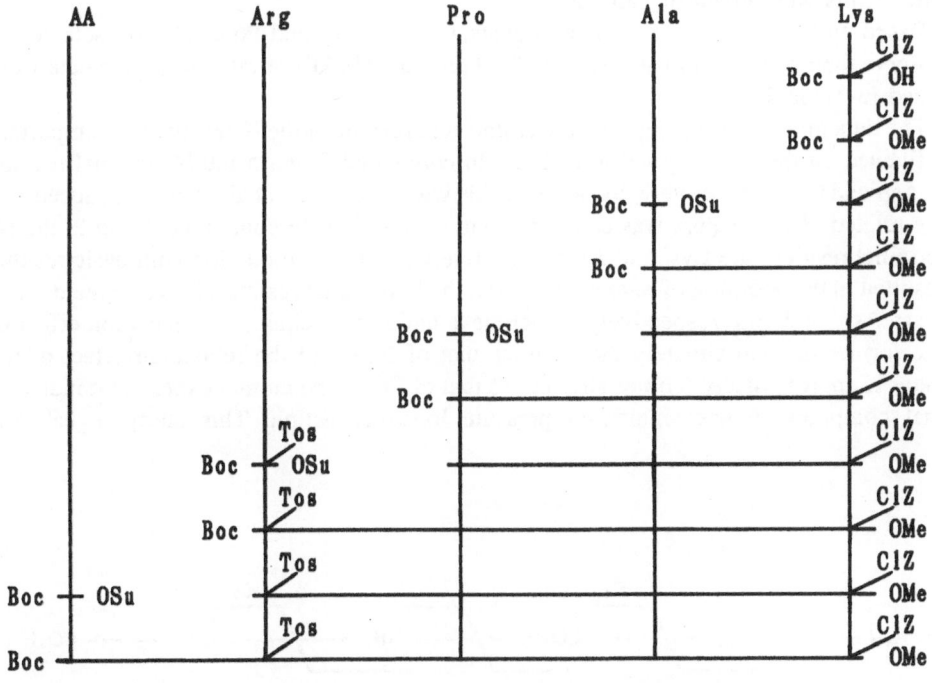

Fig. 1. The synthetic route, wherein AA represent the corresponding protected L-Lys, Arg, Asp, Asn, Glu, Glu, Ala and Gly.

Table 1 *The yields, amino acid analysis and FAB-MS for 1–16*

Compound		1	2	3	4	5	6	7	8	9	10	11	12	13	14	15	16
Total yield(%)		45	47	43	45	46	49	41	43	60	58	61	60	59	56	62	60
FAB-MS[M+H]⁺		599	627	586	585	622*	599	542	528	1203	1218	1105	1022	1119	1036	979	965
amino acid	Ala	1.01	1.01	1.03	1.03	1.02	1.02	2.02	1.02	1.02	1.03	1.02	1.03	1.03	1.02	2.03	1.03
analysis	Arg	1.03	2.03	1.04	1.02	1.03	1.01	1.02	1.02	1.03	2.03	1.03	1.01	1.02	1.01	1.02	1.01
	Asn	–	–	–	0.98	–	–	–	–	–	–	–	0.97	–	–	–	–
	Asp	–	–	0.99	–	–	–	–	–	–	–	0.97	–	–	–	–	–
	Glu	–	–	–	–	0.98	–	–	–	–	–	–	–	0.97	–	–	–
	Gln	–	–	–	–	–	0.99	–	–	–	–	–	–	–	0.98	–	–
	Gly	–	–	–	–	–	–	–	0.99	–	–	–	–	–	–	–	0.97
	Lys	1.99	0.98	0.97	0.99	0.98	1.00	0.97	0.98	1.97	1.02	1.01	0.97	1.03	1.02	0.97	0.98
	Pro	1.03	1.02	1.03	1.01	0.99	1.03	1.01	1.02	1.02	0.98	1.03	1.01	1.02	0.98	1.02	1.03

* [M + Na]⁺

defined. In this study the three fragmentations were observed. As an example the N_2-C_2-fragmentation obtained for compounds 9–16 were given in Table 2. It was shown that the structures of all compounds are correct.

After the confirmation of the purity by TLC or HPLC (98-99%) P6A derivatives were used for bioassay on rat aortic strips treated with NE to observe the relaxation actions. The results were given in Table 3.

Based on the isolated animal experiments, Gln^1-P6A(6) and P6A (7) were selected to observe their effects on mean arterial blood pressure (MAP) of rats (n= 4), the data were listed in Table 4.

The data given above suggested the amino acid residue at the N-terminal has important influence on the relaxing action of P6A. In compound 3 and 5 the N-terminal residue is Asp^1 and Glu^1 respectively, the amino scids with acidic side chain were introduced and the related effect of P6A was converted into contraction. In compound 1 and 2, the N-terminal residues are Lys^1 and Arg^1 respectively and the introduction with basic residue resulted in the abolition of relaxation action; the N-terminal residues for compound 4 and 6 are Asn^1 and Gln^1 respectively, which have amide side chain, the relaxation effect of the former is approximately the same as that of P6A and the relaxation effect of the latter is approximately 6 times stronger as that of P6A. The animal experiment indicates that compound 6 has significant pressure lowering action. This study reveals the

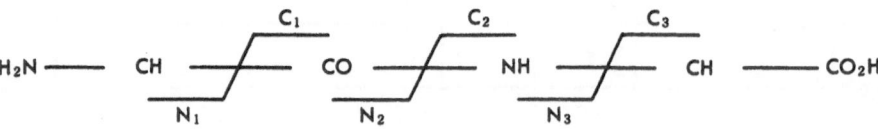

Fig. 2. *The fragmentation of peptides.*

structure relaxation effect relationships for P6A derivatives and gives more excellent antithrombia thus provides theoretical and experimental evidences for SAR and antithrombotic drugs researches.

Table 2 *The FAB-MS data for compound 9–16*

Compound	AA represent	deprotected Boc	$Boc\!\!\nearrow\!\!100$ $\xrightarrow{-Boc}$ AA	$\xrightarrow{835}$ Arg(Tos)	$\xrightarrow{524}$ Pro	$\xrightarrow{427}$ Ala	$\xrightarrow{356}$ Lys(ClZ) —OMe
9	Lys(ClZ)	1103	368	679	776	847
10	Arg(Tos)	1117	382	693	790	861
11	Asp(OcHex)	1005	269	580	677	748
12	Asn	921	186	497	594	655
13	Glu(OcHex)	1019	283	594	691	762
14	Gln	936	201	511	608	679
15	Ala	879	143	454	551	622
16	Gly	864	129	440	537	608

Table 3 *The effects of P6A derivatives on rat aortic strips treated with NE^a*

| Compound | relaxing extent for contraction strip with different dose(X±SD%) | |
	$10^{-6}M$	$10^{-5}M$
1	4.37 ± 5.10	6.70 ± 6.96**
2	0.00 ± 0.00	1.31 ± 5.29***
3	−3.20 ± 5.99***	−6.87 ± 3.04***
4	5.96 ± 6.95	16.00 ± 3.84
5	−2.40 ± 5.14**	3.60 ± 9.77
6	29.70 ± 4.16***	33.1 ± 6.59
7	5.72 ± 2.14	22.57 ± 8.35

[a] n=6, the contractive altitude of strip by $NE(10^{-9}M)$ was defined as 100%; the negative values represent contraction effect; compared with P6A, ** $P<0.05$, *** $P<0.01$.

Table 4 *Effects of Gln^1-P6A & P6A on rat MAP*

| compound and dose (mmol/kg) | Δ MAP(Kpa) at different time(min) after administration(X ± SD%) | | | | |
	1	5	10	20	30
P6A(7) (0.022)	0.93±0.28	2.05±0.40	1.12±0.49	0.59±0.36	0.43±0.21
Gln^1-P6A(6) (0.020)	1.25±0.21	2.57±0.36	2.14±0.55	0.52±0.32	0.39±0.27

M. Zhao et al.

Acknowledgement

The authors thank the Health Ministry of China for financial support.

References

1.   Mehta, J.L., J. Cardiovasc. Pharmacol., 13(1989)803.
2.   Zhao, M., Peng, S.Q. and Tang, C.S., HuaXue TonBao., 12(1990)29.
3.   Fan, G., Zhao, M. and Peng, S.Q., J. Beijing Med. Univ., 23(1991)358.

166

"Classical" and "non-classical" endothelin analogues

**Roberto de Castiglione, Mauro Galantino, Luisa Rusconi, Fabio Corradi,
Rita Perego, Cinzia Cristiani and Fabrizio Vaghi**
Farmitalia Carlo Erba Srl, R. & D., Via Giovanni XXIII 23, 20014 Nerviano, Italy

Introduction

In previous papers on endothelin-1 (ET-1) analogues we presented a systematic approach based on single-point substitution of the non-cysteinyl residues by alanine and the corresponding D-amino acids[1,2], followed by a series of modifications spanning C-terminal amidation, singlepoint deletion and single or multiple replacements with amino acids of either L- or D-configuration[3]. The immediate object of the research was a basic knowledge of the structural requirements for receptor binding and biological activity, with the ultimate goal of identifying ET-1 selective receptor antagonists.

In the present paper we examine the effects of C-terminal extension, peptide bond nicking or reduction, and truncation of large portions of the sequence with or without amino-acid substitutions.

Synthesis

All the analogues or their precursors were obtained by SPPS, using either the N^α-Boc/side-chain benzyl-based protection strategy on PAM resin (peptides 1-8) or the N^α-Fmoc/side-chain t-butyl-based protection strategy on PepSyn KA resin (peptides 9-15). Other side-chain protecting groups were, respectively, His(Dnp)/Trp(For) and Cys(Trt). Cysteine mono-t-butyl ester and cysteine monoamide were used for the preparation of compounds 12 and 13. The reduced peptide bond of analogues 3-6 was directly introduced as the suitable dipeptide synthon, obtained by conventional methods in solution.

After appropriate removal of protecting groups and cleavage from the solid support, disulfide formation was carried out by air oxidation in 0.1 M Tris·HCl buffer (pH 8-8.5) at different urea concentrations (2-8 M). Analogues 3–6 were obtained in two isomeric forms, corresponding to the (Cys$^{1-15}$, Cys$^{3-11}$) disulfide pairing of native ET-1 and to the inverted one (Cys$^{1-11}$, Cys$^{3-15}$). On the basis of the previous results[1-3], for each isomer pair the native-like structure should be attributable to the compound with the higher HPLC retention time.

Analogues 7 and 8 were obtained from ET-1 by treatment with CNBr in 70% HCOOH (and quenching with ammonia) and Lys C-endopeptidase, respectively, while analogue 9 was prepared from [Met^{4,10}, Ala^7]ET-1[3] by reaction with a large excess of CNBr in 70% TFA (to avoid the disulfide scrambling observed with formic acid). In such conditions, concomitant introduction of a CN group occurred, probably on the Tyr residue. The carbamoyl derivative 11 formed as side-product from the 8M urea employed to dissolve the linear intermediate in the cyclization step.

Table 1 *HPLC retention times (RT), and relative binding affinity (RBA) and contractile activity (RCA) of ET-1 analogues (ET-1=100)*

N°	Modification	RT[a]	RBA[b]	RCA[c]
1	Gly^{22}	19.27	2.1	0.3
2	$Val^{22},Asn^{23},Thr^{24}$	19.10	4.4	0.9
3	$Leu^{17}\Psi(CH_2NH)Asp^{18}$	21.02	0.16	–
4	$Leu^{17}\Psi(CH_2NH)Asp^{18}$	20.81	0.76	–
5	$Ile^{19}\Psi(CH_2NH)Ile^{20}$	19.77	0.4	–
6	$Ile^{19}\Psi(CH_2NH)Ile^{20}$	19.39	0.12	–
7	$Hse-NH_2^7$-nicked	17.81	25.0	4.8
8	Lys^9-nicked	18.47	0.8	0.3
9	$Hse>^4$,des (5-10), $Tyr(CN)^{13}$	19.76	5.0	0.23
10	$Ala^{3,4,5}$,des(6-11)	22.31	0.03	–
11	$H_2NCO-Cys^1,Ala^{3,4,5}$,des(6-11)	22.38	<0.0001	–
12	des(1-14),$Cys(Cys-OH)^{15}$	16.16	<0.0001	–
13	des(1-14),$Cys(Cys-NH_2)^{15}$	16.28	<0.0001	–
14	des(1-7),$D-Cys^{11}$	19.83	<0.0001	–
15	des(1-15)	16.38	<0.0001	<0.03

[a] HPLC retention time on a reverse-phase C_{18} Vydac 218 TP54 column with a linear gradient of 15→85% buffer B at a flow rate of 1.5 ml/min and monitored at 225 nm. Buffer A: 0.05% TFA in H_2O/CH_3CN 95/5; buffer B: 0.05% TFA in H_2O/CH_3CN 40/60.

[b] Binding on subconfluent cultures (1-3 × 10^5 cells/cm^2) of h-VSM cells incubated in serum supplemented with 0.1% BSA at 37°C for 2 h; IC_{50} for ET-1=0.47±0.24·10^{-9} (n=19).

[c] Contraction of the rabbit vena cava: EC_{50} for ET-1=2.89 ± 0.4·10^{-10} (n=8).

– not determined.

All analogues were purified by RP-HPLC and their structures confirmed by amino-acid analysis, FAB mass spectrometry and (for the nicked compounds) partial amino-acid sequencing. Structures of nicked and truncated analogues are reported in Fig. 1.

Bioassays

All the analogues were screened for binding affinity on subconfluent cultures (1-3 × 10^5 cells/cm^2) of the human vascular smooth muscle (h-VSM) cell line A 617 incubated in serum supplemented with 0.1% BSA at 37°C for 2 h. A few of them (particularly those displaying some receptor binding) were tested also for contractile activity on strips of rabbit vena cava without endothelium in the presence of 1×10^{-6} M indomethacin. Both tissues possess ET-1 selective receptors (ET_A receptor subtype). Results are expressed as molar potencies relative to ET-1=100, and are calculated from the corresponding IC_{50} or EC_{50} (Table 1).

Results and Discussion

C-Terminal extension (analogues 1 and 2) greatly reduced binding and even more strongly lessened contractile activity. However, compound 2, corresponding to Big ET-

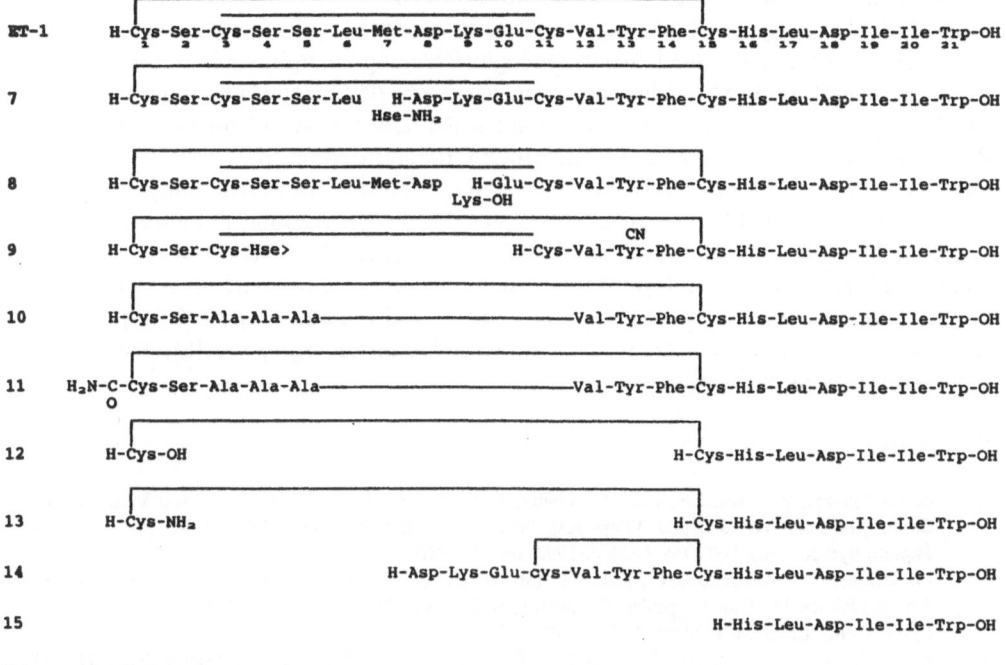

Fig. 1. Structures of ET-1 and its nicked and truncated analogues.

1(1–24), was at least twice as active as the Gly-extended derivative 1, confirming other reports [4] on C-terminal elongated forms of ET-1, even though the Gly[22] residue is not found in big-endothelins. The analogue was originally synthesized to test the enzymatic conversion of a glycine-extended peptide into the corresponding des-glycine amidated derivative by the peptidyl α-amidating monooxygenase isolated from rat medullary thyroid CA-77 cell lines[5].

Peptide analogues containing reduced peptide bonds often display antagonistic properties. Unfortunately, the two pairs of analogues 3–6, with a reduced peptide bond in positions 17–18 and 19–20, had very low binding affinity, and were not tested for either agonistic or antagonistic activity.

All other analogues were synthesized to evaluate the role of the constrained portion included between the two disulfide bridges. Data from the literature were rather contradictory on the importance of the integrity of the inner loop. CNBr cleavage at Met[7] did not affect the binding properties or the biochemical response (PI hydrolysis) on rat cerebellar and atrial preparations[6], whereas cleavage at Lys[9] by lysyl endopeptidase lowered the contraction of porcine coronary strips approximately 200 fold[7]. In our hands, the homoserine lactone formed by CNBr treatment was found to undergo intramolecular aminolysis, probably because of a favorable constraint. The resulting [Hse[7]]ET-1 analogue was more or less as active as the parent compound[3]. Therefore we blocked the reconstitution of the inner loop by quenching the reaction with excess of ammonia, obtaining the amidated derivative 7. The results with compounds 7 and 8 showed that the integrity of the inner loop was important for full biological activity, but was nonessential, as confirmed also by analogue 9 lacking six amino-acid residues in the loop. Nicking at the Lys^9 residue, however, was more detrimental than removal of the (5–10) sequence and even more than nicking at the Met^7 residue. Complete removal of the inner loop, and reclosure of the outer loop through insertion of an $(Ala)_3$ sequence, gave analogue 10 with only 0.03% of the binding affinity of the parent compound. The carbamoyl derivative 11, as well as analogues 12 and 13 (lacking the entire peptide sequence between Cys^1 and Cys^{15}) and analogue 14 (with the deletion of the first six amino-acid residues and a $D\text{-}Cys^{11}\text{-}Cys^{15}$ disulfide bridge) did not bind the ET_A receptor at all. Compound 15, corresponding to ET(16–21) and reported to be a full agonist at ET_B receptors while being inactive or weakly active at ET_A receptors [8], showed no binding affinity in our test.

References

1.   de Castiglione, R., Tam, J.P., Liu, W., Zhang, J.-W., Galantino, M., Bertolero, F. and Vaghi, F., In Smith, J.A. and Rivier, J.E. (Eds.) Peptides: Chemistry and Biology (Proceedings of the 12th American Peptide Symposium), ESCOM, Leiden, 1992, pp. 402–403.

2.   Galantino, M., de Castiglione, R., Tam, J.P., Liu, W., Zang, J.-W., Cristiani, C. and Vaghi, F., In Smith, J.A. and Rivier, J.E. (Eds.) Peptides: Chemistry and Biology (Proceedings of the 12th American Peptide Symposium), ESCOM, Leiden, 1992, pp. 404–405.

3.   de Castiglione, R., Galantino, M., Giordano, P., Cristiani, C. and Vaghi, F., In Schneider, C.H. and Eberle, A.N. (Eds.) Peptides 1992 (Proceedings of the 22nd European Peptide Symposium), ESCOM, Leiden, 1993, pp. 681–682.

4.   Nishikori, K., Akiyama, H., Inagaki, Y., Ohta, H., Kashiwabara, T., Iwamatsu, A., Nomizu, M. and Morita, A., Neurochem. Int., 18(1991)535.

5.   Tamburini, P.P., Young, S.D., Jones, B.N., Palmesino, R.A. and Consalvo, A.P., Int. J. Peptide Protein Res., 35(1990)153.

6.   Fleminger, G., Bousso-Mittler, D., Bdolah, A., Kloog, Y. and Sokolovsky, M., Biochem. Biophys. Res. Commun., 162(1989)1317.

7.   Kimura, S., Kasuya, Y., Sawamura, T., Shinmi, O., Sugita, Y., Yanagisawa, M., Goto, K. and Masaki, T., Biochem. Biophys. Res. Commun., 156(1988)1182.

8.   Maggi, C.A., Giuliani, S., Patacchini, R., Rovero, P., Giachetti, A. and Meli, A., Eur. J. Pharmacol., 174(1989)23.

Type 1 DNA topoisomerase-like activities of trichosanthin and momorcharin

Zu-he Qi[a], Shi-de Liu[a], Xi-lin Zhu[a], Xian-rong Yang[b] and Hui-yun Liu[b]
[a]Institute of Basic Medical Sciences, Chinese Academy of Medical Sciences, Beijing 100005, China
[b]Chinese Medicinal Material Research Centre, Chinese University of Hong Kong, Hong Kong

Introduction

Trichosanthin(TCS) and momorcharin(MMC) are ingradients of Chinese medicinal materials, Tian Hua-fen and Ku Gua-zi respectively, both having the characteristics of ribosomal inactivating proteins. They have been purified and used clinically. Especially trying to use them for the treatment of AIDS is very attractive. In studying their anti-viral activities, the DNA was used as the substrate. We were the first to find that the suppercoiled DNA could be relaxed by them [1]. Further investigations are described as follows.

Results and Discussion

The procedures used to identify the activities of TCS and MMC including agarose electrophoresis separation and electron microscopy observation of treated DNA are published elsewhere. Fig.1 a indicates the degradation of the single strand DNA by TCS and MMC in PB buffer, pH6.2. The smear band is shown by the arrow. Meanwhile the relaxation and catenation of the duplex DNA occurred, which are denoted by the

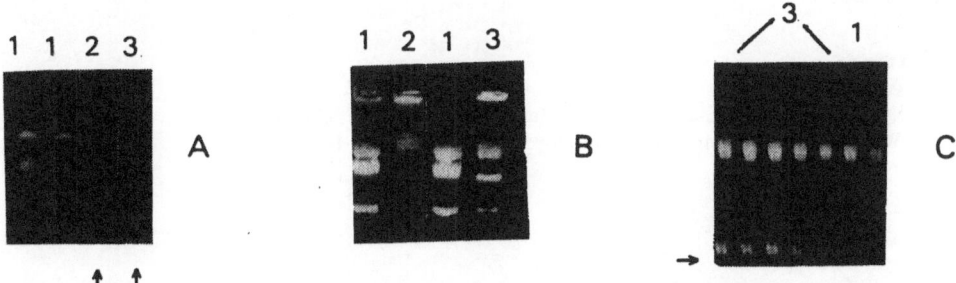

Fig. 1. The electrophoretographs of DNA before and after treatment by TCS and MMC. a, degradation of M13mp8 ssDNA; b, relaxation and catenation of M13mp8 dsDNA; c, reconstruction of supercoiled DNA. 1. control DNA; 2. DNA/TCS; 3.DNA/MMC.

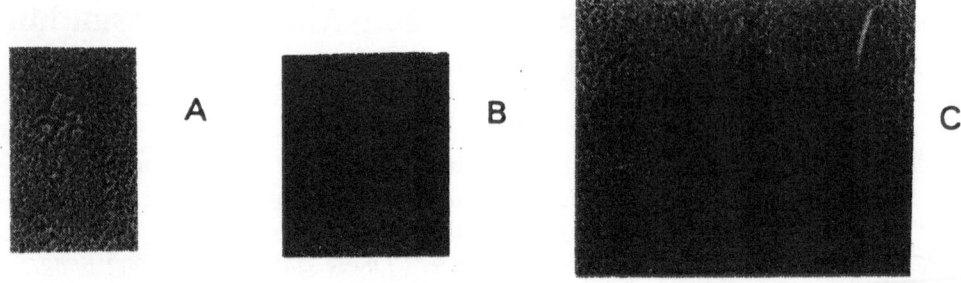

Fig. 2. *The electron micrographs of dsDNA before and after treatment by MMC. a, control DNA; b, relaxation of DNA; c, catenation of DNA.*

mobilities of the more slowly migrating bands in the electrophoretogram, Fig.1 b. These activities are ATP independent, but decatenation can also happen during addition of 1% SDS just before running on the gel. On the other hand, the supercoiled DNA is able to reform when the reaction is going on under the optimal conditions with 25-200 mM Tris·HCl, pH6.8-6.9, 1mM Na_2EDTA, 5mM DTT, BSA and incubation at 67.5°C. Na^+ and Mg^{++} ions are found to be beneficial to the reaction. The arrow indicates the reconstructed DNA band in Fig.1c. In order to provide direct proof of the interconversion of DNA configuration, The relaxation and the catenation of the molecules are shown in the electron micrograph (Fig.2).

It has been known that the type 1 DNA topoisomerases catalyze the relaxation of supercoiled DNA by the formation of transient single strand breaks and can also knot and catenate duplex DNA rings of unrelated sequence providing that at lease one of the component rings is nicked [2]. in considering all the actions of TCS and MMC on ssDNA and dsDNA their activities are the same as type 1 DNA topoisomerases.

Acknowledgment

This work was supported by a grant of National Natural Sciences Foundation of China No.39070218.

References

1. Qi, Z.H. and Yan, J., Acta Academiae Medicinae Sinicae, 10(1988)3012.
2. Wang, J.C., The Enzymes (edited by Boyer, P.D.) Acad.Press, 14(1981)331.

Purification and characterization of mammalian neurotoxins from the venom of the spider *Selenocosmia huwena*

Song-ping Liang[a], Dong-yi Zhang[a], Xin Pan[a], Qun Chen[b] and Pei-ai Zhou[b]
[a]Department of Biology, Hunan Normal University, Changsha 410006, China
[b]Department of Biology, Peking University, Beijing 100871, China

Introduction

The spider *Selenocosmia huwena* was recently identified as a new species of genus *Selenocosmia*. It is distributed in the hilly areas of south China. This spider is rather aggressive and venomous. The venom from the spider *S. huwena* contains a mixture of

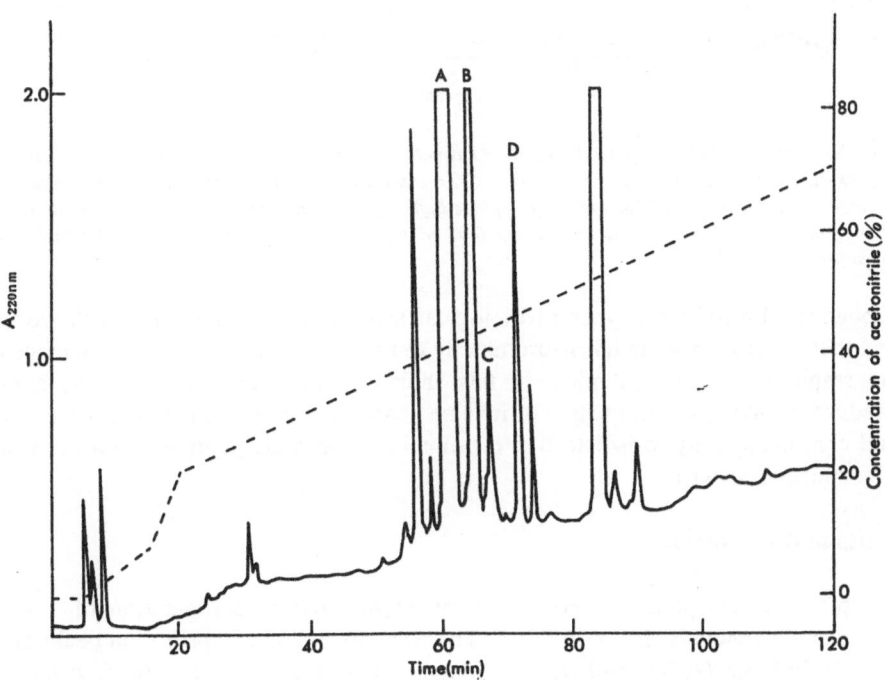

Fig. 1. Reverse-phase HPLC of S. huwena venom on C₄ column. 1 mg crude venom from S. huwena was applied to a Delta-pak C₄ 300 A column (30×0.46cm) and a linear gradient from 0 to 70% acetonitrile in a constant 0.1% trifluoro acid over 120 min was used. The flow rate was 0.7 ml/min; The column temperature was 40°C.

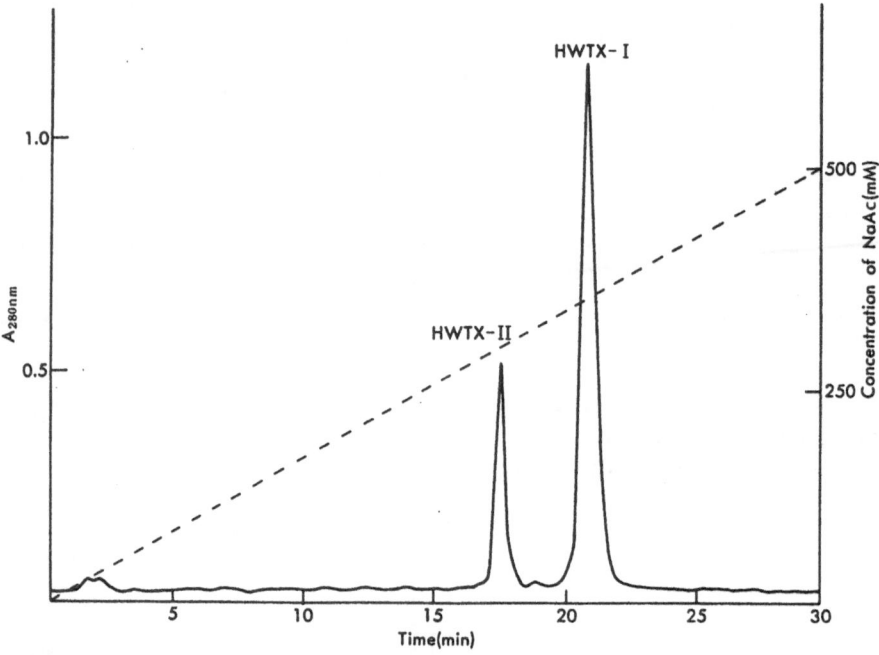

Fig. 2. Ion-exchange HPLC of partially purified huwentoxins. About 0.5 mg of fraction A (Fig.1.)obtained by C$_4$ RP-HPLC was applied to a Shim-pak WCX-1 ion-exchange column (0.4×5cm). The column was previously equilibrated with 0.02M sodium phosphate buffer, pH6.6. The elution was performed with a linear gradient from 0-50% 1M NaAC in a constant 0.02 M phosphate buffer, pH 6.6 over 30 min, at room temperature.

compounds with different types of biological activity. In our previous work we have found that the crude venom has neurotoxicity to mice. It can cause paralytic damage and rapid respiratory failure in mice. In this short communication we describe a rapid procedure involving a combination of reverse phase and ion-exchange high performance liquid chromatography to isolate five mammalian neurotoxic peptides from the venom of the spider *S. huwena*.

Results and Discussion

A typical reverse phase C$_4$ 300A column chromatographic separation of 1 mg *S. huwena* crude venom is shown in Fig. 1. There are about 23 u.v. absorbtion peaks at 220 nm. Four fractions (A,B,C and D) were found toxic to mice. The fraction A was further purified through ion-exchange HPLC on a WCX-1 column. Two peaks were obtained and both are toxic (Fig.2). The major peak (R.T.=21.5min.) was directly applied to a Nova-pak C$_{18}$ column, and finally a pure toxic peptide named huwentoxin-I was eluted out with a linear gradient of acetonitrile. The smaller peak (R.T.=17.5min) in Fig. 2, and the other three toxic fractions (fraction B,C and D) of C$_4$ HPLC were also further

174

purified by RP-HPLC on the Nova-pak C_{18} column and were named huwentoxin-II, III, IV and V respectively. The homogeneity of the toxins were confirmed by SDS-PAGE. All the five huwentoxins can induce paralytic damage and are lethal to mice.

The LD_{50} of huwentoxin-I in mice is 0.70 mg/kg. With 1×10^{-5}g/ml huwen toxin-I, the neuromuscular transmission of the isolated mouse phrenic nerve diaphragm preparation was irreversibly blocked in 13.4±1.3 min (X ± S.D., n=5). However, the conductivity of the phrenic nerve and the twitch response to direct stimulation were unaffected. The isoelectric point of huwentoxin-I determined by IEF-PAGE is 8.95. The reduced and S-carboxymethylated huwentoxin-I was hydrolysed with HCl and MSA and the amino acid composition was determined by the Pico-Tag method. Each molecule of the toxin contains 33 amino acids including 6 Cys, 6 Lys, and 1 Leu. The N-terminal residue determined by DABITC/PITC method and the C-terminal residue determined by carboxypeptidase A were Ala and Leu respectively. The complete amino acid sequence of huwentoxin-I analysed by Milligen model 6600 Solid-phase prosequencer is:
NH_2-Ala-Cys-Lys-Gly-Val-Phe-Asp-Ala-Cys-Thr-Pro-Gly-Lys-Asn-Glu-Cys-Cys-Pro-Asn-Arg-Val-Cys-Ser-Asp-Lys-His-Lys-Trp-Cys-Lys-Trp-Lys-Leu-COOH

Acknowledgement

This work is supported by National Natural Science Foundation of China under contract No.39070213.

References

1. Wang, J.F., Peng, X.J. and Xe, L.P., Acta Sci. Nat. Univ. Norm. Hunan., 16(1993)65.
2. Liang, S.P., Zhang, D.Y., Pan, X., Chen, Q. and Zhou, P.A., Zoological Research., 14(1993)65.
3. Liang, S.P. and Laursen, R.A., Anal. Biochem., 188(1990)366.

Antiplatelet and vasodilation effects of RGDS

Ming Zhao, Shi-qi Peng, Meng-shen CAI, Chao-shu Tang and Chang-ling Li
National Laboratory of Natural and Biomimetic Drugs, Beijing Medical University, Beijing 100083, China

Introduction

Platelet aggregation plays an essential role in normal hemostasis and is dependent on the interaction of the membrane glycoprotein II_b/III_a (GP II_b/III_a) complex with plasma adhensive glycoproteins, including fibrinogen (Fgn), von Willebrand factor (vWF), and fibronectin (FN).

Studies of Fgn binding to GP II_b/III_a have identified two distinct amino acid sequences in the Fgn molecule that mediate its attachment to the GP II_b/III_a receptor. One of the sequences is Arg-Gly-Asp(RGD), which occurs in the Fgn Aα chain. This sequence is also present in FN and vWF and appears to mediate their binding to GP II_b/III_a [1-3]. In addition to monoclonal antibodies, the binding function of GP II_b/III_a can be blocked by synthetic small molecular peptides containing the sequence RGD (RGDS, Arg-Gly-Asp-Ser). Small peptides related to RGD may provide informations for the researching of peptide antithrombus drugs. Among them the endogenous RGDS obviously has its importance.

In an attempt to demonstrate the biological activities, RGDS was synthesized and used for antithrombus bioassay. The studies are summarized as follows.

Results and Discussion

Platelet aggregation studies

Platelet-rich plasma was prepared by centrifugation of normal rabbit blood anticoagulated with sodium citrate at a final concentration of 0.38%. The platelet counts were adjusted to 200000/µl by the addition of autologous plasma. Platelet aggregation studies were conducted in an aggregometer using turbidimetric technique. The agonists used were platelet-activating factor (PAF) and adenosine diphosphate (ADP) (final concentration, 5 µM). The effect of RGDS on PAF or ADP-induced platelet aggregation was studied. RGDS dissolved in normal saline (NS) was added to 300µl platelet-rich plasma, the mixture stirred at 37°C in an aggregation cuvette. In control experiments, NS alone was added. The data were listed in Table 1.

Vasodilation studies

Immediately after decapitation, rat aortic strips were prepared and put in perfusion bath with 5ml of warmed (37 °C), oxygenated (95% O_2/5% CO_2) Krebs solution (pH 7.4). The aortic strips were connected to tension transducers and relaxation- contraction curves of these preparations were registered. Administration of 10^{-9} M noradrenaline (NE)

Table 1 *Inhibition effects of RGDS on platelet aggregation*

Inducer	final concentration of RGDS(M)	inhibition (%)
control		0
ADP	2.4×10^{-7}	37
	4.8×10^{-7}	68
	9.6×10^{-7}	87
control		0
PAF	4.8×10^{-7}	12
	9.6×10^{-7}	67

Compared with control, $P<0.01$, $n=7$.

induced hypertonic contraction of vessel strips. When the contraction arrived at maximal level, NE was washed out and the vessel strips were stabilized for 30 min. After renewal of the solution, NE (10^{-9}M) was added. When the hypertonic contraction value of aortic strips reached the peak, RGDS was administrated. The relaxing extent is represented by the percent of peak value. The related data were listed in Table 2.

Cyclic GMP accumualtion studies

In parallel experiments, rat aortic strips were equilibrated with Krebs-Ringer bicarbonate solution. Tissues were exposed to RGDS and processed as described above and then frozen in clamps precooled in liquid nitrogen. Cyclic GMP (cGMP) levels were assayed as previously described. Frozen tissues were homogenized in 6% trichloroacetic acid and centrifuged. Supernatant fractions were extracted with ether and a portion of the extract was acetylated and radioimmunoassayed for cGMP. The kit used was provided by Peninsula Lab. The data were listed in Table 3.

The data obtained here proved that RGDS inhibited PAF- and/or ADP-induced platelet aggregations obviously. This observations are consistent with the findings that RGDS appeared to delay thrombus formation. It was shown that RGDS was more sensitive to ADP than to PAF, but its inhibition effect on PAF was still significant. This study revealed that RGDS had vasodilative action, and its relaxing effect on rat aortic strips should not be neglected. Both the vasodilation and antiplatelet effect may be used for

Table 2 *Effects of RGDS on rat aortic strips*

Compound (M)	relaxing extent for contraction strips (X±SD%)
control	3.00 ± 3.4
RGDS(10^{-5})	8.08 ± 5.0 *
RGDS(10^{-4})	9.65 ± 1.5 *

* Compared with control, $P<0.01$, $n=6$.

Table 3 *Effects of RGDS on cGMP accumulation*

Group	cGMP (pmol/mg) (X ± SD)
normal strips	1.89 ± 0.3
strips with 10^{-5} M RGDS	4.68 ± 1.9 *
strips with 10^{-4} M RGDS	6.14 ± 2.8 *

* Compared with group of normal strips, $P<0.05$, n=6.

the design of new antithrombotic agents. The cGMP accumulation studies suggested that the enhance of cGMP accumulation may be one of the mechanisms by which RGD exerts its bioactivities.

References

1. Plow, E.F., McEver, R.P., Coller, B.S., Woods, V.L., Marguerie, C.A. and Ginsberg, M.H., Blood, 66(1985)727.
2. Plow, E.F., Pierschbacher, M.D. and Ruoslahti, E., Proc. Natl. Acad. Sci. USA, 82(1985)8057.
3. Gartner, T.K. and Bennett, J.S., J. Biol. Chem., 260(1985)11891.

Studies on squash family protease inhibitors

Xiao-ming Chen[a], Min-hua Ling[a], Hai-yan Qi[b], Zhi-wei Xie[a] and Cheng-wu Chi[a]

*[a]Shanghai Institute of Biochemistry, Chinese Academy of Sciences,
Shanghai 200031, China
[b]The Fourth Military University, Xian 710032, China*

Introduction

Squash family protease inhibitors are small peptides consisting of 27-32 residues with three disulfide bonds. Due to the low molecular weight and the low antigenicity, they are regarded as ideal targets for protein structure-function study and are of importance in clinical application if they could be chemically or genetically converted into elastase or cathepsin G inhibitors. Hence this family inhibitors have been intensively investigated for recent years [1-4]. In this paper, we describe the characteristics of several new members of the squash family protease inhibitors both at protein and gene levels. The total synthesis and the protein engineering of one member were also reported.

Results and Discussion

Amino acid sequences of Trichosanthes trypsin inhibitor(TTI), towel gourd trypsin inhibitor(TGTI) and Hami melon trypsin inhibitor(HMTI)

TTI has been isolated and purified with immobilized trypsin and HPLC from the roots of *Trichosanthes kirilowii*, a chinese medical herb. Two components of TTI, both consisting of 27 residues were found, being the smallest protein inhibitors so far known. They are different only in one residue located at the position 9, being Lys or Gln respectively[5]. TGTI and HMTI have also been purified with the same strategy [6]. TGTI was isolated from the juice of towel gourd. TGTI-I consists of 28 amino acid residues, while both TGTI-II and HMTI are 29 residues. The reactive sites of all these inhibitors are located at Arg-Ile peptide bond near the N-terminus. The three disulfide bonds are assembled in a knotty structure by an unique connecting form (I-IV, II-V and III-VI). The determined sequences of the above inhibitors show high homology with those of other squash family inhibitors in disulfide bonds, reactive site and C-terminal sequence (Fig. 1).

cDNA and genomic sequences of the towel gourd trypsin inhibitor

Though the primary and second structures of squash family inhibitors have been intensively studied, no cDNA and genomic structures of any of this family have been reported. In order to understand the molecular characteristics of the squash family inhibitors at gene level, we constructed a cDNA library of towel gourd. Instead of screening with a DNA probe, PCR was used to amplify the coding gene of the inhibitor directly from the cDNA library. Two overlapping cDNA fragments of inhibitor TGTI-II

```
                              5        10       15       20       25       30
                              •
LCTI-I  (Towel gourd)       I C P R I L M P C S S D S D C L A E C I C L E N - G F C G
LCTI-II (Towel gourd)     G I C P R I L M P C K T D D D C M L D C R C L S N - G Y C G
TTI-I   (Trichosanthes)       C P R I L M P C K V N D D C L R G C K C L S N - G Y C G      (8)
TTI-II  (Trichosanthes)       C P R I L M P C Q V N D D C L R G C K C L S N - G Y C G      (8)
HMTI    (Hami melon )     V G C P R I L M K C K T D D D C L L G C K C L S N - G Y C G
MCTI-I  (Bitter gourd)    E R R C P R I L K Q C K R D S D C P G E C I C M A H - G F C G    (18)
MCTI-II (Bitter gourd)    R I C P R I W M E C K R D S D C M A Q C I C V D - - G H C G      (18)
MCEI-II (Bitter gourd)    R I C P L I W M E C K R D S D C L A Q C I C V D - - G H C G      (18)
CMTI-I  (Squash)          R V C P R I L M E C K K D S D C L A E C V C L E H - G Y C G      (10)
CMTI-III (Squash)         R V C P R I L M K C K K D S D C L A E C V C L E H - G Y C G      (10)
CMTI-IV (Squash)      H E E R V C P R I L M K C K K D S D C L A E C V C L E H - G Y C G    (10)
CPGTI-I (Zucchini)        R V C P K I L M E C K K D S D C L A E C I C L E H - G Y C G      (10)
CPTI-II (Summer squash)   R V C P K I L M E C K K D S D C L A E C I C L E H - G Y C G      (10)
CPTI-II (Summer squash) H E E R V C P K I L M E C K K D S D C L A E C I C L E H - G Y C G  (10)
CSTI-I  (Cucumber)        M V C P K I L M K C K H D S D C L L D C V C L E D I G Y C G V S  (10)
CSTI-IV (Cucumber)        M M C P R I D M K C K H D S D C L P G C V C L E H I E Y C G      (10)
EETI-II (Ecballium elaterium)  G C P R I L M R C K Q D S D C L A G C V C G P N - G F C G S P  (18)
CVTI-I  (Citrullus vulgaris) G R R C P R I Y M E C K R D A D C L A D C V C L Q H - G I C G  (9)
BDTI-I  (Bryonia dioica)  R G C P R I L M R C K R D S D C L A G C V C Q K N - G Y C G      (9)
```

Fig. 1. *Amino acid sequence comparison of squash family inhibitors reactive site.*

```
                                                            CCGGGTCGACCC

ACGCGTCCGATCAAAAGCAAATAAGATTCAAAAGTCGTGTGTGAAAAAAATAATGGAGTGGAA

GAAAGTTGCTCTGATGGCAATGGTGGG ATG TTG CTG ATG GCG AGT GTT GCA GAG
                                Met Leu Leu Met Ala Ser Val Ala Glu
              -25                         -20                 -15
TCG AGC GGC GTG GTG GAG GTG ATT GAA CTG ATT TCC GAC GGA GGG AAT
Ser Ser Gly Val Val Glu Val Ile Glu Leu Ile Ser Asp Gly Gly Asn
        -10                     -5              *1
GAT TTA CCA AGA AAG ATC ATG AGC GGT CGC CAT GGA GGA ATT TGC CCA
Asp Leu Pro Arg Lys Ile Met Ser Gly Arg His Gly Gly Ile Cys Pro
5                       10                  15                  20
AGA ATC CTT ATG CCC TGC AAG ACC GAC GAT GAC TGC ATG CTT GAC TGT
Arg Ile Leu Met Pro Cys Lys Thr Asp Asp Asp Cys Met Leu Asp Cys
                    25
CGC TGC CTG TCC AAC GGC TAT TGT GGT TGA ACAAACTATGTTCTGTCGGAATC
Arg Cys Leu Ser Asn Gly Tyr Cys Gly

TCTCTGTGTGTGCTGTTTGTCTTTTCTCCTTCTTCTATGAAAATAAAGGTTGTCTCTTTGATC

ACTCTTCCTCTATGTGGGAGTGTATGTGATTAATTGAAACATCTGTGTAATGTAATCTTTTCT

ATGTTCTATGAAGCTCTACTATCTTCTTTGAGTAAAAAAAAAAAAAAAAAAAAAAAAAAAAAAAA

AAAA
```

Fig. 2. *The nucleotide and deduced amino acid sequence of TGTI-II cDNA. The putative polyadenylation signal AATAAA is underlined.*

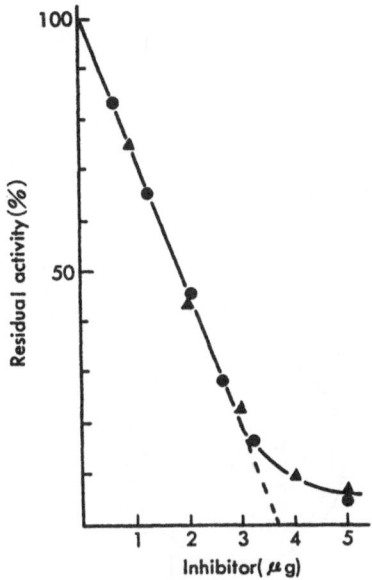

Fig. 3. Antitrypsin activity of the natural (-●-) and synthetic (-▲-) TTI.

were amplified. A full-length cDNA sequence with 481 bp coding for TGTI-II was then completed (Fig. 2). The open reading frame encoded a TGTI-II precursor with 63 amino acid residues. The deduced amino acid sequence of 29 residues for the inhibitor is consistent with that determined by primary structure analysis. In the leading peptide, the first 21 amino acid residues having a hydrophobic core are most likely to be a signal peptide. The following propeptide with 13 residues displays a hydrophilic feature. It may be involved in directing the correct folding of TGTI. Using the total DNA as a template, a 100 bp DNA fragment corresponding to mature inhibitor TGTI-II was also amplified and elucidated. It shows that the genomic DNA sequence of the inhibitor was identical with its cDNA sequence. There is no intron in the gene structure of TGTI-II [6].

Chemical synthesis of TTI and its analogue

As a small molecule, TTI can be regarded as an ideal target for structure-function study either by chemical synthesis or by protein engineering. TTI and its analogue [Ala^6]-TTI have been synthesized on the Applied Biosystem 430A peptide synthesizer. After removal of all protecting groups, the synthetic inhibitors were reduced with DTT, absorbed on a DEAE Sepharose CL-6B column and reoxidized during elution with a concentration gradient of urea and bithioreagent, DTT. The active inhibitor was further purified by affinity chromtography on an immobilized trypsin column and HPLC. The synthetic TTI showed the same inhibitory activity towards trypsin and the same retention time on HPLC as the natural one (Fig. 3) Its peptide sequence was also proved to be correct. The Ala-substituted analogue [Ala^6]-TTI also show compatible trypsin inhibitory

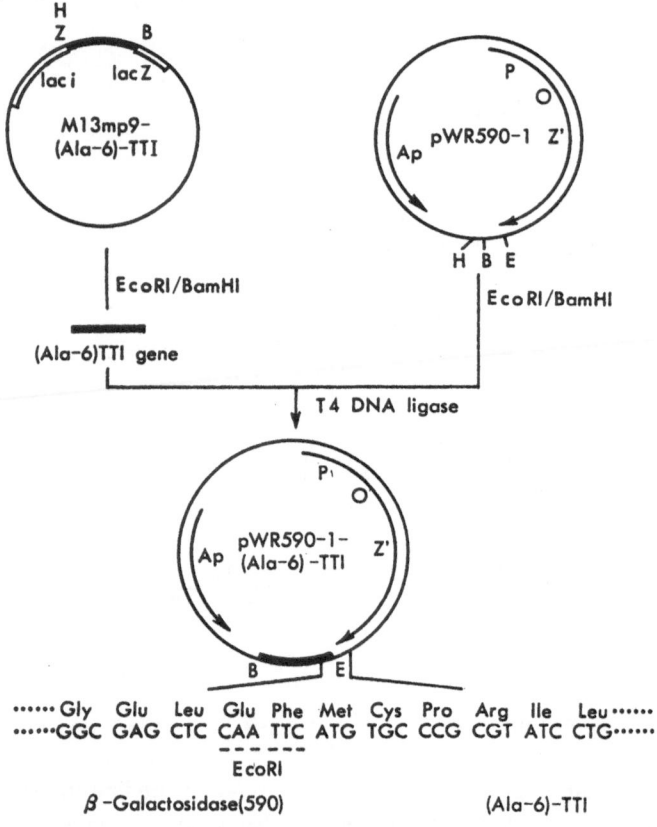

Fig. 4. *Construction of expression plasmid pWR590-1-[Ala⁶]-TTI. Abbreviations: E, EcoRI; B, BamHI; H, HindIII.*

activity with natural TTI. It implies that the replacement of Met with Ala at position 6 gives no detectable effect on the antitrypsin activity of TTI [7]. The result would provide the possibility of using genetic engineering to express [Ala⁶]-TTI in a form of fusion protein and the porduct could then be cleaved by CNBr.

Expression of [Ala⁶]-TTI gene in E.coli and S.cerevisiae

Based on the above mentioned results, the gene coding for [Ala⁶]-TTI was synthesized chemically. The synthetic gene was cloned into plasmid pWR590-1 and expressed in *E. coli* as a fusion protein composed of β-galactosidase fragment of 590 amino acid residues and [Ala⁶]-TTI, with Met as a connecting residue (Fig.4). After cyanogen bromide cleavage and reduction of the fusion protein, followed by refolding with trypsin-Sepharose 4B as a matrix and affinity chromtography on the immobilized enzyme, the fully active [Ala⁶]-TTI was obtained. The trypsin inhibitory activity and amino acid composition of the recombinant [Ala⁶]-TTI were consistent with those of the natural one. The [Ala⁶]-TTI gene was also cloned into the yeast secretion vector pVT102U/α, which has the mating α-factor secretion system to direct the expression of fused foreign products (Fig. 5). The recombinant vector was expressed in *S. cerevisiae*. The secreted

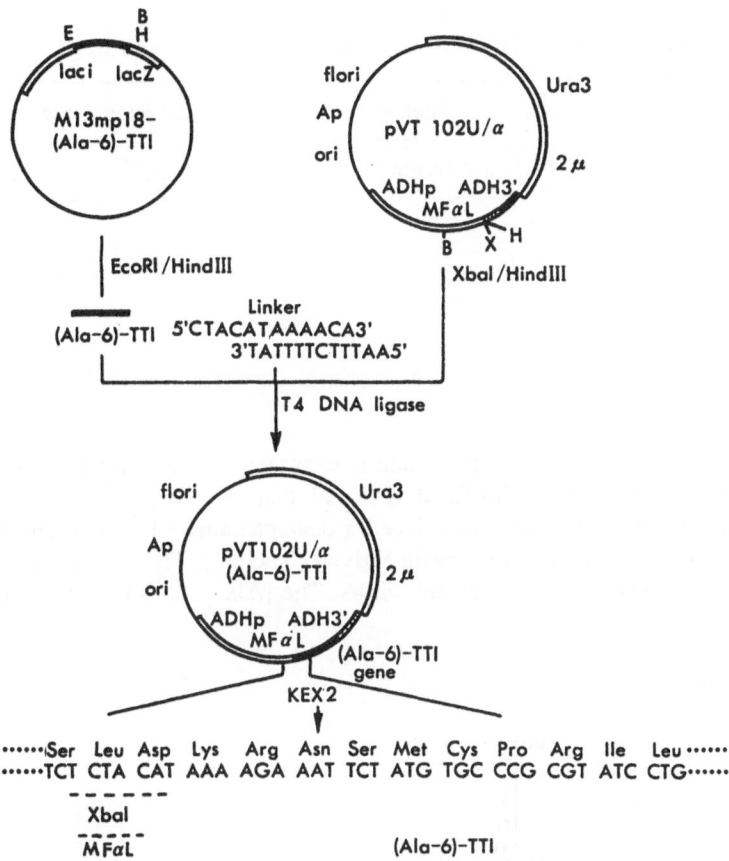

Fig. 5. Construction of plasmid pVT102U/α-[Ala⁶]-TTI. The protein and nucleotide sequences of the fused region was given.

[Ala^6]-TTI was purified and found to be correctly processed at the junction between the α-factor leader peptide and [Ala^6]-TTI downstream. Of the two expression systems, the latter is more advantageous in high yield (>2mg/L), easy purification and needlessness of disulfide refolding. Hence the yeast expression system is more suitable one for protein engineering [8].

Protein engineering of TTI

The inhibitor can be regarded as a nonproductive substrate of respective protease with a very low dissociation constant and also a very low catalytic constant. The specificity of inhibitor is mainly determined by its reactive site P1 and corresponds to the substrate specificity of the target protease. With a successful yeast expression system of TTI, protein engineering can be employed to study the structure-function relationship of TTI. The genes of TTI mutants, [Leu^3, Ala^6]-TTI, [Phe^3, Ala^6]-TTI, [Ala^3, Ala^6]-TTI, [Ala^3, Ser^4, Ala^6]-TTI, were constructed by PCR and site-directed mutagenesis and then cloned into pVT102U/α and expressed in *S. cerevisiae*. The secreted TTI mutants in the

183

Table 1 *Inhibitory activity of TTI and its mutants toward 3 serine proteases*

Reactive site	Inhibitory activity against serine protease		
	trypsin	chymotrypsin	elastase
Arg-Ile	+++	−	−
Phe-Ile	−	−	−
Leu-Ile	−	−	+
Ala-Ile	−	−	++
Ala-Ser	−	−	+++

supernatant were concentrated by rotating evaporation and purified by using gel-filtration, ionexchange and HPLC. It revealed that the simple replacement of Arg reactive site with Leu, Phe or Ala induced a disappearance of antitrypsin activity and also did not show any antichymotrypsin activity (Table 1). The P1 Leu, Ala subsituted mutants became pancreatic elastase inhibitors. The [Ala³, Ala⁶]-TTI showed a stronger

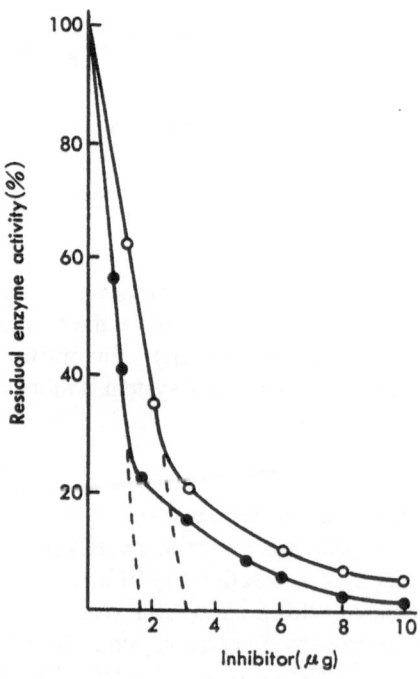

Fig. 6. Antielastase activity of the TTI mutants. [Ala³, Ala⁶]-TTI (-O-); [Ala³, Ser⁴, Ala⁶]-TTI (-●-).

elastase inhibitory activity than the [Leu^3,Ala^6]-TTI. Moreover, the [Ala^3, Ser^4, Ala^6]-TTI turned out to be a more potent elastase inhibitor with a typical eqimolar inhibition than the [Ala^3, Ala^6]-TTI (Fig.6). It demonstrated that the specificity and activity of the inhibitor is not only   attributed to the reactive site P1 residues, but also to its neighbouring residues, such as P1' and others.

References

1.   Otlewski, J., Whatley, H., Polanowski, A. and Wilusz, T., Biol. Chem. Hoppe-Seyler., 368(1987)1505.
2.   Dryjanski, M., Otlewski, J. Polanowski, A. and Wilusz, T., Biol. Chem. Hoppe-Seyler., 371(1990)889.
3.   Bode, W., Greyling, H.J., Huber, R., otlewski, J. and Wilusz, T., FEBS Lett., 242(1989)285.
4.   Le-Nguyen, D., Nalis, D. and Castro, B., Int. J. Peptide Protein Res., 34(1989)492.
5.   Qian, Y.W., Tan, F.L. and Chi, C.W., Science in China (Series B), 33(1990)599.
6.   Ling, M.H., Qi, H.Y. and Chi, C.W. J., Biol. Chem., 1992 (in press).
7.   Huang, Z.F., Wu,M.L. and Chi, C.W., Science in China (Series B), 33(1990)1192.
8.   Chen, X.M., Qian, Y.M., Chi, C.W., Gan,K.D., Zhang, M.F. and Chen, C.Q., J.Biochem., 112(1992)111.

The function of vasoactive protected L-arginine

Ming-di Gu, Shi-qi Peng, Xue-min Yu, Meng-shen Cai, Ting Yu, and Chao-shu Tang

National Laboratory of Natural and Biomimetic Drugs, Beijing Medical University, Beijing 100083, China

Introduction

In order to probe the biological behaviors of the N^G-protected-L-arginine and related oligopeptides[1–3], in the present study N^G-NO_2-L-Arg-OH (compound II), HCl•N^G-NO^2-L-Arg-OCH_3 (compound III), HCl•N^G-NO^2-L-Arg-N^G-NO_2-L-Arg-OCH_3(compound IV) and N^G-Tos-L-Arg-N^G-Tos-L-Arg-OH(compound V) were synthesized. After the confirmation of the purity (98-99%) by TLC and/or HPLC, the synthetic compounds were used for bioassay on rat aortic atrips treated with noradrenaline (NE), mean arterial blood pressure (MAP) of rats and cGMP accumulation in rat aortic strips.

Results and Discussion

The results are given in Table 1–6.

Table 1 *Vasodilation effect of compounds I-V*

Comp.	relaxing extent (X±SD%)
I.	3.88 ± 1.05
II.	61.3 ± 15.4
III.	66.3 ± 20.7
IV.	31.9 ± 13.1
V.	15.5 ± 3.6

n=6; dose 1×10^{-4}M; II-IV compared with I, P<0.01; compound I=Arg-OH. Strips with endothelium.

Table 2 *Vasodilation effect of compounds II and III*

Comp.	relaxing extent (X±SD%)
I	2.88 ± 1.20
II	3.10 ± 2.50
III	75.2 ± 24.7

n=6; dose 1×10^{-4}M; compound III compared with I, P<0.01; strips without endothelium.

It was shown that compound I(1×10^{-2}M, 10^{-3}M and 10^{-4}M) has no effect on rat aortic strips; Though the effect of compound II and compound III on the strips with endothelial cells are approximately equal, in the case of strips without endothelial cells they give different responses; compound III increases the MAP significantly; compound IV exerts biphasic effect; at dose of 0.016 mmol/kg, compound V decreases the MAP of hypertensive rats significantly and the action was lasted for 20 min; the effects of

Table 3 *Vasodilation effect of compound V*

Comp.	Relaxing extent (X±SD%)
I	1.20 ± 1.38
Va	20.08 ± 8.90
Vb	8.57 ± 1.93

Va: compound V with endothelium; Vb, without endothelium; n=6; dose, $1×10^{-4}$M; Va and Vb compared with I, P<0.01; After the occur of the relaxing action of Va and Vb, $2×10^{-3}$mol of compound III was administered and the contraction action was 25.08 ± 18.09%, P<0.01.

Table 4 *The effect of I-III on cGMP accumulation*

Comp.(dose, M)	Normal	III(10^{-5})	III(10^{-4})	II(10^{-5})
cGMP (X±SD)	2.84±0.08	1.34±0.74	0.69±0.70	1.38±0.71

Comp.(dose, M)	II(10^{-4})	I(10^{-3})	III+I	II+I
cGMP (X±SD)	0.75±0.34	2.34±0.63	2.07±0.71	2.23±0.77

Table 5 *The effects of compounds I, III, IV, V on rat MAP*

Comp. (mmol/kg)	MAP (X±SD%) at different time(min)					
	10	20	30	40	50	60
I (0.029)	96±2.9	97±2.6	96±2.3	97±2.4	96±2.5	96±2.6
III (0.059)	128±4.3**	134±10.6**	138±3.8	138±8.2**	139±2.6	138±6.6**
III (0.015)	120±18.0**	124±20.2**	127±22.9	132±20.5**	138±16.8	135±168**
IV (0.009)	107±6.2**	108±17.9	117±5.6*	118±13.2	114±1.3	117±9.7
IV (0.004)	92±9.2	94±10.0	85±7.0**	83±5.6	89±3.2	95±6.7
V (0.006)	98±16.6	88±16.2	81±14.8	81±13.8	74±11.1	70±12.9

The original MAP was defined 100%; compared with compound I, *P<0.05, **P<0.01, n=5.

compound II and compound III on cGMP accumutation in rat aortic strips may be responsible for their biological behavior as endothelium derived relaxing factor EDRF-antagonist.

Table 6 *The effects of compound V on the MAP of normal & hypertensive rats*

Comp. (mmol/kg)	Original MAP(mmHg)	MAP (X ± SD%) at different time(min)				
		5	10	20	30	50
a.						
I(0.13)	104±7	98.0±10.6	94.2±7.4	96.4±7.7	94.7±9.4	96.4±7.7
V(8×10⁻³)	110±5	98.4±16.6	87.8±16.2	86.5±16.3	82.3±12.4	84.6±7.2*
V(0.016)	108±6	85.5±17.2	82.8±18.8	81.2±9.8*	81.1±7.7*	80.7±6.6**
b.						
I(0.13)	162±10	101.2±7.0	96.6±8.1	96.2±4.6	96.1±7.7	98.9±6.4
V(8×10⁻³)	174±14	98.6±19.4	93.2±12.9	90.7±9.9	94.4±18.3	96.2±7.7
V(0.016)	166±17	90.4±9.2*	86.6±7.5*	88.2±6.4*	90.2±12.4	92.1±10.4

n=5; the original MAP was defined 100%, compared to compound I, * $P<0.05$, ** $P<0.01$; at the doses of 0.032 mmol/kg and 0.064 mmol/kg compound I had no depressor actions. a, normal rat group; b, hypertensive rat group.

Acknowledgement

We thank Beijing Natural Science Foundation for financial support.

References

1.   Furchgott, R.F., Nature, 288(1980)373.
2.   Gu, M.D., Peng, S.Q. and Tang, C.S., Hua Xue Tong Bao(in Chinese), 10(1991)32.
3.   Gu, M.D., Peng, S.Q. and Tang, C.S., Prog. Natl. Sci., 3(1992)231.

188

Studies on the interconversions of the neuroexcitotoxin β-N-oxalo-L-α, β-diaminopropionic acid from Panax species and its isomer α-N-oxalo-L-α, β-diaminopropionic acid

Yi-cheng Long, Fei-ning Ye, Yun-hua Ye and Qi-yi Xing

Department of Chemistry, Peking University, Beijing 100871, China

Introduction

β-N-Oxalo-L-α,β-diaminopropionic acid(β-N-ODAP) is a neuroexcitotoxic and hemostatic nonprotein amino acid[1] first isolated from *Lathyrus sativus* seeds in 1963[2]. In our previous article[3], we have reported that β-N-ODAP is found in most Panax species such as *P. ginseng* and *P. notoginseng*, and also observed that β-N-ODAP can very easily undergo interconversions with its α-isomer. α-N-ODAP is nearly physiologically inactive and probably an artifact. In this work we will discuss the intercoversions of β-N-ODAP and α-N-ODAP in detail. Experiments were performed as follows:

Studies of the interconversions of β-N-ODAP and α-N-ODAP are carried out by anion exchange HPLC(AEHPLC) developed in our lab. Column: Partisil SAX 10 μm, 250 × 4.6 mm; Detection: UV at 215 nm, Pye Unicam LC3 UV detector; Mobile phase: 2 ml/min, 15mM KH_2PO_4, Altex 110 pump.

Stability experiments: 0.3 mg/ml aqueous solutions of β-N-ODAP and α-N-ODAP are adjusted to various pH by 0.2 N HCl and 0.2 N NaOH, and then placed in water bath at 0°C, 15°C and 74°C. After every 20 h, the concentrations of β-N-ODAP and α-N-ODAP are analysed by AEHPLC.

Intercoversion kinetic experiments: three tubes of 0.3 mg/ml β-N-ODAP solutions are respectively adjusted to pH 2.0, pH 7.0 and pH 11.0 by 0.2 N HCl and 0.2 N NaOH, and the interconversions are then monitored by analysing the concentrations of β-N-ODAP and α-N-ODAP by AEHPLC at 30 minute intervals.

Results and Discussion

AEHPLC of β-N-ODAP and α-N-ODAP

Typical chromatogram of β-N-ODAP and α-N-ODAP. Please see our previous article[3].

Stabilities of β-N-ODAP and α-N-ODAP

Dry samples of β-N-ODAP and α-N-ODAP can be kept unchanged at room temperature, but their aqueous solutions are not stable. Analytical data show that both pure β-N-ODAP and α-N-ODAP in solution can convert to a mixture of β-N-ODAP and α-N-ODAP, indicating that the two isomers can reversibly interconvert to each other.

Table 1 *Interconversions of β-N-ODAP and α-N-ODAP*

Item	pH=2	pH=7	pH=11
k_1 (min^{-1})	2.81×10^{-3}	2.37×10^{-3}	9.66×10^{-3}
k_2 (min^{-1})	2.99×10^{-3}	3.95×10^{-3}	27.2×10^{-3}
K	0.940	0.601	0.355

Interconversion kinetics of β-N-ODAP and α-N-ODAP

Let k_1, k_2 and K (k_1/k_2) are the conversion rate constant of β-N-ODAP to α-N-ODAP, that of α-N-ODAP to β-N-ODAP and the interconversion equilibrium constant respectively. The values of k_1, k_2 and K obtained using kinetic method by AEHPLC monitoring are shown in Table 1, from which the following conclusions can be drawn:
1) Both bases and acids catalyse the interconversion but bases are more effective than acids;
2) The β- to α- or α- to β- conversion reaches equilibrium under various conditions. When the interconversion equilibria have been established the β-content is higher in basic solution than that in acidic solution, and α-content is higher in acidic solution than that in basic solution;
3) k_2 is always larger than k_1 indicating that β-isomer is more stable than α-isomer.

Acknowledgement

This work was supported by the National Natural Science Foundation of China.

References

1.   Rao, S.L.N., Biochem. Pharmacology, 16(1967)218.
2.   Roy, D.N., Nagarajan, V. and Gopalan, C., Curr. Sci., 32(1963)116.
3.   Long, Y.C., Ye, F.N., Ye, Y.H. and Xing, Q.Y., Chinese Chemical Letters, 7(1992)517.

Somnogenic effects of isometric oxidized glutathione (IGSSG) and oxidized glutathione (GSSG) isolated from roots of panax ginseng(pg-r)

Shi-yi Liu[a], Wen-yuan Zhang[a], Yi Zhang[b], Bin Lin[b], Yun-hua Ye[b], Liu Yang[b] and Qi-yi Xing[b]

[a]Shanghai Institute of Physiology, Chinese Academy of Sciences, Shanghai 200031, China

[b]Department of Chemistry, Peking University, Beijing 100871, China

Introduction

It is known that endogenous sleep factors have been pursued from the early decades of this century by Pieron (1910). Ever since, Delta-Sleep-Inducing Peptide (DSIP), Sleep-Promoting Substances (SPS), Factor S and Sleep-Promoting Neuropeptides (SPNP) are the four main approaches [1]. SPS isolated from water dialyzates of the brainstem of approximately 5000 rats of sleep-dep-rived rats was first reported in 1974 [2]. Until now, only two active components (SPS-A, uridine; SPS-B, GSSG) were identified. GSSG (Gamma-glutamylcysteinylglycine disulfide) has been regarded as the new, active and more purified sleep factor of SPS and was reported only very recently in 1990 [3].

It is known that roots of panax ginseng (PG-R) can potentiately reduced metabolism especially for elderly people ("Compendium of Materia Medica", 1596), so it is fundamentally stimulative and facilitatory. But in contrast to known stimulants (e.g. amphetamine), PG-R is "basically stimulative, but not insomnious". Nevertheless, no sleep-inducing ingredients have been identified or disclosed from PG-R until now.

Results and Discussion

GABA (γ-Aminobutyric acid) and several γ-glutamyl oligopeptides were isolated and identified from the water extract of PG-R. PG-R with 50% methanol was deproteinized and separated by cation exchange chromatography and gel filtration through Sephadex G-25[4]. Small quantity of acidic peptides were detected and isolated by the RP-HPLC. Two analogs of γ-glutamyl oligopeptides were identified from the water extract of PG-R for the first time: Oxidized γ-glutamylcysteinylglycine disulfide (GSSG) (< 5/100000) and Isomeric oxidized γ-glutamylglycylcysteine disulfide (IGSSG) (<1/100000).

It is of interest that both GSSG and IGSSG exhibited somnogenic effects. The physiologic effects of both of them on delta and sigma indices were evaluated after mesodiencephalic intraventricular infusion in 23 adult rabbits of either sex. Results show that both of them exerted some somnogenic effects, but they were less significant when compared to Asp^5-α-DSIP [1]. It is beyond expectation that the delta- and sigma-enhancing effects of IGSSG were potent than that of GSSG after mesodiencephalic

Table 1 *Effects of two γ-glutamyl oligopeptides (IGSSG, GSSG, 50 µg/rabbit,i.c.v.) isolated from roots of panax ginseng (PG-R) on delta and sigma indices compared to the controls (based on mean value of percent change between pre-and postinfusion) in rabbits*

	Delta index (M ± SE)	Sigma index (M ± SE)
Control	74 ± 6	114 ± 14
	(N=7)	(N=7)
IGSSG	147 ± 9 **	209 ± 14 **
	(N=8)	(N=8)
GSSG	107 ± 9 *	170 ± 23 *
	(N=8)	(N=8)

* $P < 0.05$, ** $P < 0.001$.

intraventricular infusion (50 µg/rabbit, i.c.v.) in rabbits (Table 1).

The present study demonstrates that endogenous or endogenous-mimetic substances with sleep-inducing inclusive can also be isolated from plant or herbal origin [1] and SPS may serve as one vivid example. It is known that both SPS-A and SPS-B can be isolated from herbal origin. Beside uridine which can be isolated from *Ganoderma capense*, the present study further demonstrates that GSSG can be isolated from PG-R. It is especially interesting that IGSSG is more potent than that of GSSG.

Acknowledgements

The study was supported by grants from the National Natural Science Foundation of China (9389007) and Chinese Academy of Sciences (875001).

References

1. Liu, S.Y., in Endogenous Sleep Factors, Inoue, S. and Krueger, J.M., Ed. SPB, The Hague 1990, pp.267.
2. Nagasaki, H., Iriki, M., Inoue, S. and Uchizono, K., Proc. Jpn. Acad., 50(1974)241.
3. Komoda, Y., Honda, K. and Inoue, S., Chem. Pharm. Bull., 38(1990)2057.
4. Yang, L., Ye, Y.H. and Xing, Q.Y., in Peptides: Biology and Chemistry: Proceedings of the Chinese Peptide Symposium 1990, Ed. by Du Yu-cang et al, Science Press: Beijing, 1991, pp.30.

Session VI
Peptide conformation and mimetics

Chairs: Chen-lu Tsou
Institute of Biophysics, CAS
Beijing, China

Ching-i Niu
Shanghai Institute of Biochemistry, CAS
Shanghai, China

Murray Goodman
University of California
San Diego, California, U.S.A.

and

Arthur M. Felix
Hoffmann-La Roche Inc.
Nutley, New Jersey, U.S.A.

The Insulin A and B chains contain sufficient structural information to form the native molecule

Chen-lu Tsou

National Laboratory of Biomacromolecules, Institute of Biophysics,
Chinese Academy of Sciences, Beijing 100101, China

Introduction

It was shown some years ago that oxidation of the reduced A and B chains of insulin gives reasonably good yield of the native hormone and it was suggested that among all the possible isomeric structures, the native insulin structure is the most stable. In spite of the confirmation of the above results in different laboratories, the related suggestion that the insulin A and B chains contain sufficient information to form the native structure have not met with general recognition as is evidenced by the treatment given to this problem in a large number of biochemistry and molecular biology textbooks (see for example, ref. 1). Recent results obtained in this laboratory have produced further evidence to show that the A and B chains contain sufficient structural information for the formation of the native disulfide bonds of insulin.

Results and Discussion

The expected yield of insulin by random joining of the chains?

If the formation of disulfide linkages is completely random, the expected yield of native insulin will be the reciprocal of the total number of possible ways of joining the thiols. Almost all authors dealing with this problem either considered only the A_1B_1 isomers the total number of which is 12 or used the formula for the calculation of the number of total isomers of a single chain protein giving, in both cases, the expected yield of a few percent. However, if the joining of the disulfides is indeed completely random, the formation of multichain products, A_xB_y must also be considered where x and y can be zero or any integer. The total number of different products approaches infinity and the expected yield of the protein with native disulfides should be close to zero. Experimentally, the oxidation of the reduced chains in 6 M guanidine hydrochloride gives indeed a very low yield of insulin as expected.

Can insulin be obtained from the A and B chains in yield well over that expected by random joining of the thiols?

Although the yield of insulin from its chains was considered good enough to make the commercial production of insulin worthwhile, this has not been generally accepted in most current textbooks (see, for example, ref. 1 and for a more detailed discussion, see ref. 2). The C peptide was considered essential in providing the information in the correct folding of the molecule.

For proteins with a single peptide chain, it is necessary to employ a dilute protein solution in order to minimize the formation of oligomers. However, for insulin, a dilute solution would decrease the opportunity of joining the A and B chains and favour the formation of products containing a single chains only. In order to increase the probability of the formation of A_1B_1 isomers, solutions of high concentrations have to be employed which would unavoidably favour the formation of other oligomeric products. A compromise has to be made in protein concentration and it is therefore not surprising that the yield actually obtained is not as good as that for proteins with a single peptide chain.

Do the A and B chains contain sufficient information to form insulin?

In connection with the above mentioned assertion that the structural information for the correct joining of thiols to form insulin is stored in the C peptide segment, earlier workers failed to obtain the native hormone from "scrambled" insulin through thiol-disulfide exchange reactions catalyzed by protein disulfide isomerase although under similar conditions scrambled proinsulin gave a 25% yield of the native molecule and treatment of insulin itself by this enzyme led to the destruction of native insulin structure. It was also reported that some insulin was obtained from the scrambled molecule in the presence of added C peptide suggesting structural information conferred to the whole molecule by this peptide.

We have now shown that at a temperature of 4°C, insulin can indeed be obtained from scrambled insulin or the S-sulfonates of the chains with a yield of 25-30% [3]. Furthermore, the addition of C peptide has no effect on the yield of native insulin from the scrambled molecule. In agreement with the results of earlier authors, the yield of the native hormone is low at higher temperatures. However, HPLC analysis of the products shows that the decrease in insulin is accompanied by increases in products containing only one of the chains but not inactive products containing both chains.

Insulin chemically crosslinked by reaction with carbonyl bis(L-methionyl p-nitrophenylester) through the A_1-α and B_{29}-ϵ amino groups containing native disulfide linkages can be obtained from the scrambled molecule or from the S-sulfonate derivative with very good yields. With a dilute solution of the hexa-S-sulfonate derivative, yield better than 90% can be obtained. The regenerated product is shown to have the native disulfide bridges by treatment with CNBr to regenerate insulin and by giving identical finger prints of its pepsin hydrolysate with that for crosslinked insulin [3]. Similar yields were obtained for proinsulin and for cross-linked insulin from the respective hexa-S-sulfonate derivative.

The above results indicate strongly that insulin A and B chains do contain sufficient information themselves to form the native molecule. It is already known that the reoxidation of both proinsulin and reduced insulin crosslinked by different reagents regenerates the native disulfide bonds with approximately the same yield. It is most unlikely that the crosslinking reagents, being all small molecules and different chemically, can all provide the same structural information. It is also highly suggestive that among the known proinsulin sequences from a number of species, the C peptide segment is the least conservative and has a mutation rate approximately 8 times that of the A and B chains.

The structure of the A and B chains and their interaction.

It has been reported in the literature that the separated insulin A and B chains are without secondary structures. However, we have now shown by CD and FTIR studies that they do contain significant amounts of α-helix and other secondary structures especially the B chain. Moreover, upon mixing of the A and B chains, the CD and FTIR spectra show considerable difference as compared to a simple sum of the spectra of the individual chains. The CD spectrum of the mixed chains shows increase in negative ellipticity at both 220 and 208 nm and the FTIR spectrum in D_2O shows increases of the peaks at 1651.6, 1633.5 and 1661.8 cm^{-1} and a large decrease at 1621.4 cm^{-1}. The above results suggest increased helix contents and decreases in the content of extended chains suggesting further folding upon mixing of the chains. The increased helices upon mixing of the chains is presumably stabilized through interaction of the two chains.

All the peak positions of the FTIR spectrum of the mixed chains shift to higher wave numbers with an average of 2.3 cm^{-1} as contrasted to a sum of the spectra of the separated chains. No such positive shifts was observed upon mixing of the intact A and C-terminal truncated B chains. The peak shift to higher wave numbers has been explained by different extents in H-D exchange. The pairing of the intact chains could produce folding of the chains to form a hydrophobic core especially when the chains are in high concentrations under the conditions employed for the FTIR measurements to make the isotopic exchange more difficult to approach completion.

It is therefore concluded that the insulin A and B chains indeed contain sufficient structural information so as to recognize and interact with each other to form a more native-like structure thus making the correct joining of the chains possible.

Is insulin molecule the most stable among the isomers?

Thermodynamically, relative concentrations of the products formed during the action of protein disulfide isomerase or during the oxidation of the reduced chains depend on their relative stabilities or relative energy contents.

The HPLC profile of the products formed by oxidation of the reduced insulin chains in the presence of denaturants contains 3 major peaks of approximately equal height containing, by amino acid analysis, respectively, either A or B chain alone or both chains with the ratio of B/A slightly greater than 1. This peak is biologically inactive and contain presumably isomeric products of A_xB_y. After unscrambling with protein disulfide isomerase at 4°C, the peaks containing either chain alone are unchanged but the peak containing both chains is greatly reduced in height and replaced by a new major peak which has been identified to be insulin by coelution with an authentic sample, amino acid analysis and biological assay.

The yield of insulin decreases at higher temperature, but by HPLC analysis the decrease in native insulin is accompanied by increases in products containing only one of the chains but not by any increase in inactive products containing both chains. Under non-denaturing conditions, in no case so far studied, the inactive product containing both chains is the major product formed and exceeds the amount of the active hormone. As the total number of the inactive A_xB_y isomers can be very large indeed whereas the peak containing the native hormone represents a single species, it follows that as far as the A_xB_y isomers are concerned, insulin is the most stable.

What is the role of the C-peptide segment in proinsulin?

If the C-peptide segment is not providing information for the formation of native disulfides in proinsulin during biosynthesis as most textbooks asserted it did, what is it doing there? It is now known that during renaturation process, the yield of insulin with 2 peptide chains, like molecules with one peptide chain such as ribonuclease, or proinsulin, also decreases with increasing temperature. For insulin, this is accompanied by increases in products containing only either A or B chain alone. No such separation of the chains is possible for ribonuclease A, proinsulin or for crosslinked insulin and consequently the role of the C-peptide could be just to bring and keep the insulin A and B chains together so as to favor the formation of the native, especially the inter-chain disulfides.

Conclusions

1. The A and B chains contain significant amounts of ordered secondary structures which are very likely the basis of the correct pairing of the chains and this in turn leads to furhter increase in the helix contents of the chains.

2. C-peptide does not increase the yield of insulin from scrambled insulin. The S-sulfonates of proinsulin and chemically crosslinked chains give similar yields of molecules with native disulfide bonds in the presence of protein disulfide isomerase indicating that C-peptide does not provide structure information for the formation of native disulfides.

Acknowledgements

The author wishes to express his gratitude to his colleagues taking part in various stages of this work and to the China National Natural Science Foundation for Grant No. 0388006.

References

1.	Darnell, J., Lodish, H. and Baltimore, D., in "Molecular and Cell Biology". 2nd Ed., Scientific American Books, New York, 1990, pp. 64.
2.	Wang, C.-C. and Tsou, C.-L., Trends Biochem. Sci., 16(1991)279.
3.	Tang, J.-G. and Tsou, C.-L., Biochem. J. 268(1990) 429.

The expansion of core group interaction that drives protein folding: A hypothesis based on the cytochrome c fragment complex

Hiroshi Taniuchi and Alice Fisher
Laboratory of Chemical Biology, National Institute of Diabetes and Digestive and Kidney Diseases, National Institutes of Health, Bethesda, MD 20892, U.S.A.

Introduction

Unlike intact RNase A, reduction and reoxidation of RNase (1-120) that lacks the four C-terminal residues generates non-native S-S bonds in most populations [1]. This implies that the residue-residue interaction that stabilizes RNase A is perturbed throughout the ordered structure for RNase-(1-120) [2]. Such perturbed interaction might be long range in the three-dimensional structure [2]. To learn about such long range interactions, we have studied cytochrome c fragment complexes [3].

Results and Discussion

Initially horse two-fragment complexes [4.5] were prepared. Type I complex contains a cleavage site of the polypeptide chain between residues 24 and 25 and Type II between residues 38 and 39. Type IV lacks residues 26 to 38. Type V contains two overlapping fragments, heme fragment containing residues 1-38, (1-38)H and apoprotein (1-104).

Type I can be formed from (1-25)H and (23-104) or (1-104) [4]. It most closely resembles native cytochrome c [6]. Residues 39 to 55 are disordered for Type II [4]. Type IV is ordered only in the right channel region at 25°C [3]. This hydrophobic channel is formed by contact of the N-terminal helix with the C-terminal helix [7]. Apocytochrome c essentially lacks helical structure [8,9]. Therefore, the covalently attached heme is required for the right channel to fold. Ferrous type V complex assumes two forms, Type I and II which interconvert to each other without dissociation [10]. Ferric Type V assumes Type II form [3]. Type I,II and V complexes show the 695 nm absorption band at 25°C [4], a band indicative of the Met 80-S-heme Fe^{3+} bond [11]. Type IV barely exhibits this band at 25°C [3,4]. When the temperature is lowered to 6°C, it shows a significant intensity of the band [3]. Addition of fragment (28-38) to Type IV complex generates the 695 nm band at 25°C [5].

We define a core domain as a structural region consisting of a hydrophobic core and the surrounding shell that folds and unfolds as a unit. A core domain can be formed from either contiguous or discontiguous segments of the polypeptide chain by folding and assembly [3]. Core domain 1 is assigned [3] to the region containing the right channel and a part of the heme of Type IV complex at 25°C (Fig.1). It folds and unfolds through a single energy barrier [6]. Core domain 2 is assigned [3] to the region

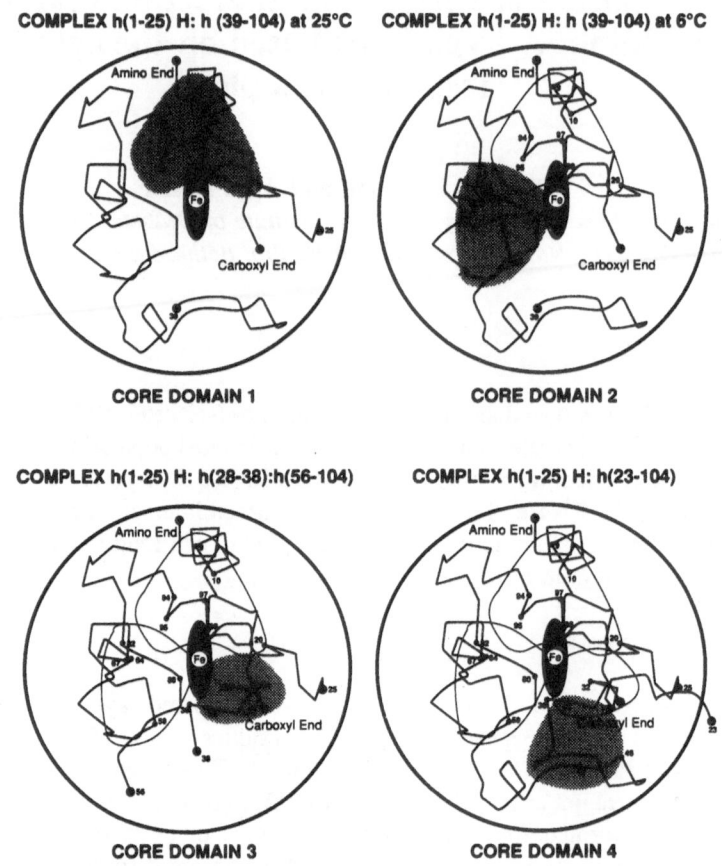

Fig. 1. *Schematic drawings of four core domains of the cytochrome c complex. The four core domains are highlighted one by one by the shaded regions. Only core domain 1 operates for Type IV (1-25)H.(39-104) at 25°C. Core domain 2 of Type IV complex folds upon lowering the temperature to 6°C. Core domain 3 forms when fragment (28-38) is added to Type IV (1-25)H.(39-104) or (1-25)H.(56-104) [5]. Core domain 4 is formed by folding of residues 39-55. Type I complex (1-25)H.(23-104) contains domains 1-4. Type II (1-38)H. (39-104) contains core domains 1-3. It is not clear whether the heme- Trp 59-Leu 64 core belongs to domain 1 or 2. It is shown here with domain 2 for simplicity. h stands for horse cytochrome c. The diagrams are based on the illustration of Dickerson and Geis [17].*

containing the core on the left side of the heme including the Fe-S bond [7] that folds in type IV complex when the temperature is lowered to 6°C (Fig. 1). Core domain 3 is assigned [3] to the region containing the core on the right side of the heme (Fig. 1). Core domain 4 is assigned [3] to the structure containing the core formed between the heme and residues 39 and 55 (Fig.1) that unfolds in Type II.

Kinetic studies [6] suggest that core domain 1 folds first before the folding of core domain 2. Removal of eight N-terminal residues of a Type I complex abolishes the 695 nm band [3]. This indicates that the N-terminal region (core domain 1) interacts with the

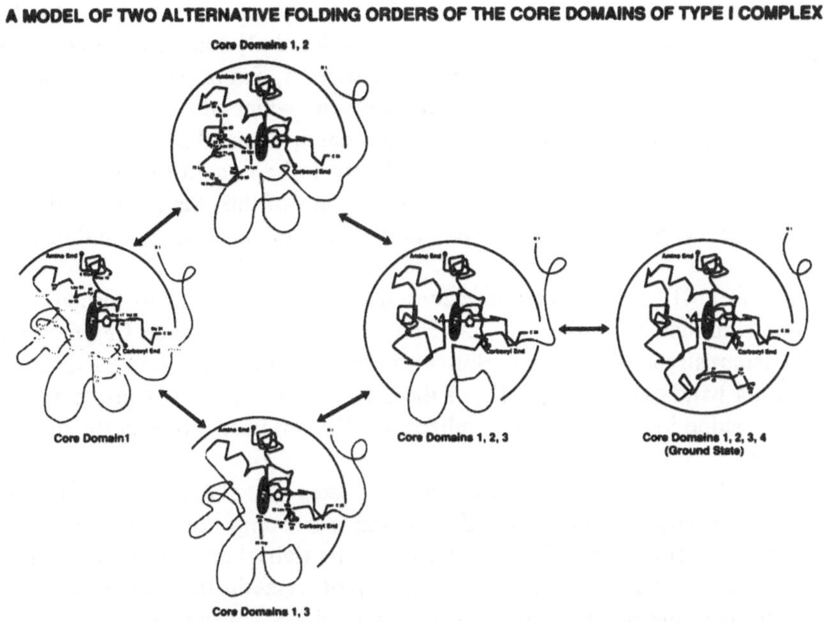

A MODEL OF TWO ALTERNATIVE FOLDING ORDERS OF THE CORE DOMAINS OF TYPE I COMPLEX

Fig. 2. *A model of the folding order of the core domains of horse Type I complex h(1-25)H. h(1-104). Core domain 1 folds first. This intermediate is compactly folded and mimicked by a major population of Type IV h(1-25)H. h(39-104) lacking the 695 nm band at 25°C [6]. In one of two alternative orders (lower), the second step is the folding of core domain 3. This is followed by folding of core domain 2. Core domain 4 folds last. In the alternative order (upper), core domain 2 folds at the second step. Core domain 3 folds next. Core domain 4 is last again. The heme-Trp-59-Leu 64 domain is depicted with domain 2 for simplicity (see the legend to Fig.1).*

Fe-S bond of core domain 2 (the distance, about 12Å, [12-13]). There is evidence that domain 1-3, domain 2-3 (the 695 nm band stabilization) and domain 2-4 or 3-4 (the 695nm band stabilization) interactions exist [3]. Based on the foregoing, a model of two alternative orders of folding of core domains for Type I complex was constructed ([3], Fig. 2). This model is consistent with the kinetic studies [6] of the generation of the 695 nm band [3].

Type I, II and V complexes were prepared from heme- and apofragments or apoproteins of horse, tuna, Candida and yeast iso-1-cytochromes c [3]. The complexes were characterized by measurements of the 695 nm band and Soret band, immunological studies, cytochrome b_2 assay and tryptophan fluorescence quenching [3]. The results have shown [3] that the corresponding heme and apo-fragments of Type I, II and V are completely exchangeable [3]. The only exception is that two specific Type II complexes showed no 695 nm band [3]. The simplest explanation of such exchangeability is to assume that the core domains and their folding order are universal for cytochromes c[3]. Furthermore, such exchangeability would only be possible if the core structures of the parent cytochromes c are homologous.

Despite such core homology, the thermal transition mid-point, Tm of the 695 nm band increases in the order of yeast iso- 1, Candida, tuna and horse cytochromes c [3]. This

201

transition involves neither core domain 1 unfolding nor much exposure of the heme to solvent [3]. The region that is presumed to be disordered [3] contains residues which are buried or partially buried in cytochrome c [7]. However, Tm of the 695 nm band differs among the above cytochromes c even in cases in which there is no difference in hydrophobicity of the residues. On the other hand, since the core domain-domain interaction stabilizes the 695 nm band, it is likely to be related to the residue-residue interaction that stabilizes the 695 nm band. In support of this, the apparent dissociation constant, K_D, of Type I complex decreases in the order of yeast iso-1, Candida and horse [3].

O'Hern et al [14] have shown that addition of a formyl group to buried Trp 59 of cytochrome c abolishes the 695 nm band. The studies with the three-fragment complex [15] have demonstrated that substitution with norvaline of buried Leu 32 decreases Tm of the 695 nm band and increases K_D of the substituted fragment. Thus, it appears that the residue-residue interaction that stabilizes the 695 nm band exists in the hydrophobic core.

K_D was measured for all combinations of complexes [3]. ΔG° was calculated from K_D [3]. There is a large variation of K_D among the complexes [3]. The heme and apofragments contain residues that are buried or partially buried in native cytochrome c [7]. However, K_D does not vary in a majority of cases in the manner expected if a difference in hydrophobicity of such residues caused a K_D change. K_D varies depending on which species of heme fragment binds to which species of apofragment [3]. Analysis of this interdependency using the ΔG° data indicates that two substituted residues, one in the heme fragment and the other in the apofragment interact with each other in the complex [3]. Furthermore, going from Type IV complex to Type I, II or V the number of residues involved in such an interaction distinctly increases [3]. Evidence suggests that this is unlikely to be an interaction between the exterior residues that make contact with each other [3]. We speculate that the difference between Type IV complex and Type I, II or V may be due to the fact that more than one core domains are operation in Type I, II and V (Fig.1). The residue-residue interaction between the two fragments may be associated with the unification of the ordered cores that occurs as core domains 1-3 or 1-4 fold. In support of this, an increase of K_D qualitatively correlates with a decrease of Tm of the 695 nm band as studied with selected complexes [3]. This suggests that such a residue-residue interaction involves the core residues [3].

Based on this reasoning, analysis of the K_D data and comparison of the amino acid sequences, 5 core residues have been tentatively assigned to be involved in such residue-residue interactions for Type I complex [3]. They are at positions 9, 20, 64, 95 and 98[3].

In summary, we hypothesize that the residue-residue interaction in the ordered core would be generated upon folding of core domain 1 of Type I complex and would expand throughout the ordered cores as core domains 2, 3 and 4 fold. This hypothetical residue-residue interaction apparently is not hydrophobic[3]. It is long range in the three-dimensional structure [3,15,16] and sensitive to substitution of the residues involved [3,15].

References

1. Taniuchi, H., J. Biol. Chem., 2465(1970)5459.
2. Taniuchi, H. and Fisher, A., in Bioengineered Molecules: Basic and Clinical Aspects (Verna, R., Blumenthal, R. and Frati, L., Eds.), Raven Press, NY, 1989, pp.1-10.
3. Fisher, A. and Taniuchi, H., Arch. Biochem. Biophys., 296(1992)1.
4. Parr, G.R., Hantgan, R.R. and Taniuchi, H., J. Biol. Chem., 253(1978)5381.
5. Juillerat, M., Parr, G.R. and Taniuchi, H., J. Biol. Chem., 255(1980)845.
6. Parr. G. and Taniuchi, H., J. Biol. Chem., 257(1982)10103.
7. Dickerson, R.E., Takano, T., Eisenberg, D., Kallai, O.B., Samson, L., Cooper, A. and Margoliash, E., J. Biol. Chem., 246(1971)1511.
8. Stellwagen, E., Rysavy, R. and Babul, G., J. Biol. Chem., 247(1972)8074.
9. Fisher, W.R., Taniuchi, H. and Anfinsen, C.B., J. Biol. Chem., 248(1973)3188.
10. Juillerat, M.A. and Taniuchi, H., J. Biol. Chem., 262(1987)13440.
11. Schechter, E. and Saludjian, P., Biopolymers, 5(1967)788.
12. Takano, T. and Dickerson, R.E., J. Mol. Biol., 153(1981)95.
13. Bernstein, F.C., Koetzle, T.F., Williams, G.J.B., Meyer, E.F., Jr. Brice, M.D., Rodgers, J.R., Kennard, O., Shimanouchi, T. and Tasumi, M., J. Mol. Biol., 112(1977)535.
14. O'Hern, D.J., Pal, P.K. and Myer, Y.P., Biochemistry 14(1974)382.
15. Juillerat, M.A. and Taniuchi, H., J. Biol. Chem., 261(1986)2697.
16. Poerio, E., Parr, G.R. and Taniuchi, H., J. Biol. Chem., 261(1986)10976.
17. Dickerson, R.E., Sci. Amer., 226(1972)58.

The conformational restriction of synthetic peptides to the α-helix, reverse turn and loops with covalent hydrogen bond mimics

Lin-chang Chiang, Edelmira Cabezas, Julio Calvo and Arnold C. Satterthwait

The Scripps Research Institute, La Jolla, CA 92037, U.S.A.

Introduction

There is an important need for synthetic methods for constraining peptides to the shapes they occupy in native proteins because shape often determines bioactivity. Conformational mimics can be used as frameworks for identifying determinants of protein structure, for structure-function studies and may find practical use as ligands, drugs, and synthetic vaccines.

Our approach is based on replacing structure-defining bydrogen bonds formed between backbone peptide bonds (NH...O=CRNH) with covalent mimics. on average, every other amino acid in globular proteins engages in this bond and different hydrogen bonding patterns define different coformations including secondary and irregular sturctures. We have developed synthetic procedures for replacing these hydrogen bonds with C=N double bonds either in the form of an amidinium link (N-CR=NH(+)CH$_2$CH$_2$) or a hydrazone link (N-N=CHCH$_2$CH$_2$) and obtained evidence that these links stabilize predetermined peptides as common secondary structures in water.

Results and Discussion

The amidinium link has been substituted for the penultimate hydrogen bond of a small loop A to give a constrained loop B.

Loop B and various analogues were synthesized by selectively activating and N-terminal thiolamide with methyl iodide and freeing a protected C-terminal diaminoethane for rapid reaction with an intermediate thiolimidate to give loop B [1]. X-ray crystallography shows that the amidinium link folds loop A, SarAlaAlaGly, and two analogues, SarProAlaGly and GlyAlaAlaGly, into a Type 1 reverse turn [2] while enforcing the characteristic (i, i+3) hydrogen bond shown for Loop B. NOE's, amide proton-C$_\alpha$ proton coupling constants ($^3J_{HN\alpha}$) and temperature coefficients indicate that the Type 1 conformation is maintained in dimethylsulfoxide and water at ambient temperatures. For example, temperature coefficients for the hydrogen bonded amide proton are ≈0 (DMSO) and ≈1 ppb/deg(10%D$_2$O/90%H$_2$O). Additionally, each of the amino acids in the loop have been substituted with glycine and other amino acids with little apparent effect on conformation as judged by the small changes in $^3J_{HN\alpha}$ [3].

The hydrazone link has been substituted for the (i, i+4) hydrogen bond characteristic of alpha helices and for the (i, i+n) hydrogen bonds found in loops and beta-hairpins

(Fig. 1). This chemistry makes use of N-aminoglycine, which can be inserted at any position in a peptide chain, and a dimethylacetal derivative that caps the N-terminus. Cyclization is achieved with acid that activates the acetal for rapid reaction with the free amino group on N-aminoglycine to give the cyclic hydrazone. These reactions have been carried out in solution [4] and with modifications to the protocol on solid supports. Although yields of the cyclic peptides vary with size and amino acid sequence, loops with 3-18 predetermined amino acids have been prepared in purified form (>95%) utilizing these procedures.

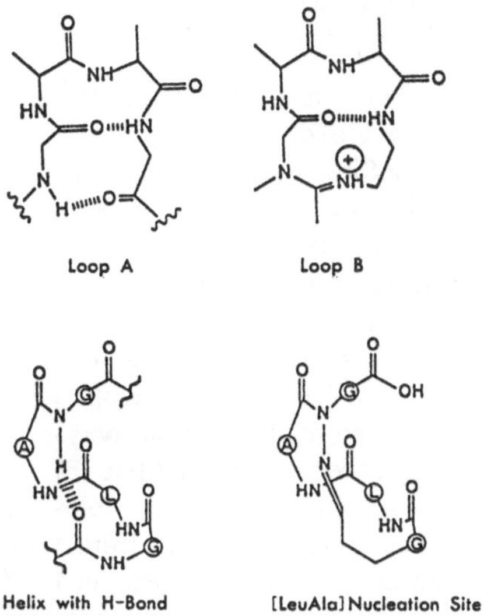

Peptides with an (i, i+4) hydrazone link are stabilized as full length alpha helices in water at ambient tempera-

Loop A   Loop B

Helix with H-Bond   [LeuAla] Nucleation Site

Fig. 1.

tures as demonstrated in extensive NMR experiments. The most detailed information is provided by the observation of sequential and short proton-porton distances for [LeuAla]NucSite-AEAAKA-NH$_2$ and deuterated analogues (summarized in Fig 2.) These data together with an analysis of chemical shifts, $^3J_{HN\alpha}$ coupling constants, and temperature coefficients of this test peptide as well as a corresponding peptide from which the covalent mimic has been removed (formylGLA GAEAAKA-NH$_2$) provide strong evidence that the mimic converts the peptide from disordered conformers to a substantial population of full length alpha helices. The observation of (i, i+3) NOE's that span amino acids in NucSite and the first helical turn of the appended peptide and throughout the full length of the appended peptide argue strongly for the formation of an undistorted alpha helix [5,6].

Changes in the amino acid content of NucSite, in amino acid sequence and length of the appended peptide can significantly impact the extent of alpha helix formation in water and are providing insights into the determinants of structure. From these ongoing studies, it appears that predetermined peptides with predictable alpha helical tendencies can be converted to alpha helical conformers using the hydrazone mimic. For example, [LeuAla]NucSite-ALFQKEKMAKA-NH$_2$, which includes an epitope from pfs66, a synthetic malaria vaccine candidate [7], forms quantities of full length alpha helices similar to that of the test peptide as suggested by similar decreases in temperature coefficients (≈2 ppb/deg) of the amide protons for the common -AKA-NH$_2$ C-termini.

Covalent hydrogen bone mimics are finding use in the conformational restriction of peptide antigens and synthetic vaccine candidates[1]. An example is provided by the use of the hydrazone mimic for constraining epitopes on the V3 loop of the gp120 surface protein of HIV-1. The V3 loop is predicted to form a beta-hairpin and an alpha helix within the confines of a large disulfide loop. The predicted hairpin of the MN strain has been constrained to give N=CHCH$_2$CH$_2$CH$_2$CO-HIGPGRAFG-NCH$_2$CO-GNH$_2$ and the predicted helix nucleated to give [LeuAla]NucSiteARQAHC(Acm)NIS RAKA-NH$_2$. NMR experiments with both constrained peptides are consistent with the formation of significant fractions of hairpin and alpha helix in water, pH 5, at ambient temperatures.

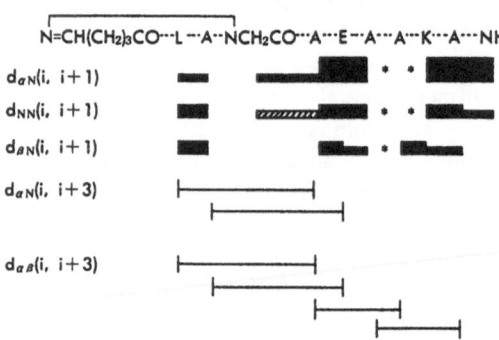

Fig. 2. Summary of NOE's observed in a ROESY spectrum for 20 mM [LeuAla]NucSiteAEAAKA-NH$_2$ in 10% D$_2$O/H$_2$O, pH2.9, 20°C. The hatched bar indicates an NOE between N=CH, the hydrogen bond mimic and the Ala NH. The asterisk indicates resonance overlap.

The constrained HIV peptide loop is antigenic, having found use as a cocrystallizing agent for an anti-HIV Fab [8] and as a panning reagent for the isolation of human monoclonal Fabs from a combinatorial Fab library generated from an asymptomatic HIV infected individual [9]. Both the loop and helix are currently being tested as immunogens.

In summary, the amidinium link and the hydrazone link have been found to be effective mimics for structure defining main chain hydrogen bonds folding peptides into common secondary structures including the alpha helix, reverse turns and loops. Approximately 50% of the amino acids in a globular protein fold into these conformations. Predictive algorithms, analyses of mutational frequencies and precise structural information for proteins from X-ray crystallography and NMR spectroscopy provide many conformational targets. It is likely that this chemistry can be used to fold many correspoding peptides with good prospects for identifying and improving their biological activities.

References

1. Satterthwait, A.C., Chiang, L-C., Arrhenius, T., Cabezas, E., Zavala, F., Dyson, H.J., Wright, P.E. and Lerner, R.A., WHO Bull., 68(Suppl)(1990)17.
2. Parge, H. and Chiang, L-C., unpublished.
3. Chiang, L-C., Cabezas, E., Noar, B., Arrhenius, T., Lerner, R.A. and Satterthwait, A.C., In Giralt, E. and Andreu, D. (Eds.) Peptides 1990 (Proceedings of the 21st European Peptide Symposium), ESCOM, Leiden, 1991, p. 465.
4. Arrhenius, T. and Satterthwait, A.C., In Rivier, J.E. and Marshall, G.R. (Eds.) Peptides: Chemistry, Structure and Biology (Proceedings of the 11th American Peptide Symposium), ESCOM, Leiden, 1990, p. 870.
5. Wuthrich, K., Billeter, M. and Braun, W., J. Mol. Biol., 180(1984)715.
6. Dyson, H.J. and Wright, P.E., Annu. Rev. Biophys. Biophys. Chem., 20(1991)519.
7. Amador, R., Moreno, A., Valero, V., Murillo, L., Mora, A.L., Rojas, M., Rocha, C., Salcedo, M., Guzman, F., Espejo, F., Nunez, F. and Patarroyo, M.E., Vaccine, 10(1992)1797.
8. Stura, E., unpublished.
9. Barbas, C., Collet, T.A., Roben, P., Binley, J., Amberg, W., Hoekstra, D., Cababa, D., Jones, T.M., Williamson, A., Pikington, G.R., Haigwood, N.L., Satterthwait, A.C., Sanz, I. and Burton, D.R., J. Mol. Biol., In press.

Topographical considerations in computerized peptide design

Victor J. Hruby, Aleksandra Misicka, Ernesto Nicolas, K.C. Russell, Guigen Li, Gregory Nikiforovich, Ding Jiao and Katalin Kövèr

Department of Chemistry, University of Arizona, Tucson, AZ 85721, U.S.A.

Introduction

Most naturally occurring peptide hormones and neurotransmitters are small, conformationally flexible molecules, the fragments of much larger protein precursors. Yet it is clear from modern structure-function studies [1,2] that their biological activities are highly dependent on their conformational, stereoelectronic and dynamic properties. Questions of potency, receptor selectivity, efficacy, stability and bioavailability all relate back to the peptide's specific chemical properties [3]. It rational design of peptides that meets all the desired criteria presents a major challenge, but the use of conformational and topographical constraint has provided a more rational approach to the peptide ligand design problem [4,5]. A wide variety of methods for local and more global backbone conformational constraints have been developed [4,6] that can provide more or less stable secondary structures(ϕ,ψ and ω angles relatively fixed). Within this context, we have been examining methods to bias or fix side chain conformations (topographical constraints) at the χ_1,χ_2, etc. torsional angles as an approach to examine systematically the 3-dimensional requirements for peptide-receptor(acceptor) interactions [5].

Results and Discussion

Our initial studies concentrated on the development of aromatic amino acids (Phe,Tyr, Trp, His,etc.) because of their almost universal importance as key residues in peptide hormone/neurotransmitter/growth factor/etc.biological activities. Aromatic amino acid analogues and mimetics such as tetrahydroisoquinoline carboxylates (**1**, Tic, Fig. 1), β-methyl-substituted aromatics such as β-MePhe (**2**, X=H; M=Me), more highly constrained aromatics (such as analogue **3** in Fig. 1) and N-alkyl-substituted analogues (such as analogue **4** in Fig. 1) are among those being developed in our laboratory. In these cases, χ_1, χ_2 or χ_1 and χ_2 are biased or constrained by substitution or local backbone to side chain ring closure. The use of these amino acids, especially **2** and **3**, have required the development of asymmetric synthesis methods (for example **2**, M=CH$_3$, exists as four isomers 2S,3R;2R,3R;2S,3R;2R,3S). Asymmetric syntheses of these amino acids have been achieved [e.g.7,8] and are being further refered for simplicity and application to trypotophan, histidine, and other amino acids. We also have demonstrated using dynamic nuclear magnetic resonance spectroscopy, that amino acids of the type **3** are highly constrained about χ_2 with barriers to rotation of 15-20 kcal/mole

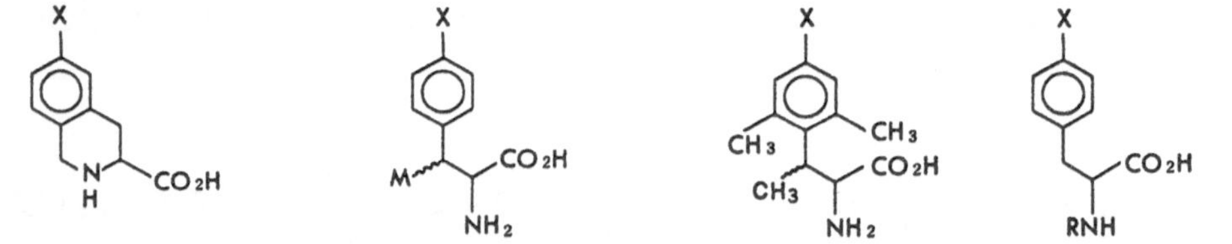

X=H,OH,CH₃,etc   M=H,CH₃,OH,etc   R=H,CH₃,C₂H₅,etc

Fig. 1. Sturctures of topographically constrained aromatic amino acids.

depending on the specific substitutions at C_α and C_β. Methods to determine precise side chain conformations about χ_1 using 1D and 2D homonuclear and heteronuclear NMR spectroscopy to obtain the necessary homon- and heteronuclear coupling constants are currently under development in our laboratory.

An early use of topographical constraint was applied to the highly δ opioid selective ligand [D-Pen², D-Pen⁵]enkephalin (DPDPE) in which the Phe⁴ residue was substituted with a Tic (Fig. 1) or a β,β-Me₂Phe residue to help determine the preferred Phe⁴ side chain conformation for biological activity[9]. More recently we have used these and related β-MePhe amino acid residues to help examine the topographical similarities that enable cyclic enkephalin analogues such as H-Tyr-D-Pen-Gly-Phe-D-Pen-OH(DPDPE) and the linear, naturally occurring deltorphins of which deltorphin I,H-Tyr-D-Ala-Phe-Asp-Val-Val-Gly-NH₂, is an example[10]. We have examined the question of topographical similarities that presumably govern the highly δ opioid receptor potency and selectivity of these structurally different compounds by a combination of synthetic biophysical and computational methods. These included the synthesis and biological evaluation of all four β-MePhe⁴-substituted analogues of DPDPE and all four β-MePhe³-substituted analogues of deltorphin I [11,12], complete conformational analysis of all four analogues of [β-MePhe⁴]DPDPE, and extensive computationally based search procedures for the low energy conformations of DPDPE, deltorphin I, and all of the β-MePhe-substituted analogues as well as several other analogues and derivatives. These studies have led to the conclusion that the DPDPEs and deltorphins, though possessing quite different backbone conformations, have considerable topographical similarity when one compares the Tyr¹ side chain, Phe³(or Phe⁴) side chain and position 2 C_α atom in three dimensional space(Fig. 2). Recent X-ray crystal structure determination of DPDPE (Flippen-Anderson et al., unpublished) provide additional evidence in support of these models. Most interestingly, the β-MePhe⁴-DPDPE analogues and the β-MePhe³-deltorphin analogues have widely different potencies and selectivities at the δ opioid receptor that largely can be accounted for by the different topographical relationships that are found in the diastereoisomeric peptide analogues. Further topographic constraints undoubtedly will provide critical further information on the conformational, topographical and stereoelectroinc requirements of the δ opioid receptor and its subtypes.

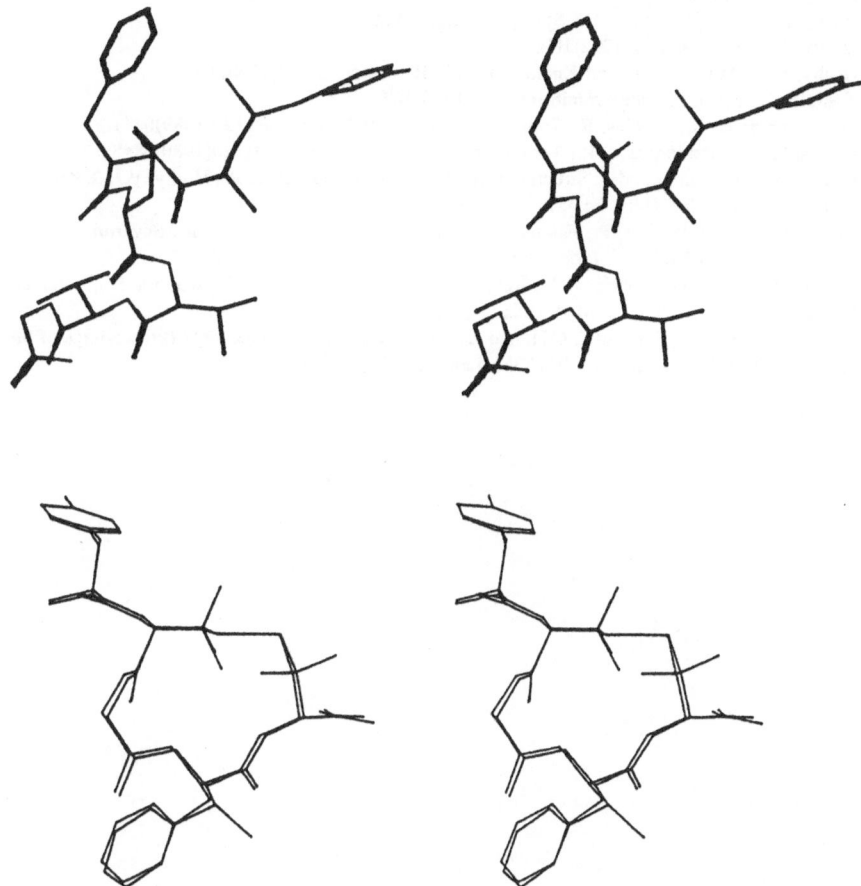

Fig. 2. Stereoviews of proposed conformations of deltorphin I(top) and [D-Pen^2, D-Pen^5] enkephalin (DPDPE and [(2S,3S)-β-MePhe^4]DPDPE (bottom, note virtually identical conformations) showing similar topographical features though different conformations. (see text for discussion).

Acknowledgements

The financial support of the National Institute of Drug Abuse, DA 06284 and DA 04248 and the U.S. Public Health Service, NS 19972 is gratefully acknowledged. In addition, we are grateful to Professors H.I.Yamamura, F.Porreca and T.F.Burks and their groups for collaboration on all biological aspects of this work.

References

1.   Hruby, V.J., in Conformational Directed Drug Design. Peptides and Nucleic Acids as Templates or Targets. (Eds. Vida, J.A. & Gordon, M.) ACS Symposium Series No.251, 1984, pp.9.
2.   Hruby, V.J. and Sharma, S.D., Curr. Opin. Biotechnol., 2(1991)599.

3. Hruby, V.J. and Hadley, M.E., in Design and Synthesis of Organic Molecules Based on Molecular Recognition, (Ed.Van Binst, G.) Springer Verlag, 1986, pp.269.
4. Hruby, V.J., Life Sci., 31(1982)189.
5. Hruby, V.J., Al-Obeidi, F. and Kazmierski, W., Biochem. J., 268(1990)249.
6. Toniolo, C., Int. J. Peptide Protein Res., 35(1990)287.
7. Dharanipragada, R., Nicolas, E., Toth, G. and Hruby, V.J., Tett. Lett, 30(1989)6841.
8. Nicolas, E., Dharanipragada, R., Toth, G. and Hruby, V.J., Tett. Lett, 30(1989)6845.
9. Hruby, V.J. et al., in Peptides: Structure and Function, (Eds. Deber, C.M., Hruby, V.J. & Kopple, K.D.) Pierce Chemical Co., 1985, pp.487.
10. Ersparmer, V., Melchiorri, P., Falconieri-Erspamer, G., Negri, L., Corsi, R., Severini, C., Barra, D., Simmaco, M. and Kreil, G., Proc. Natl. Acad. Sci. U.S.A., 86(1989)5188.
11. Hruby, V.J., Toth, G., Gehrig, C.A., Kao, L.F., Knapp, R., Lui, G.K., Yamamura, H.I., Kramer, T.H., Davis, P. and Burks, T.F., J. Med. Chem., 34(1991)1823.
12. Misicka, A. et al., In Schneider, C.H. and Eberle, A.N. (Eds.) Peptides 1992 (Proceedings of the 22nd European Peptide Symposium), ESCOM, Leiden, 1993, p. 651.

Main chain and side chain chiral methylated somatostatin analogs

Murray Goodman, Ya-bo He and Zi-wei Huang
Department of Chemistry, University of California, La Jolla, CA 92093, U.S.A.

Introduction

The cyclic hexapeptide analog of somatostatin, c[Pro^6-Phe^7-D-Trp^8-Lys^9-Thr^{10}-Phe^{11}] (I) synthesized by Veber and co-workers, shows high biological activity in inhibiting the release of growth hormone (the superscript numbers refer to the location of the residues in native somatostatin)[1]. Since its discovery this peptide (I) has been the subject of extensive structure-bioactivity studies[2]. To investigate the conformations of the main chain and side chains of this cyclic hexapeptide (I) further, we carried out syntheses and conformational analyses of a series of cyclic hexapeptide analogs of somatostatin incorporating main chain and side chain chiral methylations. Peptidomimetics α-methylphenylalanine (α-MePhe) and α-methylvaline (α-MeVal) were used to replace Phe^7 and Thr^{10} respectively in order to study the main chain conformations of residues 7 and 10. Peptidomimetics β-methylphenylalanine (β-MePhe) and β-methyltryptophan (β-MeTrp) were used to replace Phe and Trp residues at positions 7,8 and 11 in the cyclic hexapeptide (I) in order to study the side chain conformations of residues 7,8 and 11. These α- and β-methylated analogs show widely different binding potencies at the somatostatin receptor (Table 1).

Using high resolution one-dimensional and two-dimensional ^{1}H-NMR experiments and computer simulations, we have carried out conformational analyses for these somatostatin analogs containing α- and β-methylations to assess systematically the effect of chiral variations on the overall structure. This study allowed us to propose a model for bioactivity of somatostatin analogs.

The detailed discussion of the syntheses, binding studies, and conformational analyses for these α- and β-methylated analogs is published elsewhere[3].

Results and Discussion

Main Chain Methylated Somatostatin Analogs

The conformational search carried out for the parent compound(I) revealed two types of conformations:" flat" and "folded" (Figure 1). The key difference between these two conformations arises from the different backbone torsions for residue 7 and 10. The "flat" conformer shows a C_5 conformation for residues 7 and 10 while in the "folded" conformer the overall topology is folded around residue 7 and 10, assuming a C_7 conformation. Our finding is in agreement with the "flat" and "cup" shaped conformations suggested by Veber[2]. To investigate the conformational roles played by residues

Table 1 *The binding constants for somatostatin analogs*

	Analog	$IC_{50}(nM)$
I	c[Pro-Phe-D-Trp-Lys-Thr-Phe]	1
II	c[Pro-(2S,3S)-β-MePhe-D-Trp-Lys-Thr-Phe]	1
III	c[Pro-(2S,3R)-β-MePhe-D-Trp-Lys-Thr-Phe]	1
IV	c[Pro-Phg-D-Trp-Lys-Thr-Phe]	15
V	c[Pro-Phe-(2R,3S)-β-MeTrp-Lys-Thr-Phe]	<1
VI	c[Pro-Phe-(2R,3R)-β-MeTrp-Lys-Thr-Phe]	>1000
VII	c[Pro-Phe-(2S,3R)-β-MeTrp-Lys-Thr-Phe]	10
VIII	c[Pro-Phe-(2S,3S)-β-MeTrp-Lys-Thr-Phe]	>1000
IX	c[Pro-Phe-D-Trp-Lys-Thr-(2S,3S)-β-MePhe]	50
X	c[Pro-Phe-D-Trp-Lys-Thr-(2S,3R)-β-MePhe]	>1000
XI	c[Pro-Phe-D-Trp-Lys-Thr-(2R,3S)-β-MePhe]	>1000
XII	c[Pro-Phe-D-Trp-Lys-Thr-(2R,3R)-β-MePhe]	>1000
XIII	c[Pro-(S)-α-MePhe-D-Trp-Lys-Thr-Phe]	>1000
XIV	c[Pro-Phe-D-Trp-Lys-(S)-α-MeVal-Phe]	1000
XV	c[Pro-Phe-(2R,3S)-β-MeTrp-Lys-Thr-(2S,3S)-β-MePhe]	7
XVI	c[Pro-Phe-(2S,3R)-β-MeTrp-Lys-Thr-(2S,3S)-β-MePhe]	1000

7 and 10, we synthesized analogs (XIII, XIV) containing α-MePhe and α-MeVal at positions 7 and 10, respectively.

NMR studies indicated that these two analogs (XIII, XIV) maintain the backbone conformations of the parent compound (I) around D-Trp[8]-Lys[9] consisting of a β II' turn and around Phe[11]-Pro[6] bridging region consisting of a *cis* amide bond. For the α-MePhe[7] analog (XIII), the mutual strong NOEs of Trp[8]NH, α-methyl group and α-MePhe[7]C[β]H suggest the C_5 conformation for the α-MePhe[7] backbone because only in the C_5 conformation can these NOEs be detected. In the C_7 conformation, only the strong NOE of Trp[8]NH and the α-methyl group would be expected. A similar conclusion was drawn for the α-MeVal[10] residue in analog XIV. As a result, both α-methylated analogs (XIII,XIV) display a "flat" conformation. The extremely low binding potencies for these two analogs (XIII,XIV) are inconsistent with the "flat" conformation as the "bioactive conformation". Therefore, we suggest that the "folded" conformation is likely the "bioactive conformation".

Side Chain Chiral Methylated Somatostatin Analogs

As indicated by the NMR results, the backbone conformations of the parent compound (I) are maintained in all of the β-methylated analogs except for those analogs where the chirality changes at the C^α atom. For the D-Trp[8] containing analogs (I-VI,IX-XII), a β II' turn around D-Trp[8]-Lys[9] was observed. For the two L-amino acid containing analogs a β I turn was observed around Trp[8]-Lys[9]. As for the bridging region, the strong NOE between the C^α protons of Phe[11] and Pro[6] observed for the L-Phe[11] containing analogs (I-X) indicates a β VI turn with a cis amide linking Phe[11] and Pro[6]. For the (2R,3S) (XI) and (2R,3R)-β-MePhe[11] (XII) analogs which contain a D-Phe[11], the strong NOE for β-MePhe[11] C^αH-Pro[6]C[δ]H and the lack of NOE for β-MePhe[11]C^αH-Pro[6] C^αH enable us to

(A)   (B)

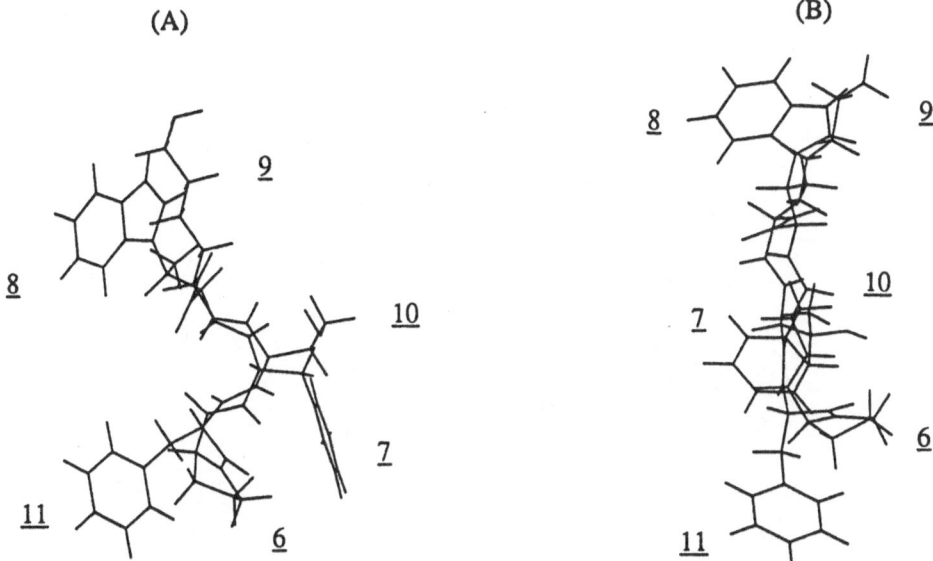

Fig. 1. Possible conformations of the active c[Pro⁶-Phe⁷-(2R,3S)-β-MeTrp⁸-Lys⁹-Thr¹⁰-Phe¹¹] (V.) (A): "folded" conformation and (B):"flat" conformation. The numbers refer to the locations of residues. This figure is taken from reference 3.

exclude the presence of a *cis* amide bond and suggest a *trans* amide bond at the bridging region. These results suggest that, unless the chirality changes at the C^α atoms, side chain β-methylation has little effect on the backbone conformations of the parent cyclic hexapeptide(I).

It is important that the β-methyl subsititution has two profound effects on the structure and bioactivity of somatostatin analogs. It specifically constrains the flexible side chains to prefer particular rotamers depending on the chiralities of the C^β atom while maintaining the backbone conformation of the parent cyclic hexapeptide. For instance, the side chain of Trp^8 in the parent compound (I) shows a large degree of flexibility with almost equal populations over the *trans* (0.44) and g^+ (0.36) rotamers. Upon β-methyl subsititution, this side chain adopts predominantly, but not exclusively, the *trans* (0.68) and g^+(0.72) rotamers for the (2R, 3S) (V) and (2R,3S) (VI) diastereomers respectively. The (2R,3S) analog(V) is more potent than the parent compound (I) while the (2R,3R) analog (VI) is less potent by more than 1000-fold. The structure of the modified side chain and the resulting enhanced binding potency provide direct evidence for the *trans* rotamer of the Trp^8 side chain as the "bioactive rotamer". A similar analysis applied to the β-MePhe¹¹ side chain suggests that the *trans* rotamer of Phe¹¹ is the "bioactive rotamer". As for the side chain at positon 7, because of the equal binding potencies shown by the (2S,3S) (II) and (2S,3R)-β-MePhe⁷ (III) analogs, despite the differences in their side chain rotameric conformations at postion 7, we hypothesize that only the aromatic group at position 7 is important for binding, but can not infer its exact

(A)   (B)

(C)   (D)

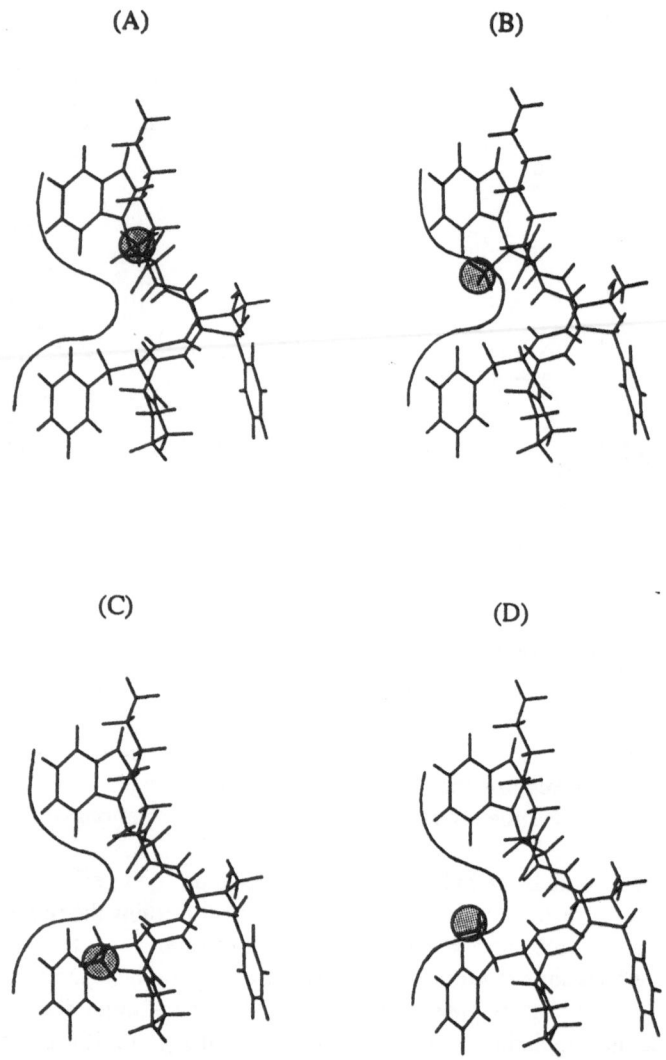

Fig. 2. *Comparison of the projections of β-methyl groups of (A): active (2R,3S)-β-MeTrp[8] analog (V), (B): inactive (2R,3R)-β-MeTrp[8] analog (VI),(C): active(2S,3S)-β-MePhe[11] analog (IX) and (D): inactive (2S,3R)-β-MePhe[11] analog (X). The methyl groups are highlighted by shaded spheres. This figure is taken from reference 3.*

topochemistry. The high binding potency shown by the Phg[7] analog (IV) fully supports our hypothesis.

Chiral side chain methylations provide a powerful means of probing the topochemical nature of the binding site. As in the cases of *(2S, 3R)* (VII) and *(2S,3S)*-β-MeTrp[8] (VIII) and *(2S,3S)* (IX) and *(2S,3R)*-β-MePhe[11] (X) analogs, comparison of the spatial

214

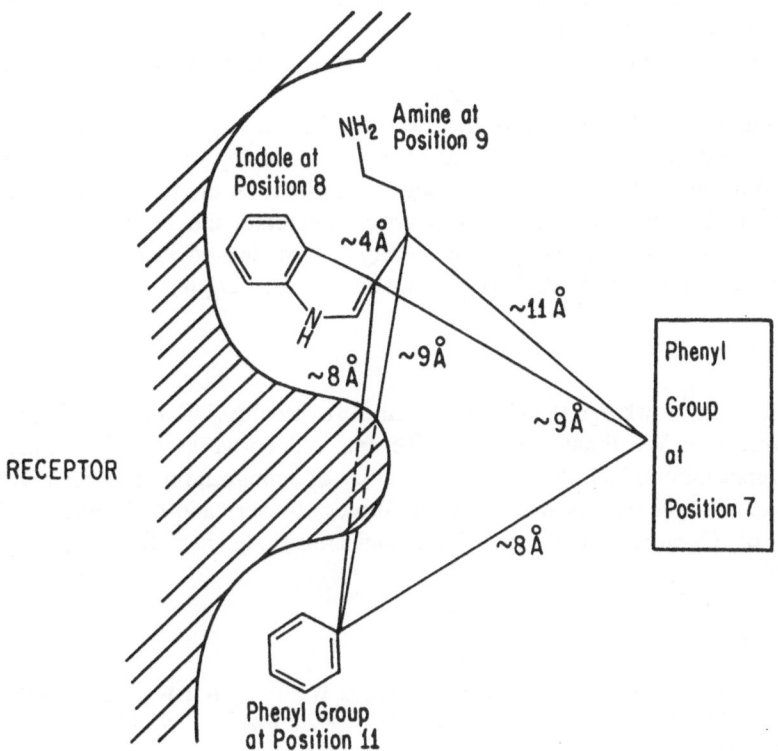

Fig. 3. Schematic model for somatostatin pharmacophores. This figure is taken from reference 3.

projections of the β-methyl groups in these analogs reveals crucial aspects of the binding "pocket" required for a peptide to interact with a receptor. As shown in Figure 2, the β-methyl group projects into a proposed binding "pocket" in the $(2R,3R)$-β-MeTrp[8] (VI) and $(2S,3R)$-β-MePhe[11] (X) analogs. We suggest that the steric hindrance created by the methyl group in this "pocket" prevents these two analogs from binding to a somatostatin receptor, which explains their loss of binding potencies.

To refine our model of the bioactivity of somatostatin analogs further, we synthesized two analogs (XV,XVI) containing β-methylations at both positions 8 and 11. The projection of the β-methyl groups in these two analogs (XV,XVI) should not have any steric effect on receptor binding. NMR studies of these analogs (XV,XVI) indicated that the side chains of β-MeTrp[8] and β-MePhe[11] assume conformations similar to those in each of the corresponding analogs containing mono β-methylation. The binding potencies of these analogs (XV,XVI) are also related to those of other analogs containing mono β-methylation. As shown in Table 1, the analog XV containing both $(2R,3S)$-β-MeTrp[8] and $(2S,3S)$-β-MePhe[11] shows binding potency intermediate between analogs V and IX containing mono β-methylation. Meanwhile, the mono β-methylation at positions 8 and 11 in corresponding analogs VII and IX lowers binding potency by 10- and 50-fold respectively. Therefore, the low binding potency shown by the analog XVI containing

215

both $(2S,3R)$-β-MeTrp[8] and $(2S,3S)$-β-MePhe[11] may be explained by the combined effect from the mono β-methylation at positions 8 and 11 in corresponding analogs VII and IX active in receptor binding. These results are fully consistent with our findings from analogs containing mono β-methylation.

To summarize our findings from these α- and β-methylated somatostatin analogs, we propose a schematic model to illustrate the important topochemical requirements for binding to a somatostatin receptor. As shown in Figure 3, the side chains of Trp[8], Lys[9] and Phe[11] in the displayed topological array consititute the most essential elements necessary for binding. We suggest that using this model, novel somatostatin agonists with peptide or nonpeptide backbone templates can be designed.

Acknowledgements

We wish to acknowledge the National Institute of Health (DK 15410) for their support of this research. We thank Dr. Terry Reisine and co-workers of the University of Pennsylvania for conducting the binding studies of somatostatin analogs. YBH and ZH would like to dedicate this work to their former advisors and colleagues in Shanghai Institute of Organic Chemistry, Hangzhou University and South China Normal University.

References

1.   Veber, D.F., Freidinger, R.M., Perlow, D.S., Palaveda, W.J. Jr., Holly, F.W., Strachan, R.G., Nutt, R.F., Arison, B.J., Homnick, C., Randall, W.C., Glitzer, M.S., Saperstein, R., Hirschmann, R., Nature, 292(1981)55.
2.   Veber, D.F., In Smith, J.A. and Rivier, J.E. (Eds.) Peptides: Chemistry and Biology (Proceedings of the 12th American Peptide Symposium), ESCOM, Leiden, 1992, pp. 3–14.
3.   Huang, Z., He, Y.-B., Raynor, K., Tallent, M., Reisine, T., Goodman, M., J. Am. Chem. Soc., 1992, in press.

Importance of intra-chain ionic interactions in stabilizing α-helices in proteins

Nian-en Zhou, Bing-yan Zhu, Cyril M. Kay and Robert S. Hodges

Department of Biochemistry and the Protein Engineering Network of Centres of Excellence, University of Alberta, Edmonton, Alberta, Canada T6G 2H7

Introduction

Electrostatic interactions are believed to play an essential role in molecular recognition, binding, catalysis, protein folding and the assembly of macromolecules. Special attention has been paid to the role of electrostatic interactions in stabilizing α-helices[1-3]. Although the interactions between Glu, Lys (i,i+3) or (i,i+4) ion pairs are thought to stabilize α-helices, the magnitude of this stabilization has been estimated in only a few instances [3,4]. Sali et al [4] demonstrated that an intra-helical surface electrostatic attraction contributed little to the stability of barnase. In contrast, Lyu et al [3] recently showed that the intra-helical solvent-exposed ion-pairs make a significant contribution to α-helical structure. O'Shea et al[5] observed intra-chain ion-pairs in the X-ray structure of the two stranded α-helical coiled-coil GCN4. In this study, we have estimated the contribution of intra-chain ionic interactions in stabilizing α-helices in a de novo designed two-stranded α-helical coiled-coil that has been demonstrated as an ideal model system for studying both intra- and inter-molecular interactions related to protein folding and stability[6-15].

Results and Discussion

The sequences of peptides used in this study are Ac-KCEALEG-KLEALEG-KAEALEG-(KLEALEG)$_2$-amide and Ac-ECEALKG-ELEALKG-EAEALKG-(ELEALKG)$_2$-amide denoted as EEK and EKE, respectively. Both peptides have identical amino acid compositions and can form two-stranded α-helical coiled-coils with the same hydrophobic residues in the hydrophobic interface and the same number of potential inter-helical ion-pairs at positions g and e of adjacent heptads(Fig.1). However, they differ in their intra-helical ionic interactions: EEK contains potential intra-helical ionic repulsions while EKE contains intra-helical ionic attractions (Fig.1). A disulfide bridge can be introduced at position 2 of the polypeptide to maintain the two helices in a parallel and in-register alignment.

The circular dichroism spectra for the two coiled-coils are very similar to each other and show the double minima absorptions at 207 and 222 nm in aqueous solution (0.1M KCl, 50 mM PO$_4$) at pH 7 and pH 3 with high ellipticity at 220 nm(−31,400±1000 deg•cm^2•dmole^{-1}). The stabilities of coiled-coils were determined by monitoring the ellipticity at 220 nm as a function of guanidine hydrochloride(Gdn•HCl) concentration

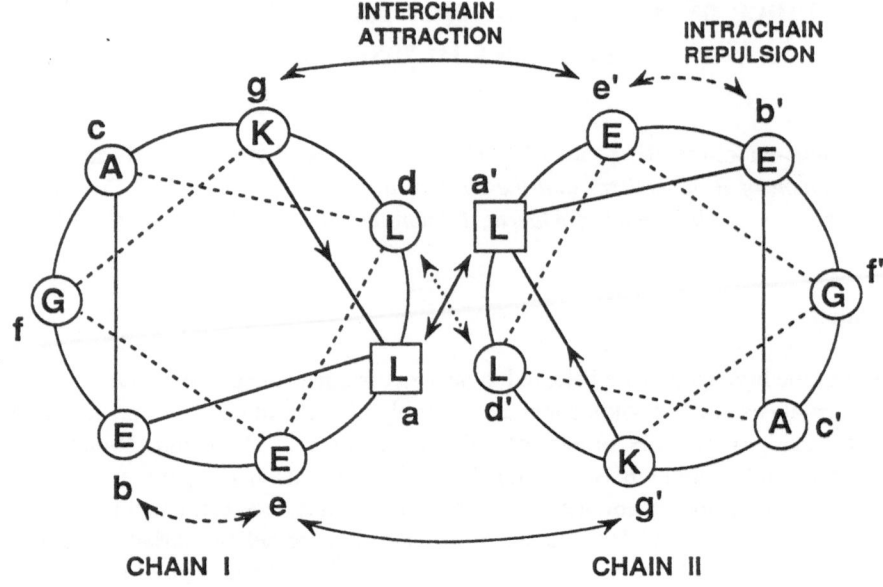

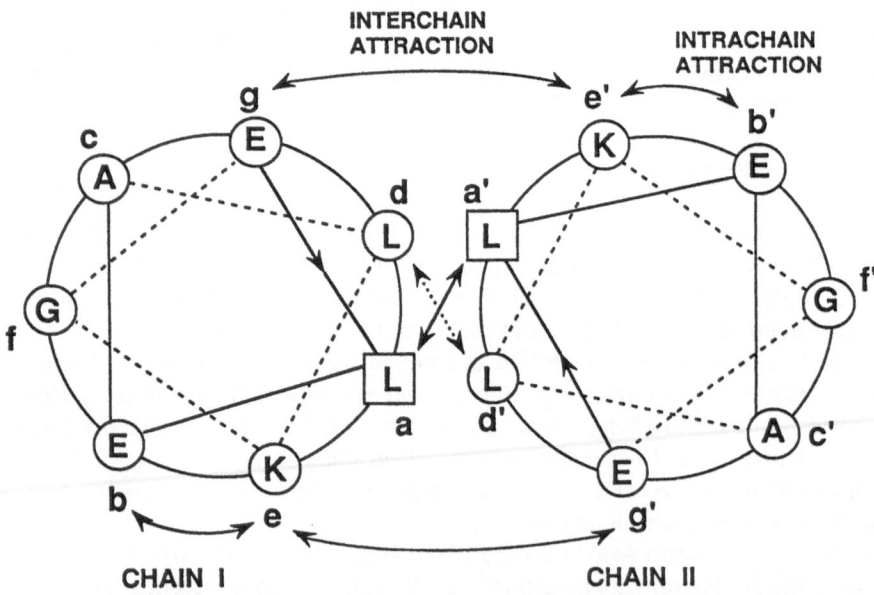

Fig. 1. Cross-sectional representation of one heptad of the model coiled-coils EEK (top) and EKE(bottom). Each heptad is designated by the letter a through g, in which the hydrophobic residues at "a" and "d" positions are responsible for the formation and stability of the coiled-coil. Electrostatic attractions are indicated as solid arrows and electrostatic repulsions are indicated as dashed arrows.

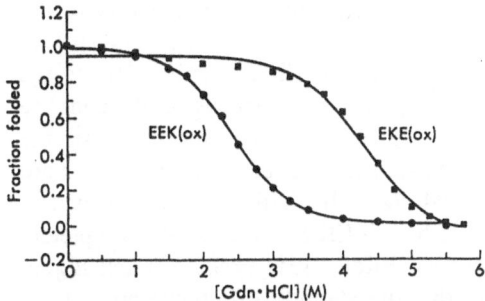

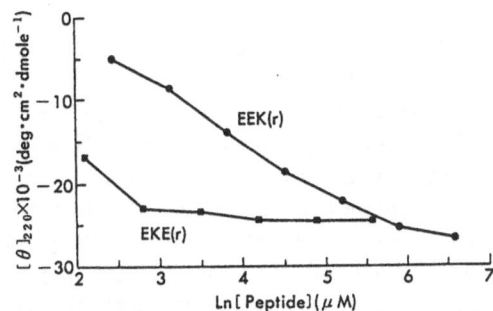

Fig. 2. *Gdn•HCl denaturation profiles of the oxidized (with an inter-helical disulfide bond) coiled-coils EEK(ox) and EKE(ox) in 0.05 M PO₄, 0.1 M KCl buffer, pH 7 at 20°C. The molar fraction of folded peptide (fn) was calculated as fn=([θ]-[θ]u)/([θ]n-[θ]u), where [θ] is the observed mean residue ellipticity at 220 nm at any particular Gdn•HCl concentration; [θ]n and [θ]u are the mean residue ellipticities at 220 nm of the native (folded) and unfolded states, respectively.*

Fig. 3. *The concentration dependence of the mean residue molar ellipticity at 220 nm for the reduced (without a disulfide bond) peptides EEK(r) and EKE(r) in 0.05 M PO₄, 0.1 M KCl buffer, pH 7 at 20°C.*

at 20°C. As shown in Fig. 2, peptide EKE was more stable than EEK. The transition midpoint determined from the Gdn•HCl denaturation profiles were 2.4 M for EEK and 4.3 M for EKE at pH 7 and 5.1 M for EEK and 5.7 M for EKE at pH 3. The differences in the free energy of unfolding($\Delta\Delta G$) between the coiled-coils EKE and EEK were calculated from the free energy values halfway between the two transition midpoints as previously described by Zhou et al [12] and the $\Delta\Delta G$ values were 2.6 kcal/mole at pH 7 and 1.2 kcal/mole at pH 3. To evaluate the net contribution of the intra-helical ionic interactions to protein stability, the effect of mutating Glu residues at positions e and Lys residues at positions g in EEK to Lys residues at positions e and Glu residues at positions g in EKE has to be taken into account. Since no ionic interactions occur in both EEK and EKE at pH 3, the $\Delta\Delta G$ value of 1.2 kcal/mole can be considered to result from sequence dependent effects. Thus the net contribution of the intrahelical ionic attractions to coiled-coil stability is 1.4 (2.6–1.2) kcal/mole. Baldwin and

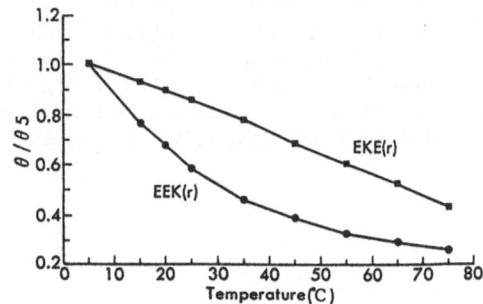

Fig. 4. *The thermal melting profiles of the reduced peptides EEK(r) and EKE(r) in 0.1 M KCl, 50 mM PO₄ buffer with TFE (1:1,v/v), pH 7. θ/θ₅ represents the ratio of the ellipticity at 220 nm at the indicated temperature to the ellipticity at 5°C.*

coworkers [1] reported that i, i+4 ionic attractions provide a greater stability to α-helices that i, i+3 interactions. In this study, there are four potential i, i+4 interactions in each α-helix of the coiled-coil. Though one ion-pair contributes little to the stability of the coiled-coil(~0.2 kcal/mole). which is in agreement with the results of Sali et al [4] on barnase, a large number of these weak interactions can contribute significantly to the overall stability as observed in this study (~1.4 kcal/ mole). Fig. 3 shows the dependence of the molar ellipticities of EEK and EKE on peptide concentration. As the concentration of peptide is increased, the monomer-dimer equilibrium is shifted toward the formation of a coiled-coil dimer which stabilizes the individual α-helices and increases the α-helical content of the peptide. Although EKE can associate completely to a coiled-coil dimer at ~ 15 μM peptide concentration, EEK needs a much higher peptide concentration (~200 μM). To confirm that the stability difference arose from intra-helical interactions, thermal denaturations of single-stranded α-helices of peptides EEK and EKE were determined in benign buffer, pH7, containing 50% TFE (Fig.4). The single-stranded α-helix of EKE is more stable than that of EEK, suggesting that the difference in stabilities of these coiled-coils is a result of differences in intra-helical ionic interactions. Thus, these results suggest that the more stable the single-stranded α-helix is the more stable will be the resulting dimer or coiled-coil, and that the single-stranded α-helix initiates the protein folding process.

References

1. Marqusee, S. and Baldwin, R.L., Proc. Natl. Acad. Sci. U.S.A., 84(1987)8898.
2. Serrano, L. and Fersht, A.R., Nature, 342(1989)296.
3. Lyu, P.C., Gans, P.J. and Kallenbach, N.R., J. Mol. Biol., 223(1992)343.
4. Sali, D., Bycroft, M. and Fersht, A.R., J. Mol. Biol., 220(1991)779.
5. O'Shea, E.K., Klemm, J.D., Kim, P.S. and Alber, T., Science, 254(1991)539.
6. Talbot, J.A. and Hodges, R.S., Acc. Chem. Res., 15(1982)224.
7. Hodges, R.S., Saund, A.K., Chong, P.C.S., St-Pierre, S.A. and Reid, R.E., J. Biol. Chem., 256(1981)1214.
8. Lau, S.Y.M., Taneja, A.K. and Hodges, R.S., J. Biol. Chem., 259(1984)13253.
9. Hodges, R.S., Zhou, N.E., Kay, C.M. and Semchuk, P.D., Peptide Res., 3(1990)123.
10. Hodges, R.S., Curr. Biol., 2(1992)122.
11. Zhou, N.E., Kay, C.M. and Hodges, R.S., J. Biol. Chem., 267(1992)2664.
12. Zhou, N.E., Kay, C.M. and Hodges, R.S., Biochemistry, 31(1992)5739.
13. Zhou, N.E., Zhu, B.Y., Kay, C.M. and Hodges, R.S., Biopolymers, 32(1992)419.
14. Zhu, B.Y., Zhou, N.E., Kay, C.M. and Hodges, R.S., Int. J. Peptide Protein Res., 1992, in press.
15. O'Neil, K.T. and DeGrado, W.F., Science, 250(1990)646.

(αMe)Val and (αMe)Phe peptides fold into turns and helices of opposite handedness

Fernando Formaggio[a], Giovanni Valle[a], Marco Crisma[a], Monica Pantano[a],
Stefano Mammi[a], Evaristo Peggion[a], Claudio Toniolo[a], Gilles Précigoux[b],
Gerlind Sulzenbacher[b], Wilhelmus H.J. Boesten[c], Hans E. Schoemaker[c]
and Johan Kamphuis[c]

[a]*Biopolymer Research Center, CNR, Department of Organic Chemistry, University of
Padova, 35131 Padova, Italy*
[b]*Laboratory of Crystallography and Crystalline Physics, University of Bordeaux I,
33405 Talence, France*
[c]*DSM Research, Bio-organic Chemistry Section, 6160 MD Geleen, The Netherlands*

Introduction

Insertion of C^α-methylated α-amino acids with defined conformational tendencies into peptides as a means of reducing flexibility has been recognized as an important approach in the construction of specially folded analogues of bioactive peptides. We and others have already described the structural propensity of Aib homo-peptides (for a recent review article see [1]). In this communication we present the results of extensive X-ray diffraction and ^1H NMR analyses of two series of N^α-blocked homo-peptides, [(αMe)Val]$_n$ (n-2,3) and [(αMe)Phe]$_n$ (n=2-5). These two residues are characterized by side-chain branching at the β- and γ-positions, respectively.

$$H_3C \diagdown \diagup CH_3$$
$$- NH - C - CO$$
$$\text{Aib}$$

$$\overset{CH_3}{\underset{}{|}}$$
$$H_3C \diagdown \overset{}{CH} - CH_3$$
$$- NH - C - CO -$$
$$\text{(αMe)Val}$$

$$H_3C \diagdown \diagup CH_2 - \hexagon$$
$$- NH - C - CO -$$
$$\text{(αMe)Phe}$$

Results and Discussion

We determined by X-ray diffraction the crystal-state molecular structures of two D-(αMe)Val and three D-(αMe)Phe homo-peptides. All residues are helical, but one of them, Phe^1 of molecule B of the (αMe)Phe homo-dipeptide, falls within the "bridge" region ($0° < \psi < 10°$) of the conformational map [2]. The average absolute values are 55.3°, 30.3°, closer to those expected 3_{10}-helix (57°, 30°) than for α-helix (63°, 42°) [3]. The two incipient 3_{10}-helices of the D-(αMe)Val homo-peptides are left-handed, but, surprisingly, four out of the six 3_{10}-helices of the D-(αMe)Phe homo-peptides, are right-handed. The two dipeptide methylamides are folded in a 1←4 C=O...H-N intramolecularly H-bonded β-turn conformation, close to types I/III [4,5]. The structure of molecules A and B of the (αMe)Phe homo-tripeptide free acid is further stabilized by an unusual oxy-analogue of the type-III β-turn (1←4 C=O...H-O intramolecular H-bond) [6]. The tripeptide methylamide and the tetrapeptide ester show two consecutive 1←4 C=O...HN intramolecular H-bonds of the type III-III. The distribution of the side-chain conformations (χ^1 torsion angle) for the five D-(αMe)Val residues is 4 (t,g$^+$) and 1(t,g$^-$), while that for the 18 D-(αMe)Phe residues is 15 gauche and 3 trans [7]. In general, the sign of χ^1 angle seems to be correlated to the handedness of the helical structure that is formed.

The preferred conformation in CDCl$_3$ solution of the longest homo-oligomers from D-(αMe)Val and D-(αMe)Phe was assessed by ^{1}H NMR. The analysis of the ROESY and DQF-COSY spectra allowed the assignments of all proton resonances. The delineation of unaccessible (or intramolecularly H-bonded) NH groups was performed with use of solvent (DMSO) dependencies of NH chemical shifts [8] and free-radical (TEMPO) induced line broadening [9] of NH resonances. Two classes of protons are clearly shown: (i) The first class [N(1)H and N(2)H protons] includes protons whose chemical shifts and line widths are dramatically sensitive to the addition of the perturbing agents. (ii) The second class (all other NH protons) includes those displaying a behavior characteristic of shielded protons (very modest sensitivity to the perturbing agents).

These ^{1}H NMR results support the view that in CDCl$_3$ solution the N(3)H and the following protons in the peptide chain are not freely accessible to the perturbing agents and therefore most probably intramolecularly H-bonded. In view of these observations it seems reasonable to conclude that the ordered secondary structure these oligomers predominantly assumed in this solvent is the 3$_{10}$-helix rather than the α-helix, which would require the NH protons involved in the intramolecular H-bonding to begin from the N(4)H proton. As additional proofs of this structural assignment for the (αMe)Phe homo-pentamer, the presence of the CH$_2$(i)→CH$_3$(i+2) interresidue connectivities and the lack of sequential NH→CH$_2$ connectivities strongly suggest that the conformation of the peptide is a right-handed 3$_{10}$-helix. In fact, model building shows that this conformation can account for all the interresidue cross-peaks found.

References

1. Toniolo, C. and Benedetti, E., Macromolecules, 24(1991)4004.
2. Zimmerman, S.S., Pottle, M.S., Némethy, G. and Scheraga, H.A., Macromolecules, 10(1977)1.
3. Toniolo, C. and Benedetti, E., TIBS, 16(1991)350.
4. Toniolo, C., C.R.C. Crit. Rev. Biochem., 9(1980)1.
5. Rose, G.D., Gierasch, L.M. and Smith, J.A., Adv. Protein Chem., 37(1985)1.
6. Toniolo, C., Valle, G., Bonora, G.M., Formaggio, F., Bavoso, A., Benedetti, E., Di Blasio, B., Pavone, V. and Pedone, C., Biopolymers, 25(1986)2237.
7. Benedetti, E., Morelli, G., Némethy, G. and Scheraga, H.A., Int. J. Pept. Protein Res., 22(1983)1.
8. Pitner, T.P. and Urry, D.W., J. Am. Chem. Soc., 94(1972)1399.
9. Kopple, K.D. and Schamper, T.J., J. Am. Chem. Soc., 94(1972)3644.

Conformational studies in solution on cyclolinopeptide A analogs. A 2D NMR study of cyclo{Pro^1-Pro-phe-Phe-Ac$_6$c-Ile-ala-Val^8}

M. Mazzeo, C. Pedone, L. Paolilio, V. Pavone, C. Isernia, F. Rossi, M. Saviano
and E. Benedetti
Department of Chemistry, University of Naples, Naples, Italy

Introduction

Cyclolinopeptide A and antamanide belong to a class of cyclic peptides well known for their competitive inhibitory activity toward cholate uptake in hepatocytes [1]. it has recently been shown that they can assume several conformational states in various organic solvents [2,3,4]. CD studies have also shown their ability to bind mono and bivalent metal ions [5,6]. Binding properties have been explained in terms of conformational flexibility, as revealed by NMR studies. The sequence Pro-Pro-Phe-Phe, common to this class of peptides, has been hypothesized as essential for biological activity.

In order to obtain a structurally rigid peptide, and to further ascertain the role of the common sequence, we have designed and synthesized a cyclic octapeptide containing a (D)-Phe residue (phe) in position 3 and a bulky aminoacid in position 5:

$$cyclo\{Pro^1\text{-}Pro\text{-}phe\text{-}Phe\text{-}Ac_6c\text{-}Ile\text{-}ala\text{-}Val^8\}$$

A D-Ala residue (ala) was also introduced in position 7, since diffraction studies on native cyclolinopeptide A have shown that the aminoacid in this position has structural features typical of D residues[2].

The conformation of this novel cyclooctapeptide has been examined by 2D NMR and molecular dynamics computer simulation.

Synthesis

The synthesis of the linear octapeptide H-ala-Val-Pro-Pro-Phe-Phe-Ac$_6$c-Ile-OH was achieved by solid phase methods with t-Boc protected aminoacids using a 430A Applied Biosystem automatic synthesizer.

Peptide cyclization was obtained in diluted solution of DMF and CH$_2$Cl$_2$ in the presence of isobutylchloroformate and N-methylmorpholine at pH=8.2. The purification, carried by a linear gradient H$_2$O (0.1% TFA): CH$_3$CN (0.1% TFA) by preparative HPLC, gave the pure cyclopeptide (yield 73%). FAB mass for MH+was 897 as expected.

NMR

NMR measurements were carried out on a VARIAN UNITY 400MHz spectrometer equipped with a SUN somputer SPARC330.

All ^{1}H spectra were recorded at 298K and referenced to internal TMS. NMR samples were prepared by dissolving 16 mg and 2.2mg of peptide in 0.75ml of CD$_3$CN (99.96% ^{2}H atom) and in 0.75 ml of CDCl$_3$ (99.96% ^{2}H atom), respectively.

Homonuclear NOESY, ROESY, TOCSY and DQCOSY and heteronuclear HMQC and HMBC in reverse detection two dimensional spectra were acquired with the States-Haberkorn method. Mixing time values were of 70 ms for TOCSY, from 50 to 400 ms for NOESY and from 30 to 200 ms for ROESY spectra. Two-dimensional experiment were typically acquired with 256 increments and 2048 data points in t$_2$ and zero filled to 1K in F$_1$.

Restrained molecular dynamics (RMD)

RMD simulations were carried out on a Silicon Graphics (Personal lris 4D25GT Turbo) workstation, employing the INSIGHT/DISCOVER program. CVFF force field was utilized for all the simulations. All clearly distinguishable NOE effects were used for the MD calculations. In each RMD simulation, performed with a time step of 0.5 fs, the molecule was equilibrated for 40 ps. After this first step, an additional 60 ps of simulation without rescaling was carried out.

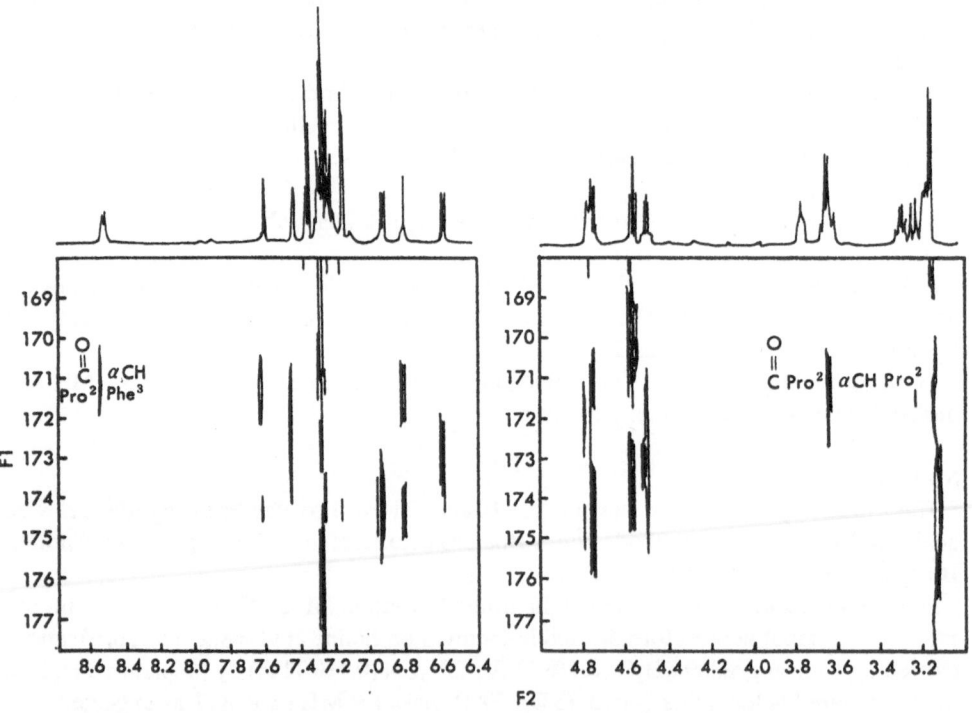

Fig. 1. *Regions of the HMBCR spectrum in CDCl$_3$ showing CO-NH and CO-αCH correlations (long-range coupling constant=7 Hz).*

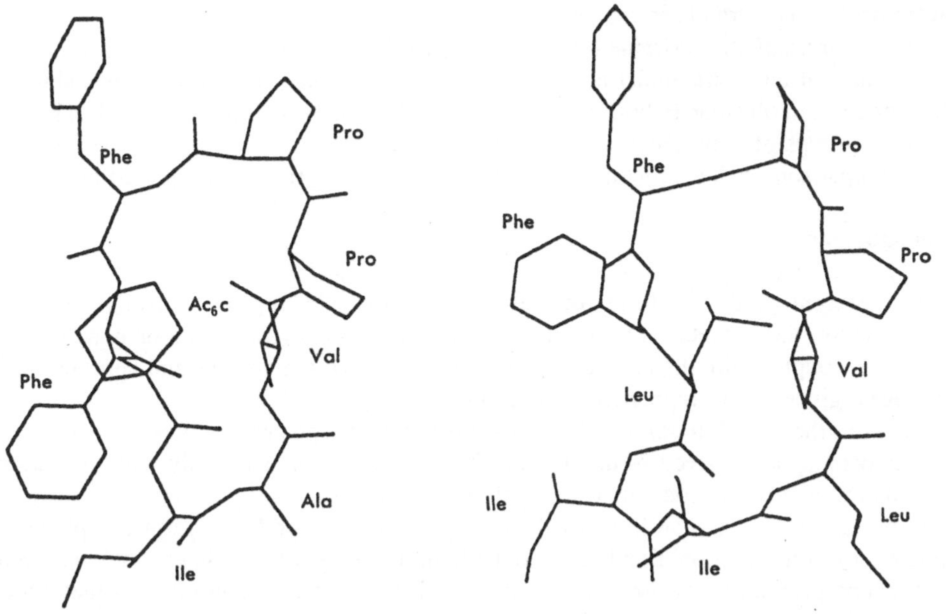

Fig. 2. *Comparison of the model obtained in CD₃CN (left) with the solid state structure of Cyclolinopeptide A (right).*

Results and Discussion

Cyclo{Pro¹-Pro-phe-Phe-Ac₆c-Ile-ala-Val⁸} was studied in chloroform and acetonitrile solutions. Sequential assignments of all spin systems was achieved by the conventional procedure using TOCSY and NOESY spectra. Selective 1D-TOCSY experiments were acquired to confirm the assignment and to measure vicinal coupling constants. Coupling constants and temperature coefficients of amide protons have been measured in both solvents.

The wide dispersion of the chemical shift of ¹³C makes heteronuclear correlation experiments valuable for assignment and, in some cases, for conformational purposes. Here the ¹H-¹³C correlation experiment (HMQC) was used to assign the Val⁸-Pro¹ and Pro¹-Pro² peptide bond isomerism unambiguously. In both solvents a *trans* and a *cis* peptide linkage for the Val⁸-Pro¹ and Pro¹-Pro² bonds were found, respectively. These results were further confirmed by NOE contacts between αCH(Val⁸)-δCH(Pro¹) and between αCH(Pro¹)-αCH(Pro²).

In chloroform solution the sequential assignment of the two prolines was also obtained acquiring an HMBC spectrum in reverse mode. In Fig. 1 two regions of the HMBC experiment, showing the CO-NH and CO-αCH correlations, are presented.

Conformational parameters, coupling constants, temperature coefficients, relaxation times T₁ and NOE effcts point to a very rigid stucture with intramolecular hydrogen bonds. The NOESYs and ROESYs spectra at diffetent mixing times were used to evalutate the NOE buildup rates and to calculate interproton distances. These distances

225

were used as constraints in molecular dynamics calculations.

On the basis of the restrained molecular dynamics simulation we have been able to detemine that only one model is consistent with NMR data. This structure is stabilized by three intramolecular H-bonds involving the NH groups of Phe^4, ala^7, and val^8 with the CO groups of Val^8, Ac$_6$c, and Phe^4, respectively.

A comparison of this structure with that of cyclolinopeptide A is reported in Fig. 2.

Conclusions

The present study confirms our expectation that, in organic solvents at room temperature, the reduction of the length of the main chain (from deca or nona to octa peptide), coupled with the insertion of sterically hindered aminoacids and appropriate D residues, gives rise to highly rigid molecule.

As far as the overall topology of the Pro-Pro-Phe-Phe segment is concerned, the model of the octapeptide derived from our data shows considerable similarity with that found in solution and in the solid state for cyclolinopeptide A.

Furthermore we can confirm that the ability of cyclic peptides to form complexes is related to conformation flexibility. In fact, preliminary data indicate that the rigid compound presently investigated is unable to bind monovalent and bivalent metal ions in organic solvents. Further studies are presently in progress to evaluate the biological activity of this molecule.

References

1.   Kessler, H., Klein, M., Muller, A., Wagner, K., Bats, J., Ziegler, K. and Frimmer, M., Angew. Chem. Int. Ed. Engl., 25(1986)997.
2.   Di Blasio, B., Rossi, F., Benedetti, E., Pavone, V., Pedone, C., Temussi, P.A., Zanotti, G., Tancredi, T., J. Am. Chem. Soc., 111(1989)9089.
3.   Tancredi, T., Zanotti, G., Rossi, F., Benedetti, E., Pedone, C., Temussi, P.A., Biopolymers, 28(1989)513.
4.   Patel, D.J., Biochemistry, 12(1973)661.
5.   Tancredi, T., Benedetti, E., Grimaldi, M., Pedone, C., Rossi, F., Saviano, M., Temussi, P.A., Zanotti, G., Biopolymers, 31(1991)761.
6.   Ivanov, V.T., Ann. New York Acad. Sci., 264(1975)221.

Conformational studies of a dodecacyclopeptide in solution

Li-bin Ma[a,b], Lu-hua Lai[a], Zhen-wei Miao[a] and Xiao-Jie Xu[a]
[a]Department of Chemistry, Peking University, Beijing 100871, China
[b]Beijing Institute of Microchemistry, Beijing 100091, China

Introduction

De novo design of proteins provides simple model systems for understanding protein structure and function, and for determinating which residues are essential to the function of a protein[1]. This approach allows the construction of model system that should be easier to study and understand than natural proteins, and finally, should allows the construction of novel proteins whose structures and properties are presents or not present in nature. This approach has been successfully applied to the study of peptides. Cyclopeptide is a useful segment for de novo design of four-helix-bundle proteins as supporting template for the helix. We have synthesized a dodecacyclopeptide and studied its conformation by the combined use of 2D-NMR and distance geometry structure calculations with DIANA program.

```
    1       2       3       4       5       6
   phe  →  Lys  →  Pro  →  Gly  →  Lys  →  Gly
    ↑                                       ↓
   Gly  ←  Lys  ←  Gly  ←  Pro  ←  Lys  ←  Phe
   12      11      10       9       8       7
```

The four lysines are protected by C-O-CH$_2$-Ph groups.

Dodecacyclopeptide was synthesized by solid phase technique. The NMR sample was prepared by dissolving dodecacyclopeptide in d$_6$-DMSO. The peptide concentration was approximately 15 mM. All data were collected at 27°C on a Bruker AM 500 Spectrometer equipped with Aspect 3000 computer. Spectral widths of 4500 Hz were used for spectra. Chemical shifts are reported in reference to internal DMSO set at 2.49 ppm. COSY spectrum was acquired in magnitude mode. 1024 data points were collected in F2 domain with 128 time increaments. A final size of 1024×512 was used to yield the spectrum. NOESY spectra were acquired in phase sensitive mode. 4 spectra were collected at mixing periods of 200 ms, 400 ms, 600 ms and 800 ms. In each case the mixing time was subjected to a random variation of about 20 ms to suppress zero quantum coherence. 256 tl points were collected with 1024 data points along the F2 dimension. A 90°-shifted sine bell was applied along both domains. The final matrix size was 1024 × 512 real points. The 400 ms spectrum was used for distance quantitation after baseline correction, which showed little effect of spin diffusion.

Results and Discussion

Sequential resonance assignment

Sequential resonance assignment was carried out by using COSY and NOESY spectrum. In COSY spectrum, only 5 cross-peaks of NH-C_αH were found from 12 amino acid residues. It shows that the structure of the cyclopeptide contains two identical parts which give rise to the overlapping of chemical shifts. In each part there is a proline in which no NH exists. So only 6 residues are individuals. Their spin systems were easily identified by COSY spectrum.

5 NOE cross-peaks of $C_\alpha H_i$-NH_{i+1} were found in NOESY spectrum for sequential assignment. The proline was assigned by the cross-peak of C_αH of Lys^2 and C_δH of Pro^3. This peak also implies that the proline exists in trans-conformation. Complete sequential assignment was shown in Table 1.

Conversion of NOE intensities into upper distance limits

The NOESY cross-peak intensities were evaluated by their integrated volumes. To calculate distnaces, relationship $V=C/d^6$ was used. C is a constant obtained by scaling the ^1H-^1H distance as 1.78 A related to the integrated volume of cross-peak between two C_β protons of Phe^2. Other cases of interproton distances from NOESY spectra were not attempted for the following reasons. The cyclo-peptide is a rigid molecule and most of NOESY cross-peaks are short and medium-range interactions. 62 distances thus obtained were used as upper distance constraints, 18 derived from intraresidue protons, 36 from neighbouring-residue protons and other 8 from non-neighbouring-residue protons.

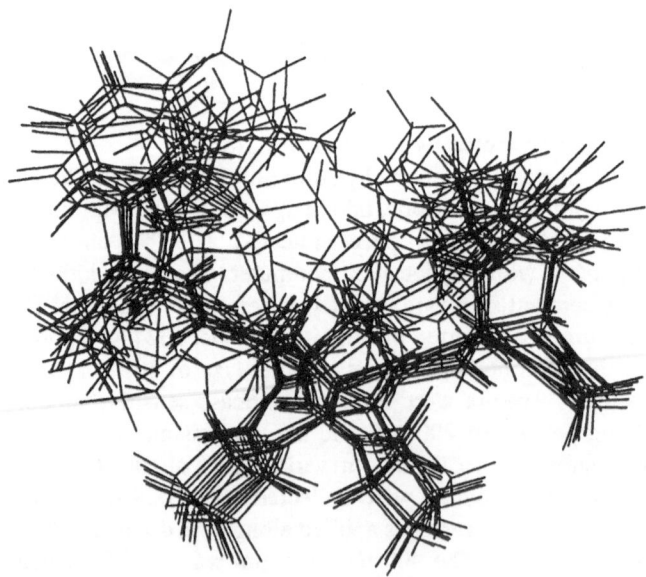

Fig. 1. The 10 structures of dodecacyclopeptide obtained from distance geometry calculation.

Table 1 *^{1}H Chemical shift for dodecacyclopeptide (ppm)*

	NH	C$_\alpha$H	C$_\beta$H	Others
Phe^1 (Phe^7)	8.56	4.35	3.00, 2.77	ring(2,6):7.04, ring(3,5,4):7.18
Lys^2 (Lys^8)	7.99	4.53	1.61	C$_\gamma$H: 1.65, C$_\delta$H: 1.32, C$_\epsilon$H:2.93, sidechainNH:7.15
Pro^3 (Pro^9)		4.16	1.82, 2.06	C$_\gamma$H:1.82, 1.96, C$_\delta$H:3.55, 3.75
Gly^4 (Gly^{10})	8.33	3.59, 3.76		
Lys^5 (Lys^{11})	7.83	4.23	1.51	C$_\gamma$H:1.16, C$_\delta$H:1.32, C$_\epsilon$H:2.93, side chain NH: 7.15
Gly^6 (Gly^{12})	8.38	4.03, 3.29		

Structure calculation with DIANA

Three-dimensional structures were calculated from the distance constraints using the DIANA program[2]. 100 structures were obtained, in which the 10 top structures with the lowest target function values were considered as final results. Figure 1 shows the 10 structures which superimpose quite well. The backbone root-mean-square deviation (RMSD) between these 10 structures was 0.29±0.06 Å and all heavy atoms was 1.24±0.19 Å. The maximal residual violation of NOE distance constraints is 1.24 Å. The results showed that the conformational structure obtained from NMR data was satisfactory.

Discription of the conformation in solution

Figure 2 shows that the structure of dodecacyclopeptide comprised of two β-turns and two γ-turns. The first β-turn involves the tetrapeptide Lys^5-Gly^6-Phe^7-Lys^8. The second

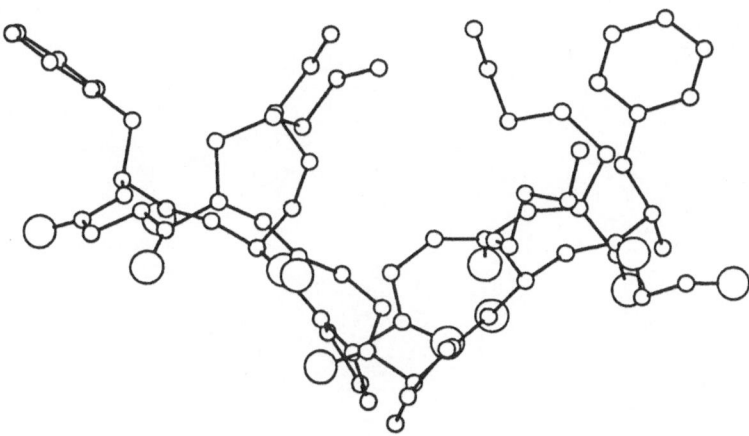

Fig. 2. One of the 10 final structures showing orientation of the side chains of four lysines.

229

β-turn involves the tetrapeptide Lys^{11}-Gly^{12}-Phe^1-Lys^2. The two γ-turns involve the tripeptides Lys^2-Pro^3-Gly^4 and Lys^8-Pro^9-Gly^{10}. All the side-chains of four lysine residues are oriented towards the same side of the cyclopeptide ring. This result is consistent with the designed structure, making the cyclopeptide an ideal template for further synthesis of four-helix-bundle proteins.

References

1.   DeGrado, W.F., Wasserman, Z.R. and Lear, J.D., Science, 243(1989)622.
2.   Guntert, P., Braun, W. and Wuthrich, K., J.Mol. Biol., 217(1991)531.

Low temperature NMR studies of an antigentic peptide of trichosanthin

Yi-bin Zhang, Yi-hau Yu and Qi-wen Wang
State Key Laboratory of Bioorganic and Natural Products Chemistry,
Shanghai Institute of Organic Chemistry, Chinese Academy of Sciences,
Shanghai 200032, China

Introduction

Based on the distribution of local hydrophilicities along the peptide china of trichosanthin, a linear peptide, AAGKIRENIPL, was synthesized and its ELISA tests by reaction with rabbit antitrichosanthin serum showed that the peptide has antigenic activities. Therefore studying its structure-activity relationships and reactive properties with the receptor will benefit the improvements of the clinic practices of trichosnthin.

Studying the structure-activity relationships of such medium sized peptides by NMR spectroscopy in solution is, in principle, favourably owing to that the highly regiospecific NMR spectra allows one to monitor the intrinsically conformational tendencies with the influence of various physico-chemical environments simulated in solution. However the mainly practical obstacles are the conformational flexibilities of the peptides and the unfavourable values of their correlation times at room temperature, which make the NOEs undetectable and consequently hamper the sequential-specific assignments and revealing of the conformationally spatial restraints. For resolving these problems we use a solvent mixture, 23% CD_3OD: 77% H_2O (V:V), to dissolve the quoted peptide and NMR can be measured at very low temperature, e.g. −16°C, in this study. Increasing the viscosity of the solution and the correlation times by low-temperature NMR measurements made a lot of intra- and inter-residue NOEs observable and helped us to complete the assignments, which is essential for further researches.

Results and Discussion

These intra- and inter-residue NOEs can also be observed even measuring at room temperature when the sample is dissolved in another solvent mixture, 80% DMSO-d_6: 20% H_2O(V:V). However comparing the fingerprint area of the DQF-COSY spectra with that of the sample dissolved in water solution we found CD_3OD/H_2O is more favourable because its cross-peak pattern is almost the same as that of the sample in water whereas very different crosspeak pattern is observed when the sample is dissolved in DMSO-d_6/H_2O. It indicates that the solvent mixture CD_3OD/H_2O has properties closer to water and the whole conformation may be changed in DMSO-d_6/H_2O. Furthermore, the low temperature measurement in CD_3OD/H_2O shifts down the water signal to about 5.4 ppm, making α proton area of the NMR spectra more visible and easier to analyze the

Table 1 *Proton resonance assignments of the quoted peptide*

Residue	Chemical shift values (ppm)			
	NH	$C_\alpha H$	$C_\beta H$	Others
Ala[1]		4.13	1.61	
Ala[2]	8.96	4.41	1.50	
Gly[3]	8.81	4.02		
Lys[4]	8.52	4.41	1.88 1.82	$_\gamma$H: 1.51 $_\delta$H:1.76 $_\epsilon$H: 3.06
Ile[5]	8.61	4.20	1.92	$_\gamma$H: 1.54 $_\gamma$CH$_3$:0.99 $_\delta$CH$_3$:0.95
Arg[6]	8.77	4.36	1.85 1.91	$_\gamma$H:1.74 1.72 $_\delta$H:3.29
Glu[7]	8.78	4.38	1.99 2.13	$_\gamma$H:2.46
Asn[8]	8.75	4.73	2.86 2.78	NH:7.26 7.97
Ile[9]	8.41	4.54	1.95	$_\gamma$H:1.21 1.60 $_\gamma$CH$_3$:1.04 $_\delta$CH$_3$:0.95
Pro[10]		4.47	2.15 2.38	$_\gamma$H:2.04 $_\delta$H:3.78 3.99
Leu[11]	8.40	4.28	1.80	$_\gamma$H:1.66 $_\delta$CH$_3$:1.03 0.99

cross peaks in this area. Further investigations of the whole conformations with the sample dissolved in the different solvent mixtures and the influences of the solution viscosity, solvent contents and measuring temperature on the conformational tendencies are in progress in our laboratory.

Secondary structure transitions of trichosanthin fragment

Hong-yu Hu, Zi-xian Lu and Yu-cang Du
Shanghai Institute of Biochemistry, Chinese Academy of Sciences,
Shanghai 200031, China

Introduction

It is generally accepted that peptide fragment of proteins adopting secondary structures in solution has enormous implicarions for initiation of protein folding[1]. In order to elucidate the folding mechanism of trichosanthin (TCS) domains and the relationship between amino acid sequence and secondary structure, we pay attention to the α-helical segment locating at 183-205 positions of native protein, i.e. N-terminal of the smalll domain, which lies in the active site of N-glycosidase[2].

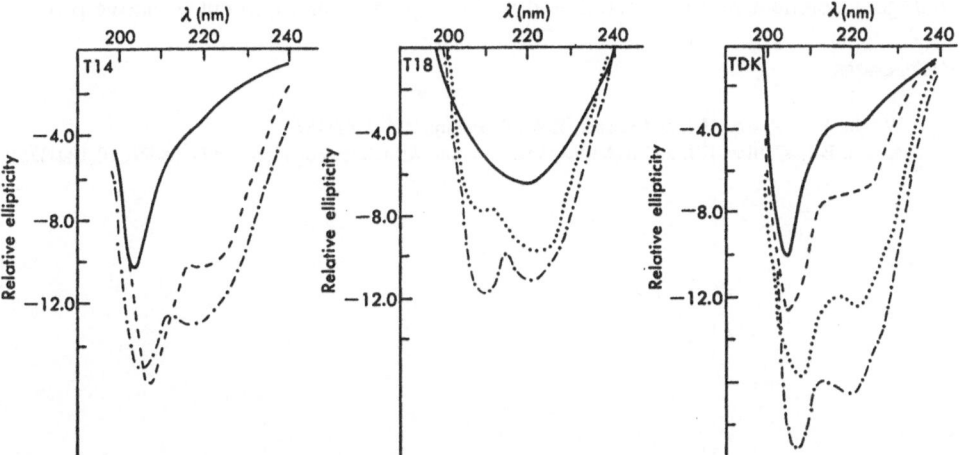

Fig. 1. CD spectra of T14(a), T18(b) and TDK(c) indicating solvent induced transitions of secondary structures. ——— H_2O, pH 10.0; ---------- 50% methanol; ·············· 50% trifluoroethanol; —·—— 50% hexafluoropropanol. Sample concentration: 0.1 mg/ml; Cell path: 2 mm; Scanning speed: 5 nm/min; Time constant: 8.0 sec.

H.Y. Hu, Z.X. Lu and Y.C. Du

Results and Discussion

We designed and synthesized three polypeptides and their structures are described as following:

     T14   SLENSWSALSKQIQ,

     T18   LAIISLENSWSALSKQIQ (anslogue of TCS position 183-200),

     TDK   DLAKISLENSWSALSKQIQ.

The CD spectra of the three peptides indicating solvent induced transitions of secondary structures are shown in Fig. 1. T14 undergoes random coil to α-helix conformation transition induced by 50% methanol, trifluoroethanol and hexa-fluoropropanaol (Fig 1a), whereas T18 adopts β-sheet to α-helical transition(Fig 1b). The hydrophobic region (LAII) of N-terminal in T18 may be possible to form a β-strand conformation. In the case of TDK peptide, it exhibits a slightly probability of α-helix in water, which is also strongly improved by alcohols(Fig 1c), suggesting that the substitution of Asp-1 and Lys-4 might apparently enhance the α-helical potential.

In summary, the N-terminal amino acid features determine the conformation of these peptides and the transitions of secondary structure are induced by low polarity of solvents, that is to say, the α-helical probability is proportional to the degree of solvent nonpolarity. Comparatively, the α-helical (other than β-strand) structure of this analogous segment might be facilitated by hydrophobic environment in native protein.

References

1.   Wright, P.E., Pyson, H.J. and Lerner, R.A., Biochem., 27(1988)7167.
2.   Katzin, B.J., Collins, E.J. and Robertus, J.D., proteins: Structure, Function and Genetics, 10(1991)251.

Studies on the conformation of HBeAg P3 peptide by NMR

Hou-li Jiang[a], Zu-han Du[b], Rui Feng[a] and Zheng-ying Chen[b]
[a]Instrumental Analysis Center,
[b]Institute of Basic Medical Sciences, Academy of Military Medical Sciences,
Beijing 100850, China

Introduction

HBeAg P3 peptide is a synthetic octapeptide with the following sequence: Leu-Glu-Asp-Pro-Ala-Ser-Arg-Asp, which is predicted to be an epitope of hepatitis B e antigen. It is estimated that the P3 peptide is hydrophilic with a β-turn propensity and low charge distribution by means of PR02 program of 'SEQNCE' soft ware package. The conformation of HBeAg P_3 peptide in water solution was studied with 2D NMR techniques. According to the distance constraints derived from the NOE's and the hydrogen bond information, the secondary structure of HBeAg P3 peptide in water has been identified.

Results and Discussion

HBeAg P3 peptide was synthysized[1]. and then purified through HPLC C_{18} reverse phase column. The pH values adjusted with DCl and NaOD were the direct meter readings without isotope correction. The experiments were done with 6 mM solution and the residual methanol was used as internal standard.

All NMR experiments were performed on a JEOL-GX400 NMR spectrometer. ^1H-^1H COSY and NOESY experiments in 90%H_2O/10%D_2O solution were carried out with 0.25M $NH_2OH.HCl$ and 0.25 M $NH_4SO_3NH_2$ added as water proton spin relaxation reagents. The huge water resonance was effectively suppressed by homo-gated irradiation and resonances of the P3 peptide which were difficult to obtain before adding the relaxation reagents were clearly observed.

Spin system recognition was accomplished by analysing the ^1H-^1H COSY spectra of HBeAg P3 peptide in D_2O solution. The backbone NH resonances of six amino acid residues were assigned on the COSY spectra of the P3 peptide in 90%H_2O/10%D_2O solution. According to the sequence specific recognition method, the chemical shifts assignment of HBeAg P3 peptide was readily accomplished (Table 1).

Deuteration of the amide protons showed that the amide proton of Asp-8 is extremely stable and might form a hydrogen bond. The stability of the six amide protons of HBeAg P3 peptide is as follows:

Asp-8>Ala-5>Asp-3, Ser-6, Arg-7>Glu-2

Table 1 *Chemical shifts of HBeAg P3 peptide in H_2O (pH=5.0)*

Residue	NH	αH	βH	γH	δH
Leu-1		4.00	1.62	1.60	0.90
			1.68		0.91
Glu-2	8.71	4.31	1.89	2.24	
			1.98	2.30	
Asp-3	8.51	4.81	2.56		
			2.75		
Pro-4		4.35	1.91	1.96	3.74
			2.26	2.00	3.80
Ala-5	8.29	4.26	1.35		
Ser-6	8.03	4.37	3.85		
Arg-7[a]	8.16	4.35	1.68	1.58	3.14
			1.86	1.62	3.16
Asp-8	8.00	4.39	2.62		
			2.71		

[a] εNH of Arg-7: 7.14.

Temperature dependence of the backbone NH protons of the P3 peptide was tested, showing that the NH of Asp-8 should have been involved in a hydrogen bond.

NOESY experiment (mixing time τm=20ms) in 90%H_2O/10%D_2O solution revealed the NOE effect between protons of the αH of Ser-6 and the backbone NH of Arg-7. This together with the hydrogen bond formation of the NH of Asp-8 indicated that there should be a Type-II β-turn involving residues 5-8 in the HBeAg P3 peptide.

Further NOESY experiments (mixing time τm=40ms, 60ms) revealed that there were NOE effects between the backbone NH and βH within each residue of Asp-3, Ala-5, Ser-6, Arg-7 and Asp-8. THe NOE information, the different temperature coefficients and the distinct stability of the amide protons could be applied in the Distance Geometry method and Restrained Molecular Dynamics calculations to further improve the structure of HBeAg P3 peptide.

Reference

1. Chen, Z.Y., Yao, Z.J., Fen, X.J., Dong, X.J., Zeng, Z., Ma, X.K., Song, X.R. and Shi, M.Q., in Peptides: Biology and Chemistry: Proceedings of the Chinese Peptide Symposium 1990. Ed. by Du, Y.C. etal., Science Press, Beijing, 1991, pp.128.

Structure–activity relationship studies of memory enhancing peptide ZNC(C)PR

Peng-liang Wang[a], Ke-yin Ma[a], Yu-cang Du[a] Lu-hua Lai[b] and Xiao-jie Xu[b]

[a]Shanghai Institute of Biochemistry, Chinese Academy of Sciences,
Shanghai 200031, China
[b]Chemistry Department, Peking University, Bejing 100871, China

Introduction

It has been proved that ZNC(C)PR is a new neuropeptide in rat brain, which has the function to facilitate memory and learning specifically. ZNC(C)PR distributes mainly in hippocampus and cerebral cortex with the primary structure as follows:

```
                Cys
                 |
  pGlu-Asn-Cys-Pro-Arg
```

To study the structure–activity relationship of ZNC(C)PR, a series of ZNC(C)PR analogs had been synthesized and their activities had been measured[1]. Also the conformations of ZNC(C)PR and [D-Arg] ZNC(C)PR had been studied[2]. But further research is still necessary to understand more details on the structure-activity relationship of ZNC(C)PR. All the calculations were performed on IRIS 4D/70GT workstation, implemented with QUANT program package from POLYGEN Company to draw molecular graphics, and CHARMM[3] was employed in conformational calculations. The algorithm[4] we proposed was employed to predict the conformations of ZNC(C)PR analogs we synthesized.

Results and Discussion

Design of ZNC(C)PR analogs

45 ZNC(C)PR analogs were constructed by considering various situations including the substitutions of one or two residues with hydrophobic or hydrophilic residues, with large side chain or small side chain residues, etc. And the conformations of these analogs were calculated. By examing the calculation results, the roles of every residue in ZNC(C)PR can be deduced theoretically. The theoretical results showed that Pro is essential to the conformation of ZNC(C)PR and cannot be substituted. While Cyt can be substituted by hydrophobic residue and the analogs will still have the similar conformation to ZNC(C)PR. It is believed that the analogs with ZNC(C)PR activity should have the similar conformational properties to ZNC(C)PR. The conformation of ZNC(C)PR is characterized with the near side chains between Asn and Arg and moderate stability. So the conformational characteristics and the lowest energies were chosen as parameters to

Table 1 *Effective ZNC(C)PR analogs duduced from prediction*

sequence	1	2	predicted activity
Asn Leu Pro Lys	−68.42	7.42	+
Asn Ile Pro Arg	−68.50	7.44	+
Asn Tyr Pro Arg	−75.66	8.08	+
Asp Leu Pro Arg	−105.39	8.33	−
Asp Leu Pro Lys	−85.75	9.42	−
Glu Leu Pro Lys	−80.23	9.64	−

Note: All the analogs in this table have the similar conformational properties of ZNC(C)PR. 1. the lowest energy, Kcal/mol; 2. distance between the C_α atoms of residue 1 and 4. "+", positive effect to enhance memory; "−", nagative effect.

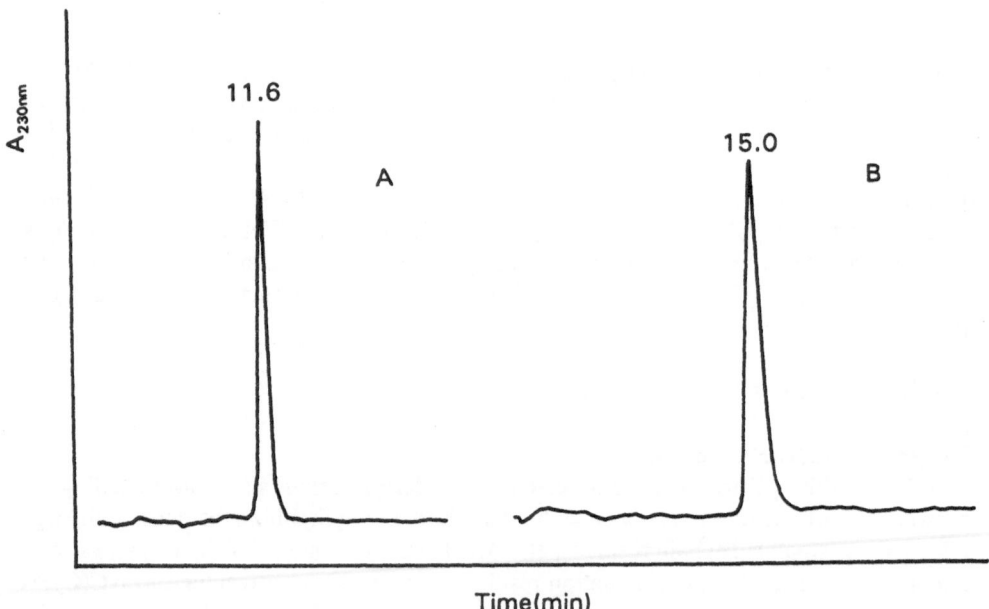

Fig. 1. *Reverse-phase HPLC of synthetic peptide. Column: DELTA PAK C18-300A(7.8×300mm). Eluent: (A) 0.1% trifluoroacetic acid; (B) 90% acetonitrile and 0.1% trifluoroacetic acid. Flow rate: 2ml/min. Detection: 230nm. Linear gradient 0-10%(B) in 2.5min and then 10-30% (B) in 15 min. Curves A and B are the profiles for NAPR and DLPR, respectively.*

Table 2 *Effects of NAPR and DLPR on the retention of passive aviodance behavior in rats*

Sample	Dose	N	Mean latency (s)	
	(µg/rat)		24 h	48 h
NAPR	Saline	20	97±11	107±12
	0.003	19	133±26	119±16
	0.01	19	131±20	158±21*
	0.03	19	147±19*	125±19
	0.1	19	108±17	107±18
DLPR	Saline	19	59±16	117±14
	0.1	18	40±15	73±12*
	0.3	18	74±30	91±14
	1.0	18	64±23	122±20
	3.0	18	72±21	119±19

Peptides were subcutaneously injected immediately after the learning trial. N: number of experimental animals.
*: $p < 0.05$, vs. saline control (Mann-Whitney U-test).

classify the analogs, so as to find out the analogs with the same properties as ZNC(C)PR, see Table 1.

Synthesis of ZNC(C)PR analogs

Two ZNC(C)PR analogs NAPR and DLPR were chosen to synthesize, NAPR was preidcted with ZNC(C)PR activity, while DLPR was predicted with inhibitory activity. By investigating NAPR, not only if it is true that Cyt in ZNC(C)PR can be substituted by hydrophobic residues, but also the importance of pGlu can be examined.

NAPR and DLPR were synthesized and purified to homogeneous material which showed a single peak in reverse-phase HPLC. Fig.1 is the HPLC profiles of NAPR and DLPR.

2D-NMR spectra of ZNC(C)PR analogs

Unexpectedly, there were no long-range NOEs in the NOESY spectrum of DLPR and ROESY spectrum of NAPR. Only a few short-range (intraresidue or neighbour residue) NOEs were observed. These results may suggest that the conformations of NAPR and DLPR are flexible.

Although no NOEs were observed in the 2D-NMR spectra, it is possible that both analogs NAPR and DLPR may have a compact structure (the side chains of the first and the last residue are closed). Because they are just tetrapeptides, the whole molecules are relatively flexible. Asn and Asp are at the N-terminal of NAPR and DLPR, even the side chains of Asn and Asp still tend to the sidechains of Arg in NAPR and DLPR, NOEs will not perform or will be very weak.

Effects of ZNC(C)PR analogs on passive avoidance behavior

The effects of NAPR and DLPR on passive avoidance behavior in rats were measured. From Table 2, we can see that NAPR induced significant facilitation of passive avoidance behavior ($p<0.05$), while DLPR showed somewhat inhibitory effect ($p<0.05$) at 48 hours. These results are agreed with our prediction.

To summarize all this work, firstly the conformations of ZNC(C)PR and its analogs were studied theoretically, analogs for experimental investigation were chosen according to the predicted conformations. This helps a lot to our research work.

Acknowledgements

The authors thank Mr. Ren-yi Liu and Mr. Li-bing Ma in assistance of bioactivity analysis and 2D-NMR experiments.

References

1. Lin, C., Liu, R.Y. and Du, Y.C., Peptide, 11(1990) 633.
2. Xu, L.F., Xu, C.J., Du, Y.C., Lin, C. and Lu, Z.X., Acta Biochem. Biophys. 22(1990) 533.
3. Brooks, B.R., Bruccoleri, R.E., Olafson, B.D., States, D.J., Swaminathan, S. and Karplus, M., J. Comput. Chem., 4(1983) 187.
4. Wang, P.L., Lai,L.H., Xu, X.J. and Du, Y.C., A fast systematic search algorithm for peptide conformational analysis, to be published.

A new structural type of zinc insulin observed in a mutant of [A21, Ser]-human insulin

Da-cheng Wang[a], Zhong-hao Zeng[a], Yong-lin Hu[a] and Jen Markussen[b]
[a]Institute of Biophysics, Chinese Academy of Sciences, Beijing 100101, China
[b]Novo Research Institute, DK-2880 Bagsvaerd, Denmark

Introduction

It has been a long time to know insulin as a polypeptide hormone and as a pharmaceutical preparation for the treatment of diabetes mellitus. However, some recent findings from X-ray analysis and spectroscopic analysis revealed that the zinc insulin hexamer displays many of the classical elements of an allosteric protein which has cast a new perspective on the insulin as a fundamentally interesting protein. The major development for the new conception of insulin mainly comes from the finding of three structure species existed in the hexameric zinc insulin which were designated as T_6, R_6 and T_3R_3 conformational state [1]. T_6 and R_6 correspond to 2 Zn insulin and phenol insulin respectively which differ each other in both the conformation of B1-B8 and the coordination mode of zinc ions. The T_3R_3 corresponds to the 4 Zn insulin and is considered as a stable intermatiate in the T_6 to R_6 conformation transition. One also knows that the chloride ion with high concentration is the lyotropic factor for the T_3R_3 structure. A series of recent reports with spectroscopic analysis provides definite evidences to show the existance of T_6 and R_6 conformations in solution. But so far there are no unambiquous data to identify the T_3R_3 structure in solution. Therefore it specially needs to gain more observation on the intermediate structure. Here we report a new structure type of such intermediate ($T^3R^t_3$) observed in the X-ray crystal structure analysis of an insulin mutant, [A21,Ser] human insulin.

[A21,Ser]-human insulin (A21S) is an insulin mutant with a high stability prepared by site-directed mutagenesis through a mutation of A21-Asn→Ser. It basically retains the biological potencies (71.4% and 76.6% of the native potancy in FFC assay and MBG assay respectively), but it is 5-10 times more stable in acidic solution (pH 3-4) than the native insulin in respect to deamidation and covalent dimer formation [2]. However, A21-Asn is conservative throughout the evolution and invariant among the insulin speices known today. Therefore, it is necessary to understand the impact of A21-substitution on the 3-D structure of insulin. This is the original aim of our crystallographic studies. It has been found that A21S had an abnomal Zn^{2+}-combination behavior in solution, i.e. the number of Zn^{2+} combined by every A21S hexamer had a nearly linear increase with the concentration of Zn^{2+}. This property made the crystallization difficult. After a systematical search, nice single crystals were gained in a system containing $ZnAc_2$, citric acid, Tris and 1,4-dioxane (in reservoir), which was different from the conventional recipe.

Results and Discussion

Structure Analysis

The preliminary crystallographic analysis shown that the space group and cell constants of A21S crystal were R3 and $a_H=b_H=80.61$Å, $c_H=37.98$Å which were similar to those of 4 Zn insulin. This indicates that A21S may adopt T_3R_3-like conformation. This result is quite interesting, because in the system of A21S crystal growth, there is neither high concetration Cl⁻ ion nor phenol which are lyotropic factors for the R-type structure known today. What does induce the formation of R-conformation in A21S crystals? Does this T_3R_3-like structure emerged in A21S possess some features different from that observed in 4 Zn insulin? These are attractive suspensions of the story.

The structure of A21S was solved by using difference Fourier method at 1.8Å resolution. After a total of 52 cycles of the refinement and 5 times of the model rebuilding, the conventional crystallographic R factor was reduced to 0.181 for data 10-1.8Å which shown as a good structure having been gained. In the final (2Fo-Fc) map, except the side chain of B1, the all residues can fit to the qualified electron density map.

Structure Model

In the final refined model of the mutant dimer, the main chains of the mutant have similar conformations with that of the native insulin and the important hydrogen bonds formed between the main-chain NH of A21 and the main-chain CO of B23 are well kept in the mutant, but the salt bond between the terminal carboxyl group at A21 and the guanidino group of Arg-B22 emerged in the native is destroied in the mutant. It indicates that the main-chain structure feature of A21 residue should play a critical role for the structure-function significance of this conservative residue.

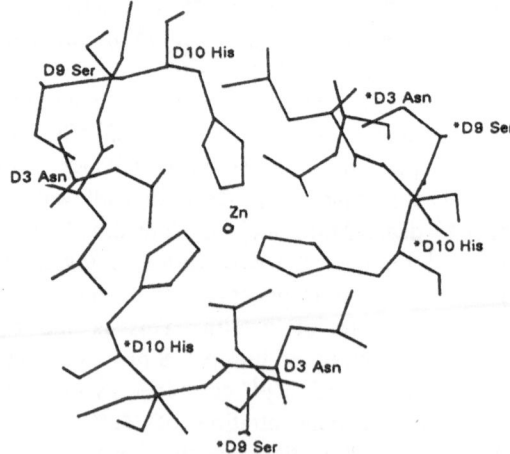

Fig. 1. Three Asn-B3 and three His-B10 of 2 molecule coordinated with Zn (II) to form an octahedral array.

242

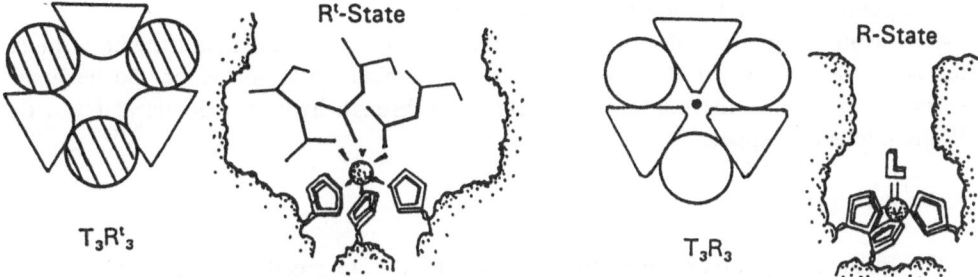

Fig. 2. Schematic diagram of $T_3R'_3$ (left) and T_3R_3 (right) structure.

The A21S hexamer adopted a general pattern of T_3R_3 structure but possesses different structure features for R_3 trimer. Compared with the R_3 structure discovered in 4 Zn insulin and phenol insulin, the R_3 trimer observed in A21S possesses the following characteristics: (1) The B1-B3 stretch has been moved about 5Å toward the Zn(II) so as the first cycle of B-chain helix emerged in 4 Zn insulin is loosed to a β-turn-like conformation and the helical conformation is only kept in a region contiguous from B4-B19. (2) The side-chain of Asn-B3 has a new position coordinated to the Zn(II), there by the coordination pattern of the Zn(II) adopts an octahedral array composed of six ligands including three Asn-B3 and three His-B10(Fig.1), other than the tetrahedral array composed of four ligands including three His-B10 and one chloride ion in conventional R_3 structure (Fig.2). Besides, both two conformations and three off 3-fold axis zinc ions emerged in 4 Zn insulin all disappeared. Consequently the coordination positions off 3-fold axis comprised of His-B10(II) and # His-B5 dose not existed at all in A21S. (3) Three solvent-accessible pockets were found on the hexamer surface of A21S, each of which bound to a 1,4-dioxane molecule via hydrogen bonding and van de Walls interactions. This should be the effector inducing the conformational change to R structure of A21S which is different from the case observed in 4 Zn and phenol insulin.

Conclusions

1. The hexameric zinc insulin structure observed in [A21,Ser]-human insulin crystal represent a new type of T_3R_3 insulin conformational state, in which the conformational pattern of subunits are basically T_3R_3, except for a nonhelical stretch of B1-B3, but the coordination mode of zinc ions in the metal chelate sites adopts a T_6-like type, namely two zinc ions are all on the 3-fold axis and both possess six ligands arranged as an octahedral array. We would like to designate it as $T_3R'_3$ conformational state.

2. In the R'_3 structure, six coordination sites of zinc ion(II) are all occupied by the residues of insulin molecule itself including three Asn-B3 and three His-B10, which has not yet been observed in other hexameric insulin structure known today. The coordinate interactions between Asn-B3 and Zn(II) should be a significant factor for stabilizing the

helical conformation of B4-B9 segment.

3. It seems likely that the $T_3R_3^t$ structure represents particularly stable intermediate in the T_3 to R_3 conformational transition, which may provide a new model for the investigation of the allosteric transition of insulin.

4. A neutral organic molecule, 1,4-dioxane, present in crystallization media is most probably the effector of R_3^t conformation, which binds to a pocket on the hexamer surface and induces the conformation transition through hydrogen bonds and van de Walls interaction appeared hereabout.

Acknowledgement

The financial supports from UNIDO grant (91/048) are greatly appreciated.

References

1. Kaarsholm, N.C., Biochemistry, 28(1989)4427.
2. Markussen, J., Protein Engineering, 2(1988)157.

Conformational analysis of dynorphin(1-13)

Qiao-lin Deng, Lu-hua Lai, Yu-zhen Han and Xiao-jie Xu
Department of Chemistry, Peking University, Beijing 100871, China

Introduction

Dynorphin-(1-13) [1] is a bioactive peptide, the sequence of which corresponds to the NH_2-terminal sequence of dynorphin. It has typical opioid activity and higher potency than [Leu]-enkephalin, normorphine and β-endorphin. In order to study its active conformation, we used Monte Carlo simulated annealing method[2] to search the preferred conformation of dynorpyin-(1-13).

The conformational analysis of peptides is difficult and challenging because of a large amount of local minima in the configuration space. So the conventional optimization methods are liable to be trapped into the local minima and difficult to find the global minimum. Monte Carlo simulated annealing method, which adopts Metropolis criteria to select the conformational changes, is a more effective method for conformational analysis of flexible molecules. It is better than the conventional molecular mechanics methods because it not only accepts all the downhill energy changes, but also some uphill moves, making it possible to overcome energy barriers and more likely to find the global minimum.

Dynorphin-(1-13) is a linear peptide consisting of thirteen amino acid residues: Tyr-Gly-Gly-Phe-Leu-Arg-Arg-Ile-Arg-Pro-Lys-Leu-Lys. There are seventy-three rotary torsion angles as the variables of the system while all the bond lengths and bond angles are kept fixed at their standard values. The conformational analysis has been carried out using program SAPEP[3] developed in this laboratory. One hundred calculations have been performed using different sets of random number seeds. All the calculations have been run on the Silicon Graphics Personal IRIS workstation.

Results and Discussion

The results show that dynorphin-(1-13) is a very flexible molecule because the final conformations are quite diverse. There are two lowest energy conformations assigned as CONF1 and CONF2 while other conformations have much higher energy. CONF1 and CONF2 have similar conformation consisting of continuous β-turn structures, close to helix conformation. The backbones of the conformations differ at both terminals. The superposition of the α-carbon traces of these two conformations shows the similarity between them (Fig. 1).

An interesting feature of these two conformations is the arrangement of side-chains. Along the backbone of the peptide, the side-chains of hydrophobic residues gather on one side, while the side-chains of hydrophilic residues except Arg^9 assemble on the opposite side. This kind of arrangement may be favourable for its binding to receptor.

Fig. 1. Stereoview of the α-carbon trace of CONF1 superposed on the α-carbon trace of CONF2.

Acknowledgement

This research was funded by the 863 High Technology Project.

References

1. Goldstein, A., Tachibans, S., Lowney, L.I., Hunkapiller, M. and Hood, L., Proc. Natl. Acad. Sci. U.S.A., 76(1979)6666.
2. Kirkpatrick, S., Gelatt, C. D. and Vecchi, M. P., Science, 220(1983)671.
3. Deng, Q. L., Han, Y. Z., Lai, L. H., Xu, X. J. and Tang, Y. Q., Chinese Chemical Letters., 2(1991)809.

The synthesis and 2D NMR study of Deltorphins

Xiao-yu Hu[a], Hong Ji[a], Zhu-ying Wang[a], Xiao-jie Xu[b], Ruo-heng Zhang[b] and Lu-hua Lai[b]

[a]Department of Biology, Lanzhou University, Lanzhou, 730000, China
[b]Department of Chemistry, Peking University, Beijing 100831, China

Intorduction

Deltorphins were isolated from the skin extract of frogs (*Genus Phyllomodusa*) in 1989. They are related to endogenous opioid peptides with high affinity and selectivity for δ opioid receptor. The amino acid sequence of Deltorphin I and II were known as Tyr-D-Ala-Phe-Asp(Glu)-Val-Val-Gly-NH$_2$. The study with cloning technique suggested that the D-Alanine in this serics are derived from L-Alanine in their precursor.

Results and Discussion

We synthesized Deltorphin I, II and their analogs with L-Ala at position 2. The synthesis was carried out with SPPS and Boc strategy. The purity of synthetic peptides was acquired in reverse HPLC analysis to be greater than 95%. The molecular weitht of these two peptides were 769d and 783d from their FAB mass spectra based on their molecular ions (M+1) peaks (770d and 784d). The sequences of Deltorphin II and [L-Ala^2]-Deltorphin II were determined by 440A amino acid sequencer to be Y-A-F-E-V-V-G.

Deltorphin II and [L-Ala^2]-Deltorphin II were studied on 500 MHz NMR spectrometer in DMSO-d$_6$. Residue types and sequential assignments were performed by means of standard 2D NMR techniques: ^{1}H-^{1}H COSY, and 2-D NOESY. It was found that NH-NH effects at F^3-E^4, E^4-V^5, V^5-V^6, and V^6-G^7, are similar between Deltorphin II and its L-Ala analog and suggested that folded conformations were present in the C-terminal. Because of the close proximity of the diagonal peaks, it is impossible to identify other significant NH-NH NOEs, particularly for the N-terminal region. In NOESY spectra, the interaction of Phe^3-C^{α}H and Val^5-NH was also found which is characteristic of 3$_{10}$ helix.

The conformations of Deltorphin II and [L-Ala^2]-Deltorphin II are almost the same, there are 3$_{10}$ helix strutures at the C-fragment tetrapeptide of both. The only difference is the configuration of Ala at position 2. The study of opioid activity of Deltorphins is in progress.

The molecular models of these two peptides were established.

Acknowledgement

We thank National Laboratory for Dynamic and Stable State of Molecular for financial support, we also thank Mr. Ma Leibin, Beijing Institute of Trace Chemistry, for his excellent work on NMR.

References

1.   Richter, K., Egger, R. and Kreil, G., Science, 238(1987)200.
2.   Kreil, G., Barra, D. and Simmaco, M., Eur. J. Pharmacol., 162(1989)123.
3.   Erspamer, V. and Melchiorri, P., Proc. Natl. Acad. Sci. U.S.A., 86(1989)5188.
4.   Amiche, M., Sagan, S. and Nicolas, P., Int. J. Protein Res., 32(1988)506.

Session VII
Peptide synthesis

Chairs: Meng-Shen Cai
Beijing Medical University
Beijing, China

and

Ruth F. Nutt
Merck Research Laboratories
West Point, Pennsylvania, U.S.A.

The design and synthesis of glucagon antagonists

R.B. Merrifield and Cecilia G. Unson

The Rockefeller University, New York, NY 10021, U.S.A.

Introduction

The bihormonal hypothesis [1] states that in diabetes the underutilization of glucose is due to a lack of insulin, while overproduction of glucose is a consequence of excess glucagon. Our aim has been to design and synthesize glucagon analogs that will be potent antagonists of the hormone and to separate the structural components required for binding glucagon to its receptors from the structures necessary for transduction of the hormonal signal. We believe such glucagon antagonists will be of therapeutic value in lowering the insulin requirements of diabetics and in controlling the disease.

Results and Discussion

A fruitful approach to the design of strong antagonists was based on the assumption that glucagon and secretin may have evolved from a common precursor. The two peptides show considerable sequence homology, but their physiological roles differ and they show no affinity for each other's receptor. Thus, during evolution the hormone structure may have passed through an intermediate stage containing elements of both glucagon and secretin that would bind to the glucagon receptor but would not transform the signal to activate adenyl cyclase in the liver membrane. It therefore would be a glucagon antagonist. To test this idea a large number of glucagon analogs containing replacements corresponding to the secretin structure were synthesized. In this way Asp^9 was identified as a critical residue for glucagon activity. When this feature was combined with the known importance of His^1 in signal transduction [2] and with the C-terminal amide, known to enhance the helical dipole of a peptide, we arrived at the analog desHis^1[Glu^9] glucagon amide [3]. It bound 40% as well as glucagon but was completely inactive (0.0001%) in the *in vitro* cAMP assay. The inhibition index, $(I/A)_{50}$, was only 12.

For a glucagon antagonist to be effective in the way we have envisioned, it is important that it not inhibit the release of insulin from pancreatic islet cells. Otherwise the benefit of reducing the insulin requirement would be lost. Both glucagon and secretin potentiate glucose-stimulated release of insulin and it was found that two of our antagonists retain such activity and do not inhibit insulin release by endogenous glucagon [4]. This is a very positive property of these antagonists.

Although *in vitro* assays of the glucagon analogs have been convenient in our search for glucagon antagonists, it is essential to know if these compounds are effective in whole animals. As expected from the membrane studies, it was shown that desHis^1[Glu^9] glucagon amide, in contrast with glucagon itself, does not cause blood glucose to rise

Table 1 *Position-9 replacement analogs*

Glucagon amide analog	Membrane binding%	Relative activity%	Inhibition index$(I/A)_{50}$	pA_2
1. Asp^9(parent)	100	15		
2. Glu^9	14	<0.004	129	5.2
3. Glu(OCH$_3$)^9	100	0.05	10	8.7
4. Nle^9	32	0.17	4	7.7
5. Lys^9	54	<0.006	30	6.4
6. Gly^9	37	0.17	219	5
7. Asn^9	42	0.17	25	6
8. D-Asp^9	1.6	<0.004	129	5.2
9. desAsp	45	1	5	7.2
10.Des-His^1-[Glu^9]	40	<0.0001	12	7.2
11.Des-His^1-[Nle^9]	125	<0.0012	19	7.6
12.Des-His^1-[Lys^9]	70	<0.0012	5	7.35
13.Des-His^1-[Gly^9]	100	<0.006	43	6.24

in rabbits and is not an agonist [5]. Even more important, it was demonstrated that this analog will inhibit the rise in glucose induced by glucagon injected into fasted rabbits. The analog therefore acts as a glucagon antagonist *in vivo*. It was also possible to demonstrate in rats made diabetic with streptozotocin that their resulting elevated glucose could be reduced by injection of a bolus of the glucagon analog and would remain at a constant low level by continuous infusion.

The antagonist, desHis^1[Glu^9]glucagon amide, was very valuable for demonstrating that glucagon in the liver can activate two separate receptors and the formation of two second messengers [5]. In pancreatic acinar tissue adenyl cyclase is stimulated by low concentrations of secretin , while higher concentrations also stimulate phospholipase C and the subsequent enzymatic hydrolysis of phosphatidylinositol bisphosphate. The second messenger products, diacylglycerol and inositol trisphosphate (Ins P$_3$) lead to activation of protein kinase C and the several cellular responses that it produces. In the hepatocyte we showed that desHis^1[Glu^9]glucagon amide is a pure antagonist toward cyclase activation but is an agonist in the production of Ins P$_3$. Thus, our data support the idea that there are two different glucagon receptors in this tissue.

To further understand the role of position 9 of glucagon, a large number of replacement analogs of that site were synthesized [6]. Every single-residue substitution for Asp^9 resulted in the retention of appreciable receptor binding, but each substitution was accompanied by severely diminished capacity to transmit the biological signal (Table 1). These results demonstrate an uncoupling of receptor binding from transduction of the signal at this locus and define a central role of aspartic acid-9 in glucagon action. The data suggest that signal transduction may be mediated by a coupled interaction between His^1 and Asp^9 of the hormone, and perhaps a third essential residue.

To examine this question further we studied the role of substitutions at position 1 in glucagon amide (Table 2). His^1 could be replaced by several aromatic derivatives such

Table 2 *Position-1 replacement analogs*

Glucagon amide analog	Membrane binding %	Relative activity %	Inhibition index $(I/A)_{50}$	pA_2
1. His[1],(Parent)	100	15		
2. Tyr[1]	155	5.6		
3. Imidazole acetic[1]	89	20		
4. Indole acetic[1]	44	8.3	72	5.7
5. Difluorobenzoic[1]	20	0.16	16	6.1
6. Pro[1]	10	<0.0004	28	7.3
7. Asp[1]	11	0.25	58	6.4
8. Arg[1]	9	0.64	158	6.2
9. His[1],Glu[9]	14	0.05		
10. Tyr[1],Glu[9]	28	<0.001	178	6.0
11. Imidazole acetic[1],Glu[9]	32	1.3	11	7.2
12. Indole acetic[1],Glu[9]	112	<0.001	7	7.3
13. Difluorobenzoic[1],Glu[9]	75	<0.008	4	8.5
14. Octanoic[1],Glu[9]	48	0.02	25	7.5
15. Succinoic[1],Glu[9]	28	<0.004	25	7.4
16. Pro[1],Glu[9]	6	<0.002	20	6.1

as tyrosine or imidazoleacetic acid[1] or could be deleted altogether with retention of good receptor binding as judged from displacement of [^{125}I]glucagon, However, substitution with certain nonaromatic residues such as Pro, Asp, or Arg had negative effect on binding. Note that relative to glucagon, glucagon amide retained full binding and a maximum activity of 100%, with a 15% relative activity. When Asp[9] is replaced by Glu[9], the same general rules apply to changes at position 1. Aromatic or long chain aliphatic residues are better than His for receptor binding, while Pro[1] reduces binding. In spite of good binding, the non-imidazole position-1 residues lead to very low relative activities and, in the absence of Asp[9], none of the position-1 residues had significant activity. The most potent antagonist to date is [2,4 difluorobenzoic acid[1], Glu[9]] glucagon amide. It showed 75% binding, < 0.008% relative activity, an inhibition index of only 4 and a very high pA_2 value of 8.5.

Finally, we have begun similar studies on the four serine residues of glucagon. Representative data are in Table 3. Individual replacements of serine at positions 2,8, and 11 by alanine modified receptor binding strength but in each case significant binding was retained. [Ala[11]]glucagon bound 5 times as tightly as the native Ser[11] molecule. In each of these three positions the change in relative activity roughly paralleled the change in binding and there was no suggestion that one of these serines played a specific role in transduction. Ser[16], however, was quite different. [Ala[16]]glucagon amide was bound 100% as well as glucagon amide, but was inactive in stimulating cAMP formation (0.9). We believe this serine residue plays an important function in glucagon action.

Thus, three specific residues have been identified that are important in transducing the hormone signal in liver membranes. The data suggest to us that this triad may function

Table 3 *Serine replacement analogs*

Analog	Membrane binding %	Relative Activity %
Ala^2	44	22
Ala^8	13	16
Ala^{11}	500	44
Ala^{16}	100	0.9

in a way analogous to that of the serine proteases, which utilize these same residues in the enzymatic cleavage of their respective substrates. we think it is conceivable that glucagon, when bound to its receptor, may assume a conformation that aligns the side chain functional groups of Asp^9, His^1 and ser^{16} so that the charge relay system can operate to catalyze the cleavage of a peptide bond with formation of an intermediate acyl-serine bond. Rapid hydrolysis would give rise to the final products. This bond may be within the receptor itself, giving rise to new amine and carboxyl end groups. One of these may represent the activated receptor, which goes on to interact with its G-protein and cause the activation of adenyl cyclase.

We have been able to show that the serine protease inhibitor, (4-amidinophenyl)-methanesulfonyl fluoride inhibits glucagon-induced cAMP formation. cAMP formation was not affected in controls in which glucagon or liver membrane receptor were pretreated singly with pA-PMSF,supporting the idea that an activated serine is present in the complex. We plan to further test this working hypothesis when the cloned glucagon receptor is isolated in purified form.

References

1. Unger, R.H., Diabetes, 25(1976)136-151.
2. Lin, M.C., Wright, D.E., Hruby, V.J. and Rodbell, M., Biochemistry, 14(1975)1559.
3. Unson, C.G., Andreu, D., Gurzenda, E.M. and Merrifield, R.B., Proc. Natl. Acad. Sci. U.S.A., 84(1987)4083.
4. Kofod, H., Andreu, D., Thams, P., Merrifield, R.B., Hedeskov, C.J., Hansen, B. and Lernmark, Å., Am. J. Physiol. 254(1988)(Endocrinol. Metab.17)E454.
5. Unson, C.G., Gurzenda, E.M. and Merrifield, R.B., Peptides 10(1989)1171.
6. Unson, C.G., Macdonald, D., Ray, K., Durrah, T.L. and Merrifield, R.B., J. Biol. Chem., 266(1991)2763.

Solid phase synthesis of cyclic peptides: effects of ring size on biological activity of growth hormone releasing factor

Arthur M. Felix, Zhi-cheng Zhao, Yung Lee and Robert M. Campbell
Roche Research Center, Departments of Peptide Research and Animal Science Research, Hoffmann-La Roche Inc., Nutley, NJ 07110, U.S.A.

Introduction

Growth hormone releasing factor, GRF(1-44)-NH_2, is produced in the hypothalamus and stimulates the release of growth hormone (GH) *in vivo* in humans and a variety of animal species[1]. GRF(1-44)-NH_2 has been shown to be clinically useful in the treatment of grwoth hormone deficient children[2]. Structure-activity studies have shown that the shortened analog, GRF(1-29)-NH_2, retains full intrinsic activity with only slightly reduced potency[3]. Linear analogs of GRF have been prepared in which replacement of Gly^{15} by Ala^{15} results in increased potency both *in vitro* and *in vivo*[4]. This effect has been attributed to enhanced α-helicity and maximization of amphiphilic character. Potent analogs of GRF(1-29)-NH_2 have been prepared which are resistant to plasma degradation and are being used for performance enhancement applications in domestic livestock (pigs and lactating cows)[5]. We have recently reported on a series of synthetic cyclic peptides which were designed to constrain the peptide toward formation of the α-helical structure through side-chain to side-chain lactamization spaced at *i-(i+4)* positions[6]. Cyclo^{8,12} and cyclo^{21,25}-[Ala^{15}]-GRF(1-29)-NH_2 have been reported to be highly potent both *in vitro* and *in vivo* [7]. These cyclic analogs of GRF are more stable to enzymatic degradation than the corresponding linear compounds since the lactam ring may stabilize the helical conformation near the NH_2-terminus and disrupt interaction with the degrading enzyme (dipeptidylpeptidase IV). In the present study two series of cyclic *i-(i+4)* homologs of GRF, one cyclized at positions (8-12) and the other at (21-25), were prepared to determine the optimal ring size for biological activity.

Results and Discussion

Cyclic homologs of cyclo^{8,12}- and cyclo^{21,25}-[Ala^{15}]-GRF(1-29)-NH_2 containing 18-20 membered rings were prepared by our previously reported solid phase procedure using the Boc/Bzl procedure together with side-chain Fmoc/OFm protection at sites to be cyclized[6]. Cyclo^{8,12}-homologs containing larger rings (21-24 membered rings) required the coupling of "spacer". Fmoc-amino acids (e.g. Fmoc-Gly-OH, Fmoc-β-Ala-OH or Fmoc-γ-Aba-OH) to the side-chain of the basic residue at position 12 by the strategy shown in Fig.1 (Left). The synthesis of cyclo^{21,25}-homologs possessing 21-24 membered rings was achieved *via* a similar strategy by coupling amino acid fluorenylmethyl ester

Table 1 *Biological activity (GH release in vitro) and receptor binding affinities for i-(i+4) cyclic analogs of [Ala^{15}]-GRF(1-29)-NH$_2$ with varying ring size*

GRF Analog	Ring Size	Relative Potency	Relative Affinity
GRF(1-44)-NH$_2$	–	1.00	1.00
GRF(1-29)-NH$_2$	–	0.71	0.91
[Ala^{15}]-GRF(1-29)-NH$_2$	–	3.81	3.60
Cyclo^{8,12} Homologs			
Cyclo-(Asp^8-[γ-Aba]-Orn^{12})-[Ala^{15}]-GRF(1-29)-NH$_2$	24	1.42	1.62
Cyclo-(Glu^8-[Gly]-Orn^{12})-[Ala^{15}]-GRF(1-29)-NH$_2$	23	1.68	1.70
Cyclo-(Asp^8-[Gly]-Orn^{12})-[Ala^{15}]-GRF(1-29)-NH$_2$	22	4.90	4.95
Cyclo-(Glu^8-[β-Ala]-Dpr^{12})-[Ala^{15}]-GRF(1-29)-NH$_2$	22	3.79	4.51
Cyclo-(Glu^8-Lys^{12})-[Ala^{15}]-GRF(1-29)-NH$_2$	21	3.43	4.27
Cyclo-(Glu^8-[Gly]-Dpr^{12})-[Ala^{15}]-GRF(1-29)-NH$_2$	21	3.80	4.29
Cyclo-(Asp^8-Lys^{12})-[Ala^{15}]-GRF(1-29)-NH$_2$	20	0.77	0.86
Cyclo-(Glu^8-Orn^{12})-[Ala^{15}]-GRF(1-29)-NH$_2$	20	1.77	1.79
Cyclo-(Asp^8-[Gly]-Dpr^{12})-[Ala^{15}]-GRF(1-29)-NH$_2$	20	1.79	1.28
Cyclo-(Asp^8-Orn^{12})-[Ala^{15}]-GRF(1-29)-NH$_2$	19	0.03	0.02
Cyclo-(Glu^8-Dpr^{12})-[Ala^{15}]-GRF(1-29)-NH$_2$	18	0.02	0.02
Cyclo^{21,25} Homologs			
Cyclo-(Orn^{21}-[γ-Aba]-Asp^{25})-[Ala^{15}]-GRF(1-29)-NH$_2$	24	2.68	1.71
Cyclo-(Orn^{21}-[Gly]-Glu^{25})-[Ala^{15}]-GRF(1-29)-NH$_2$	23	2.53	1.84
Cyclo-(Orn^{21}-[Gly]-Asp^{25})-[Ala^{15}]-GRF(1-29)-NH$_2$	22	4.14	4.48
Cyclo-(Dpr^{21}-[β-Ala]-Glu^{25})-[Ala^{15}]-GRF(1-29)-NH$_2$	22	4.05	4.18
Cyclo-(Lys^{21}-Glu^{25})-[Ala^{15}]-GRF(1-29)-NH$_2$	21	2.41	1.96
Cyclo-(Dpr^{21}-[Gly]-Glu^{25})-[Ala^{15}]-GRF(1-29)-NH$_2$	21	2.69	2.56
Cyclo-(Lys^{21}-Asp^{25})-[Ala^{15}]-GRF(1-29)-NH$_2$	20	1.33	1.75
Cyclo-(Orn^{21}-Glu^{25})-[Ala^{15}]-GRF(1-29)-NH$_2$	20	0.81	0.71
Cyclo-(Dpr^{21}-[Gly]-Asp^{25})-[Ala^{15}]-GRF(1-29)-NH$_2$	20	1.73	1.74
Cyclo-(Orn^{21}-Asp^{25})-[Ala^{15}]-GRF(1-29)-NH$_2$	19	0.04	0.03
Cyclo-(Dpr^{21}-Glu^{25})-[Ala^{15}]-GRF(1-29)-NH$_2$	18	0.03	0.02

"spacers" (e.g. H-Gly-OFm, H-β-Ala-OFm or H-γ-Aba-OFm) to the side-chain of the acidic residue at position 25 as outlined in Fig. 1 (Right). The amino acid-OFm esters were prepared in 2 steps with excellent yields from the N$^\alpha$-Boc-amino acids. Following the insertion of the "spacer" the solid phase synthesis and cyclization was completed by our general procedure[6].

Table 1 summarizes the biological activity (*in vitro*) for all the cyclo^{8,12}-[Ala^{15}]-GRF(1-29)-NH$_2$ and cyclo^{21,25}-[Ala^{15}]-GRF(1-29)-NH$_2$ homologs which also gave excellent correspondence with GRF receptor binding affinity[8]. The relationship between ring size and biological activity was remarkably similar for both the cyclo^{8,12}- and cyclo^{21,25}-homologs. The most active GRF homologs have lactam rings with 22 members. Analogs

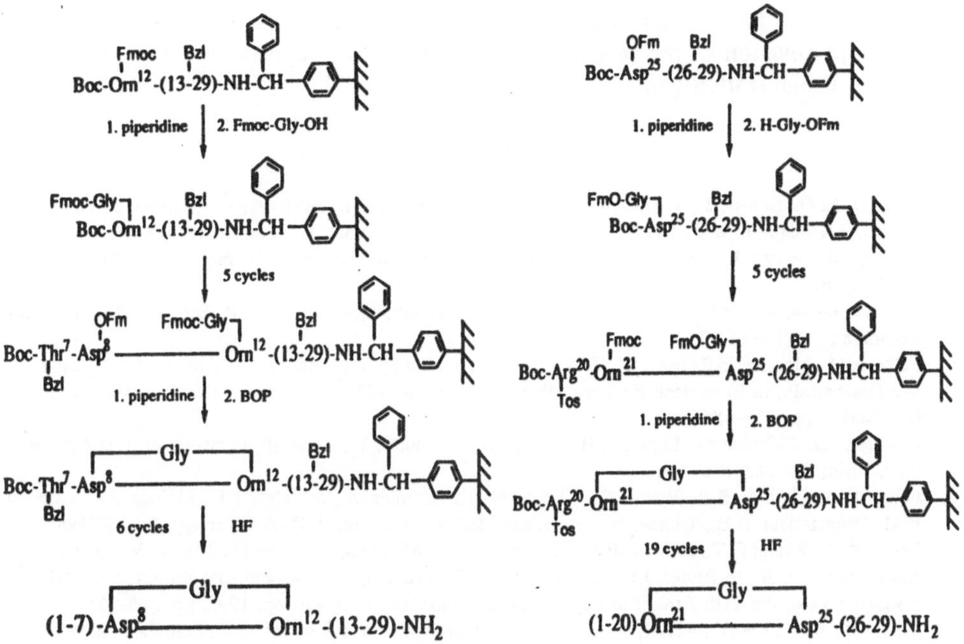

Fig. 1. (Left) Synthesis of cyclo^{8,12}-homologs of GRF containing 21-24 membered-rings exemplified by cyclo-(Asp^8-[Gly]-Orn^{12})-[Ala^{15}]-GRF(1-29)-NH$_2$. (Right) Synthesis of cyclo^{21,25}-homologs of GRF containing 21-24 membered rings exemplified by cyclo-(Orn^{21}-[Gly]-Asp^{25})-[Ala^{15}]-GRF(1-29)-NH$_2$.

with smaller (20 and 21 members) and larger (23 and 24 members) rings also retained substantial biological activity. However, GRF homologs with constrained rings containing ≤19 members have substantially decreased potencies. Isosteres of cyclo^{8,12}- and cyclo^{21,25}-[Ala^{15}]-GRF(1-29)-NH$_2$ (20, 21 and 22 membered rings) were also prepared and found to have similar biological activity for a given ring size.

Conclusions

A procedure has been developed for extending the solid phase (Boc/Bzl) strategy to include the synthesis of cyclic peptides (lactams) using "spacer" amino acids (e.g. Gly, β-Ala, γ-Aba) between the side-chains of basic (e.g. Lys, Orn, Dpr) and acidic (e.g. Asp, Glu) residues. Using this procedure, two series of cyclic *i-(i+4)* GRF analogs were designed to determine the optimum ring size for maximum biological activity. Optimal ring size for both series of cyclic GRF analogs was determined to be 22 members. Although cyclic GRF analogs containing larger and smaller rings retain substantial potencies, peptides with constrained rings (≤19 members) are biologically inactive.

Conformational analysis of the cyclo^{8,12}- and cyclo^{21,25}- homologs reveals that the preferred conformations for the biologically active peptides (20 through 24 membered

257

rings) retain substantial α-helicity. The biologically inactive homologs (≤19members) possess constrained rings which partially destabilize the α-helix. These studies demonstrate that the α-helix is an important factor for biological activity in the GRF system. Optimization of the α-helix may be achieved through the synthesis of *i-(i+4)* side-chain lactamization containing 22 members.

References

1. Thorner, M.O., Spiess, J., Vance, M.L., Rogol, A.D., Kaiser, D.L., Webster, J.D., Rivier, J., Borges, J., Bloom, S.R., Cronin, M.J., Evans, W.E., MacLeod, R.M. and Vale, W., Lancet, 1(1983)24.
2. Gelato, M., Ross, J., Pescovitz, O., Cassorla, F., Skeeda, M. and Merriam, G.R., Pediatr. Res., 18(1984)167A.
3. Ling, N., Baird, A., Wehrenberg, W.B., Ueno, N., Munegumi, T. and Brazeau, P., Biochem. Biophys. Res. Commun., 119(1984)265.
4. Felix, A.M., Heimer, E.P., Mowles, T.F., Eisenbeis, H., Leung, P., Lambros, T., Ahmad, M., Wang, C.-T. and Brazeau, P., in Proc. 19th European Peptide Symposium (Theodoropoulos,D., ed.), W.deGruyter & Co., Berlin, pp. 481–484.
5. Petitclerc, D., Pelletier, G., Lapierre, H., Gaudreau, P., Coutre, Y., Dubreuil, P., Morisset, J. and Brazeau, P., J. Anim. Sci., 65(1987)996.
6. Felix, A.M., Heimer, E.P., Wang, C.T., Lambros, T., Fournier, A., Mowles, T.F., Maines, S., Campbell, R.M., Wegrzynski, B.B., Toome, V., Fry,D. and Madison, V.S., Int. J. Peptide Protein Res., 32(1988)441.
7. Felix, A.M., Wang, C.T., Heimer, E.P., Campbell, R.M., Madison, V.S., Fry,D., Toome, V., Downs, T.R. and Frohman, L.A., In Rivier, J.E. and Marshall, G.R. (Eds.) Peptides: Chemistry, Structure and Biology (Proceedings of the 11th American Peptide Symposium), ESCOM, Leiden, 1990, pp. 226–228.
8. Campbell, R.M., Lee,Y., Rivier, J., Heimer, E.P., Felix, A.M. and Mowles, T.F., Peptides, 12(1991)569.

Synthesis of muramyl peptides

Jie-cheng Xu[a], Ming-zhu Zhang[a], Shi-yi Liu[b] and Yi Zhang[b]

[a]*Shanghai Institute of Organic Chemistry, Chinese Academy of Sciences, Shanghai 200032, China*
[b]*Shanghai Institute of Physiology, Chinese Academy of Sciences, Shanghai 200031, China*

Introduction

Muramyl peptides(MPs) is a kind of glycopeptide which is well known to exert multiple effects on immune and nervous systems as immunoadjuvant, antitumor agent, pyrogen, as well as promoter of slow wave sleep. NAcMur-L-Ala-D-isoGln (Muramyl Dipeptide, MDP) is the simplest structure required for the activities. We have found that some analogs of MDP exhibited negligible pyrogenic effect, a major co-existent side effect with immunomodulation. Following this line, some new MDP analogs were designed and synthesized (Scheme 1).

Results and Discussion

The analogs were synthesized by the conventional solution method using DCC-HOBt, except the last coupling step which was carried out with a new coupling agent, benzotriazolyloxybis(pyrrolidino)-carbonium hexafluoro-phosphate (BBC). The t-butoxycarbonyl (Boc) and benzyloxycarbonyl (Z) group was employed for α- and ε-amino protection repectively and benzyl ester for carbonyl protection. The Boc group was removed with trifluoroacetic acid (TFA) prior to each coupling. Hydrogenolysis of 2a-f gave 3a-f, which was then purified by chromatography on C18 reverse phase column with 5% MeOH-0.05% TFA as a solvent. Details of the synthesis are described in Scheme 2.

In summary, the advantages of the procedure are:

1. Tetrafluoroboric acid etherate ($HBF_4.Et_2O$) was used as the catalyst instead of sulfuric acid in the γ-esterification of D-Glutamic acid with the advantages of simple operation and high yield (about 90%).

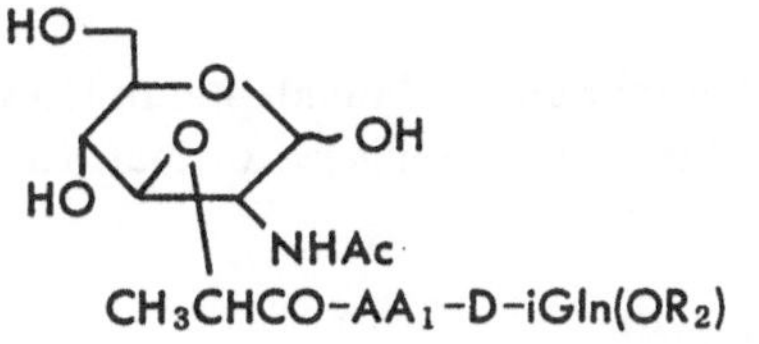

$$CH_3CHCO-AA_1-D-iGln(OR_2)$$

(3a-f)

Scheme 1. Synthetic analogues of MDP.

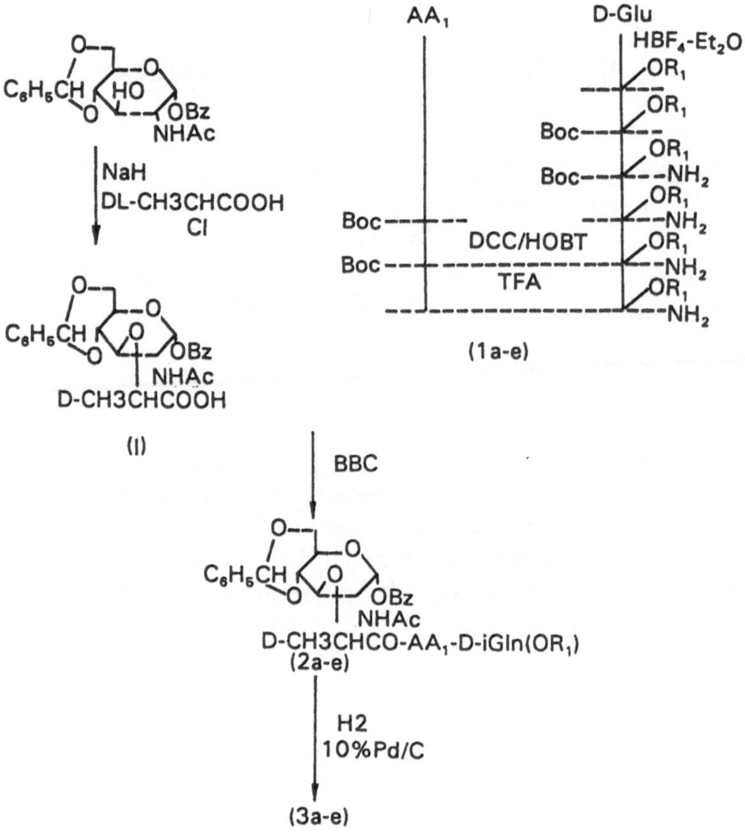

Scheme 2. Synthetic route for analogues 3a–f.

2. Racemic 2-chloropropionic acid was served to prepare the key intermediate, benzyl-2-acetamido-4,6-0-benzylidene-2-deoxy-3-0(methyl-D-1-ethylcarboxylate)-2-D-glucopyranoside(I). The resulted (R,S) isomers, after converting to their methyl esters, were then conveniently separated via column chromatography.

3. A new high efficient coupling reagent (BBC), newly developed by one of authors(Xu), was used in the coupling reaction of protected muramic acid and peptide with simple reaction condition, excellent yield and low racemization.

Acknowledgements

We thank the National Natural Science Foundation of China and State Key laboratory of Bioorganic and Natural Products Chemistry for their financial support.

References

1.   Lefoancier,P. and Lederer,E., Pure & Appl.Chem., 59(1987)449.
2.   Chen, S.Q. and Xu, J.C., Tetrahedron Lett., 33(1992)647.

260

Synthesis of peptide fragment of the principal neutralizing determinant of HIV-1$_{mn}$

Ran Zhang^a, Li-li Guo^a, Hong Zhao^a, Wei Cui^b, Chun-hua Pan^a and Guan-fu Zhu^a

^{a}Institute of microbiology and Epidemiology, Beijin 100850, China
^{b}Jilin University, Changchun 130023, China

Introduction

Neutralizing antibodies elicited by immunizing animals with gp120 envelope protein derived from a single strain of HIV-1 are highly type specific, and the principal binding site for HIV-1 neutralizing antibodies(the pricipal neutralizing determinant, PND) is within a disulfide loop in the third variable domain of the external envelope protein, gp120. Analysis of the PND from many HIV-1 isolates revealed that although variable the PND contains conserved sequence and structure. PND peptide from HIV-1$_{mn}$ isolate is similar to the consensus sequence and reacts with high percentage of the sera (80% or so). In this paper, a 23 amino acid peptide containing positions from 302 to 324 of gp120 of HIV-1$_{mn}$ was synthesized, ELISA activity was also reported.

Results and Discussion

Peptide fragment of HIV-1mn was synthesized by manual FMOC Solid-phase method, with p-benzyloxy benzyl alcohol resin as carrier. The amino acid sequence and synthesis process are as follows:

Y N K R K R I H I G P G R A F Y T T K N I I G

First Amino acid:
5 eq of Fmoc-Gly-pfp, 5 eq of HOBT, 1.5 eq of DMAP, were reacted for 1 h and two times. The coupling yeild was 85%.
DMF 6×2 min, rinsing,
20% piperidine 2×1 min, 1×5 min, 1×10 min,
DMF rinsing to neutral,
2.5 eq of Fmoc-(AA)$_{n+1}$ -pfp in 3-4 ml DMF, were reacted for 1 h, monitored with ninhydrin colour reaction.
After synthesis, the peptide was cleaved from supporting resin with 1 M TMSBr-thioanisole/TFA, purified with Sephadex G-25 and analyzed by RP-HPLC (Fig.1). For further purification, CMC-52 column chromatography was used (Fig.2.)
The protecting groups of side-chains of amino acids were: Lys(Boc), Thr(OBut), Arg(Mtr), His(Boc) and Tyr(OBut).

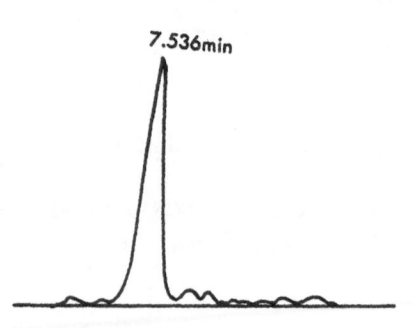

Fig. 1. HPLC of peptide on PE-Pack C18. Reservoirs A contained 0.1% aqueous TFA, B contained 0.1% TFA in acetonitrile, A linear gradient of TFA 20-50% B was developed over 30 min. Flow rate 1 ml/min.

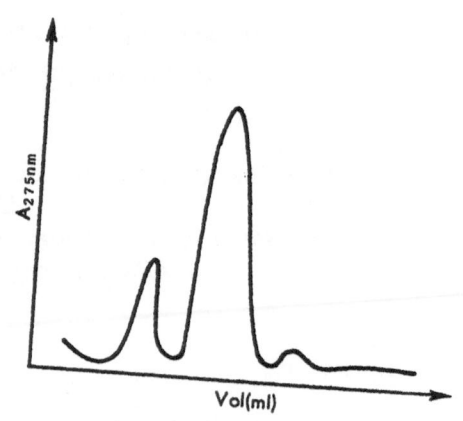

Fig. 2. Chromatogram of synthetic peptide on CMC-52. CMC-52 Column 1.5×25cm, flow rate 20 ml/h. A linear gradient of 0.02-0.5M NH_4Ac.

The result of amino acid analysis of the peptide was satisfactory. Assay of the peptide by ELISA showed that its activity as an antigen was + 0.4 (patient sera, absorbance at 490 nm) and + 0.02 (healthy sera). So, it needs to be linked to protein carriers to verify whether this low activity is due to a lack of binding between the peptide and antibody, or to a lack of attachment of the peptide to the surface of the plate.

References

1. Clement, G.J., Price-jones, M.J., Stephens, P.E., Sutton, C., Schulz, T.F., Clapham, P.R., McKeating, J.A., Mcclure, M.O., Thomson, S., Marsh, M., Kay, J., Weise, R.A. and Moore, J.P., AIDS Research and Human Retroviruses., 7(1991)3.
2. Fiflds, B. and Nobel, R.L., Int. J. Peptide Protein Res., 35(1990)161.

Synthesis and structure revision of the scorpion toxin noxiustoxin

Ruth F. Nutt[a], Byron H. Arison[b] and Jeffrey S. Smith[c]

[a]*Department of Medicinal Chemistry, Merck Research Laboratories, West Point, PA 19486, U.S.A.*
[b]*Department of Pharmacology, Merck Research Laboratories, West Point, PA 19486, U.S.A.*
[c]*Department of Drug Metabolism, Merck Research Laboratories, Rahway, NJ 07065-0900, U.S.A.*

Introduction

Noxiustoxin (NTX) is a 39-residue peptide isolated from the venom of the Mexican scorpion *Centruroides noxius HOFFMAN* with potent K^+ channel blocking properties [1]. In order to make available supplies of NTX for biological studies, the chemical synthesis of the peptide with reported sequence TIINVKCTSPKQCSKPCKELYGSSA GAKCMNGKCKCYNN was undertaken.

Results and Discussion

The chemical synthesis of the 39-peptide C-terminal acid NTX was carried out using the ABI 430A automated synthesizer and Boc/Benzyl protection for amino aicds. Symmetrical anhydride and HOBt/DCCI activation protocols were used. For each residue, double couplings were used except for Ile[2], Ile[3] and Asn[4] (three couplings) and Ser[9] and Lys[15] (four couplings). The initial synthesis, starting with commercially available Boc-Asn-O-Pam-resin, resulted in a low yield of final peptide-resin. This result was explained by the potential intramolecular cyclization involving the C-terminal Asn sidechain and the ester-resin linkage that resulted in aspartimide formation and thereby release of the first residue from the resin during peptide assembly. In order to minimize this undesirable reaction, a novel mode of synthesizing C-terminal Asn acids by the solid phase method was devised (Scheme 1).

In this approach, the C-terminal Asn was attached to the solid support via its sidechain using H_2N-MBH-resin. To minimize intramolecular cyclization, the α-carboxyl group was protected by the sterically hindered cyclohexyl ester. Thus, starting with Boc-Asp(NH-MBH-resin)-OChx, the synthesis of NTX resulted in higher yield of the 39-peptide-resin than the initial attempt. The peptide-resin was cleaved by HF-anisole (9:1) and the resulting peptide product was oxidized in air at pH 7.5 to achieve disulfide formation. The final product was isolated by gel filtration and reverse phase HPLC using a Delta Pak C_4,15μ,300Å column and gradient elution with 0.1% TFA containing aqueous CH_3CN. Final product was characterized by HPLC (95% single peak), amino

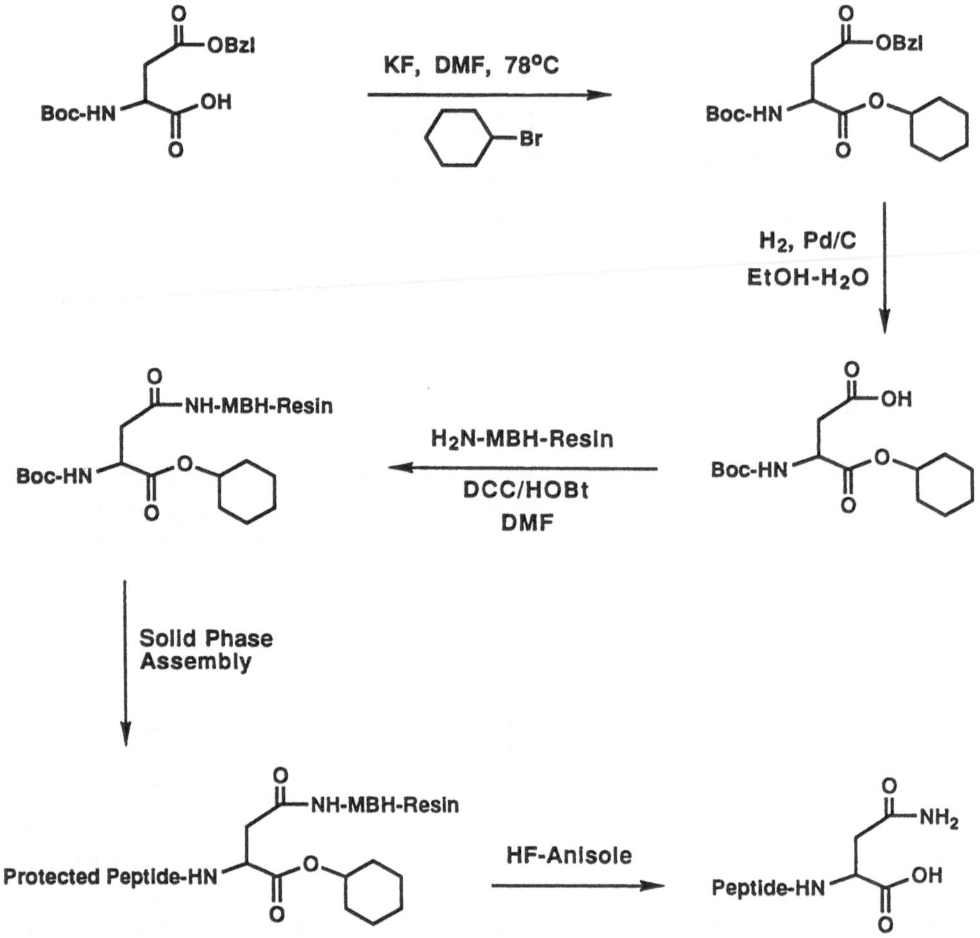

Scheme 1. Synthetic strategy used for the preparation of the 39-peptide C-terminal acid with noxiustoxin (NTX) sequence.

acid analysis after acid hydrolysis (ratios +/−5% of theory), Edman degradation and preview analysis (correct primary sequence, <1.5% preview at cycle 30), and NMR (no Met(O), spectrum consistent with structure), biological activity, and comparison with natural NTX (nNTX). Even though all characterization data of synthetic NTX (sNTX) were consistent with the published structure of nNTX, direct comparisons showed differences. A mixture of sNTX and nNTX resulted in two peaks during HPLC in the shallow gradient of 98% to 88% H_2O-CH_3CN containing 0.1% TFA over 30 min., with nNTX eluting 0.34 minutes earlier than sNTX. NMR studies showed two different sets of Tyr aromatic proton chemical shifts. In addition, sNTX exhibited 1/5 the potency as a blocker of voltage clamped Shaker locus H4 K⁺ channel expressed in Xenopus laevis oocytes [2,3] than the natural product. Upon further characterization of the two samples, it was found that only sNTX generated Asn during digestion with carboxypeptidase A. This result suggested that nNTX has a non-carboxylic acid C-terminus, possibly a carboxamide, which would be resistant to cleavage by carboxypeptidase.

To substantiate the potential structure revision of nNTX, the C-terminal amide (NTX-NH$_2$) was synthesized. Coupling of Boc-Asn-OH with H$_2$N-MBH-resin, and using synthesis and isolation protocols described above, NTX-NH$_2$ was obtained that was similar in purity to the acid. Direct comparison studies of the three products by HPLC and biological potency showed NTX-NH$_2$ to be different from the C-terminal acid sNTX and identical to the natural product nNTX.

In conclusion, the structure of the scorpion toxin noxiustoxin has been revised from the C-terminal acid to the amide and confirmed by total synthesis. In addition, a novel, generally practical synthetic approach for C-terminal asparagine peptides was developed.

Acknowledgements

The authors thank Drs. S.F. Brady, D.F. Veber for chemical and Drs. S.A. Buhrow, R.B. Stein for biochemical support, and Dr.L.D. Possani for a reference sample of nNTX.

References

1. Possani, L.D., Martin, B.M. and Svendsen, I., Carlsberg Res. Commun., 47(1982)285.
2. Iverson, L.E., Tanouye, M.A., Lester, H.A., Davidson, N. and Rudy, B., Proc. Natl. Acad. Sci. U.S.A., 85(1988)5723.
3. Swansom, R., Marshall, J., Smith, J.S., Williams, J.G., Boyle, M.B., Folander, K., Luneau, C., Antanavage, J., Oliva, C., Buhrow, S.A., Bennett, C., Stein, R.B. and Kaczmarek, L.K., Neuron, 4(1990)929.

Synthesis and function of unclassical hormone

Chao Wang, Shi-qi Peng, Meng-shen Cai, Xue-cai Qiu and Yan Zhu
*National Laboratory of Natural and Biomimetic Drugs, Beijing Medical University,
Beijing 100083, China*

Introduction

The extensive studies on steroids have established the mechanism of their actions. The peptide hormones are thought to bind on the cell membrane. With the advancement of hormone research, the relationships between steroid- and peptide-hormones were discovered. For instance, steroid hormones exert their rapid effects through bonding to specific membrane receptors[1]; some tissues in the body may be the target tissues for both steroid and peptide hormones. Many researches do demonstrated that steroid hormones increase the actions of peptide hormones[2] , furthermore, some receptors of peptide hormone (e.g. growth factors) are also present in the nucleus of the cells[3].

The interactions between steroid hormone and peptide hormone lay a foundation for the design and preparation of the compounds, called unclassical hormone here. With

Table 1 *Effect of injection of a and b on pain threshold in rats^a*

Drugs	% increase of pain threshold in rats at different times (min) (X±SD)				
	0	10	20	30	40
1	1.00±1.00	9.00±5.00	−1.00±4.00	−1.00±6.00	−3.00±8.00
2	1.84±3.11	21.15±7.08	7.00±8.49	21.45±8.49	14.34±12.84
3	68.27±33.83	35.92±12.18	55.00±18.21	57.85±17.59	22.89±23.70
4	72.21±19.63	83.61±25.87	106.35±21.0	94.46±21.37	43.98±12.69
5	114.30±16.80	89.10±18.00	88.50±17.30	80.80±21.10	60.80±14.10
6	34.30±4.70	115.00±13.20	70.30±9.80	106.20±13.90	97.20±15.20

1: Normal Saline; 2: Hydrocortisone; 3: H-Arg-Tyr-OH; 4: a; 5: H-Tyr-Arg-OH; 6: b.

^{a} Dose 0.24μ mol/10μl; 0 min point represents the data measured immediately after the injection of drug(i.c.v.); after 60 min the analgesic effect for a was 92±8.39%, for b was 51.6±4.6%; n=6; hydrocortisone group compared with normal saline group, except 10 min point ($P<0.05$) and 30 min point($P<0.01$) all of the other time points have no statistical significant; H-Arg-Tyr-OH group compared with hydrocortisone group, except 40 min point all of the time points have $P<0.01$; a group compared with H-Arg-Tyr-OH group, except 0 min all of the time points have $P<0.01$.

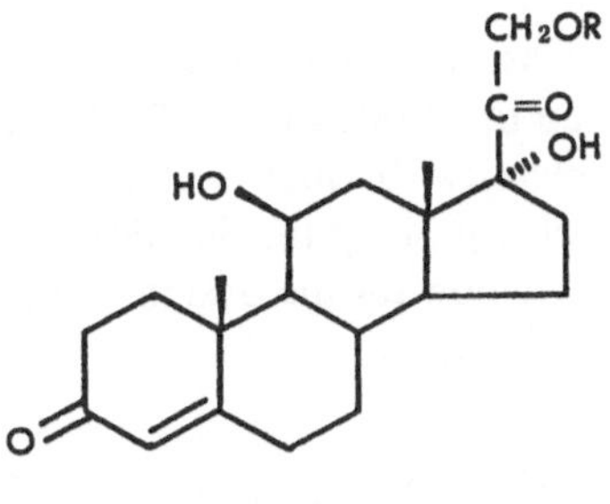

a: R=COCH₂CH₂COArgTyrOH
b: R=COCH₂CH₂COTyrArgOH

Fig. 1. Structures of linkers a and b.

these new compounds, we hope the bioactivity and duration of the peptide hormone will be enhanced and prolonged.

Results and Discussion

The dipeptides H-Arg-Tyr-OH, H-Tyr-Arg-OH which have analgesic effect were prepared with the conventional method. These peptides reacted with hydrocortisone, the corresponding linker a and b were obtained (Fig. 1). The bioassay for them was made on the analgesia model of rats to observe the increase of pain threshold in rats and the related data were listed in Tables 1 and 2.

The results demonstrate that the a is obviously better than H-Arg-Tyr-OH, after 20-30 min the analgesic effect is optimum and after 60 min of injection there was still definite analgesic effect, the duration for unclassical hormone is significantly longer than that for peptide hormone. Similarly the b is also better than H-Try-Arg-OH. After 10-40 min of

Table 2 *Dose-dependent response^a*

	% Increase of pain threshold in rats at different doses(X±SD)		
Dose	0.06μmol	0.12μmol	0.24μmol
H-Arg-Tyr-OH	8.33±8.03	25.82±5.31	36.90±8.53
H-Try-Arg-OH	12.8±1.60	32.4±3.70	72.1±3.30
a	28.57±2.92	48.64±3.01	64.64±8.89
b	28.1±2.19	40.6±2.60	82.1±4.00

[a] The data of 60 min point taken as the mean increase of pain threshold in rats at each dose; n=6; in H-Arg-Tyr-OH group 0.12μmol point compared with 0.06μmol point $P<0.01$; 0.24μmol point compared with 0.12μmol point $P<0.05$; in a group, all of the point compared with each other $P<0.01$. H-Tyr-Arg-OH group compared with hydrocortisone group $P<0.01$ for all time points, b group compared with H-Tyr-Arg-OH group $P<0.05$ for 10 min and 30 min point, $P<0.01$ for 40 min point; in H-Try-Arg-OH group, 0.24μmol point compared with 0.12μmol point $P<0.01$; in b group, 0.24μmol point compared with 0.12μmol point, $P<0.01$; b group compared with H-Tyr-Arg-OH $p<0.01$.

injection the analgesic effect was maintained, even though after 60 min the increase of pain threshold in rats is still up to 51.60±4.60%. The data mentioned above confirmed that the linkers, so-called unclassical hormones consisted of steroids and peptides not only enhance the activity but also prolong the duration of the peptides. The results first reported here lay a sound foundation for the design and preparation of new hormones.

Acknowledgement

Authors thank The National Education Committee for financial support.

References

1.   Schumacher, M., TINS, 13(1990)359.
2.   Malbon, C.C. and Hadcock, J.R., Biochem. Biophysio. Rec. Commun., 154(1988)676.
3.   Peng, Y., Prog. Biochem. Biophys. 18(1991)182.

Syntheses of iberiotoxin and its related peptides

Yun-Neng Chen, Hisaya Kuroda, Yukako Itahara, Takushi X. Watanabe, Terutoshi Kimura and Shumpei Sakakibara

Peptide Institute Inc., Protein Research Foundation, Minoh-shi, Osaka 562, Japan

Introduction

Iberiotoxin (IbTX), a minor component of the venom of Israel scorpion *Buthus tamulas*, is a 37 amino acid peptide having three intramolecular disulfide bonds[1]. IbTX exhibits 68% sequence identity with charypbotoxin (ChTX), another scorpion-derived inhibitor of the maxi-K channel (high conductance Ca^{2+}- activated K^+ channels) [2], as shown below.

```
IbTX pEFTDVDCSVSKECWSVCKDLFGVDRGKCMGKKCRCYQ-OH
ChTX pEFTNVSCTTSKECWSVCQRLHNTSRGKCMNKKCRCYS-OH
```

Despite the high degree of sequence homology, the mode of inhibition by IbTX differs greatly from that of ChTX. IbTX is a highly selective inhibitor of the maxi-K channel and does not block other types of voltage-dependent K^+ channels that are sensitive to inhibition by ChTX, suggesting that these peptides might bind at different sites on the channels and modulate channel activity by different mechanisms. ChTX has a remarkably high content of basic amino acid residues between positions 21-28, suggesting that this region might be a potential site of interaciton with the channels [2]. The major distinct region between IbTX and ChTX sequences is at position 21-24 (underlined above). In the present study, we report the syntheses of IbTX and its N-terminal biotinylated peptide (Biot-IbTX) as well as the hybrid peptide of ChTX and IbTX, in which the sequence between 21-24 in ChTX was replaced by that of IbTX, in order to elucidate the structure-activity relationships.

Results and Discussion

The synthesis was carried out by the solution procedure applying our maximum protection strategy [3]. Fully protected IbTX was synthesized by the procedure reported for ChTX [4]. The whole molecule was divided into four segments as shown in Fig. 1. All of the side-chain functional groups were protected by benzyl-type protecting groups except for Cys and Trp residues which were protected by Acm and For groups, respectively. Every segment was synthesized in the form of Boc-peptide-OPac except for the C-terminal segment which was protected by benzyl ester instead of Pac ester. The Pac ester was removed by treatment with zinc powder in AcOH except in the case of Z-(1-10)-OPac, for which a solubility problem was encountered. Therefore, a mixture of dichloromethane (DCM) and trifluoroethanol (TFE) in a ratio of 3:1 (v/v) was used for the solvent as reported previously [5]. The peptide dissolved easily in this solvent

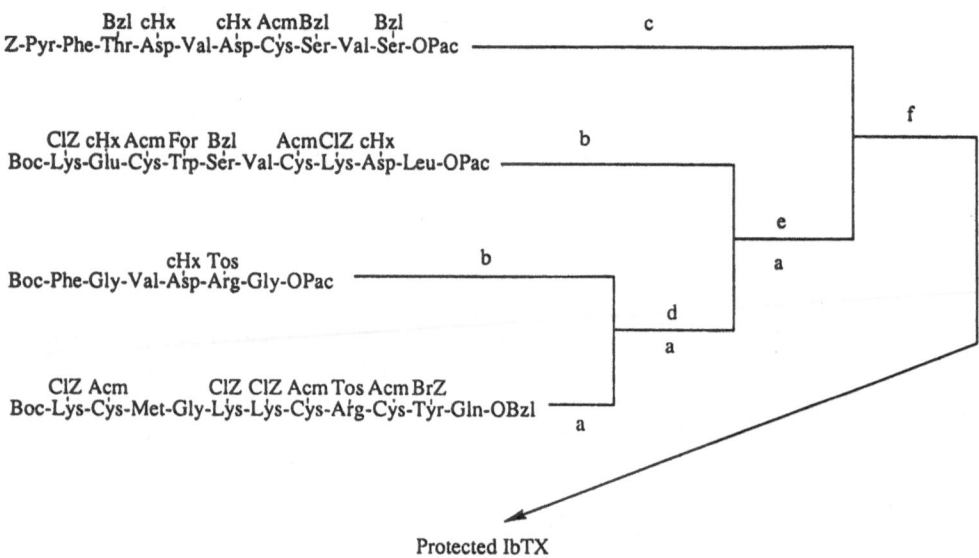

Fig. 1. Synthesis of protected IbTX. a: TFA, b: Zn/AcOH, c: Zn/DCM-TFE-AcOH, d:WSCI/HOBt, e: WSCI/
HOOBt, f: WSCI/HOOBt in CHCl₃-TFE (3:1).

Fig. 1. Synthesis of protected IbTX. a: TFA, b: Zn/AcOH, c: Zn/DCM-TFE-AcOH, d:WSCI/HOBt, e: WSCI/
HOOBt, f: WSCI/HOOBt in CHCl₃-TFE (3:1).

system, and removal of the Pac ester by zinc powder preceeded smoothly at 35°C within 1 hr after diluting with AcOH. Each segment thus obtained was coupled from the C-terminal segement using water-soluble carbodiimide (WSCI) in the presence of HOBt or HOOBt. The final coupling reaction of segment was achieved by the WSCI/HOOBt method using a mixture of chloroform and TFE (3:1,v/v) as solvents [5]. The other protected peptides were synthesized by the same procedure as above except for the synthesis of Biot-IbTX, wihich was done by coupling Biot-Gln-(2-10)-OH with NH₂-(11-37)-OBzl in the same solvent system. Each fully protected peptide thus obtained was treated with HF in the presence of anisole (9:1) at −5°C for 1 h to remove all of the protecting groups except for Acm and For groups. After evaporating HF and anisole, the peptide was treated again with HF in the presence of butanedithiol at −5°C for 30 min to remove the remaining For group (high/low HF reaction). The crude product (Fig. 2a) thus obtained was purified by CM-cellulose chromatography followed by RP-HPLC. The purified hexa-Acm peptide (Fig. 2b) was treated with Hg(OAc)₂ in 5% AcOH to remove the Acm groups. Hg ions were removed completely by adding β-mercaptoethanol followed by gel filtration on Sephadex G-25 and RP-HPLC, successively (Fig 2c). To fold and oxidize the peptide, the hexa sulfhydryl peptide was stirred in 0.1 M NH₄OAc (pH7.8) at a peptide concentration of 1×10^{-5}M in the presence of reduced and oxidized glutatione (peptide: GSH: GSSG=1: 60: 6) at room temperature for 2 days (Fig 2d). During the folding reaction, redox reagents were needed to accelearate the reaction. The principal product was isolated to homogeneity by RP-HPLC followed by ion-exchange HPLC. The deprotection and subsequent folding reaction of biotinylated peptide and hybrid peptide could be achieved under the same conditions as above. The HPLC profile

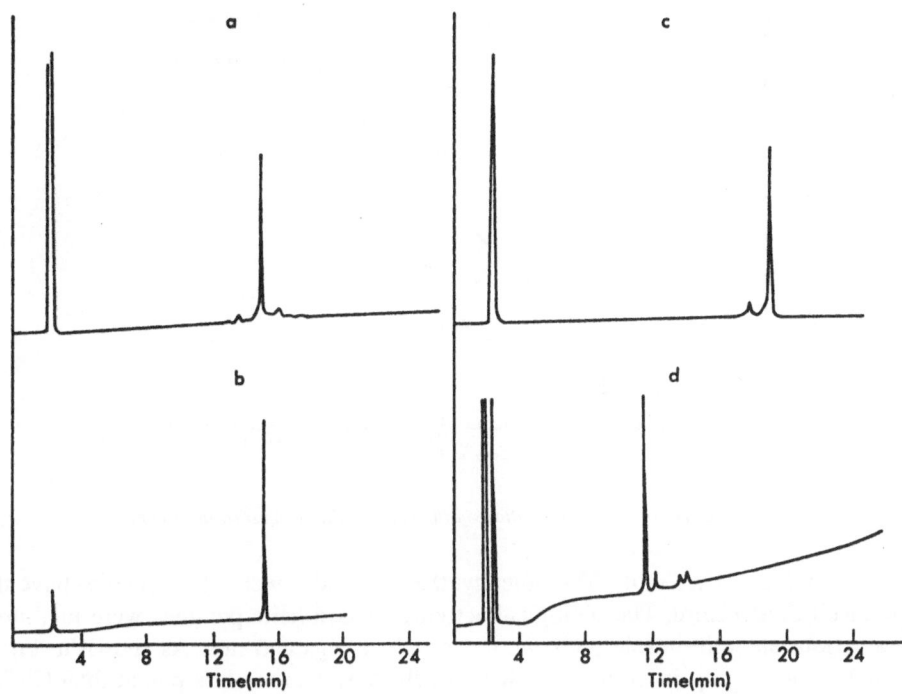

Fig. 2. HPLC profiles of IbTX. a: Crude product after HF treatment; b: Purified hexa Acm peptide; c: Hexa sulfhydryl peptide; d: After folding reaction. Column: YMC-Pak A-302(150×4.6 mm); Eluant: 10-60% MeCN in 0.1% TFA (25min); Flow rate: 1.0ml/min; Temperature: 50°C; Detection: 220nm.

of the mixture of purified IbTX, biotinylated IbTX, hybrid peptide and ChTX is shown in Fig.3; it indicated that they were completely separable and the retention time of each peptide was quite reasonable on RP-HPLC. Amino acid analysis after acid hydrolysis and molecular weight measurement by plasma desorption (PD) mass spectrometry of the synthetic peptides showed good agreement with the theoretical values. In order to determine the disulfide structure of synthetic IbTX, the purified product was treated with a mixture of trypsin and chymotrypsin. Each peak was isolated and the amino acid compositions were determined.

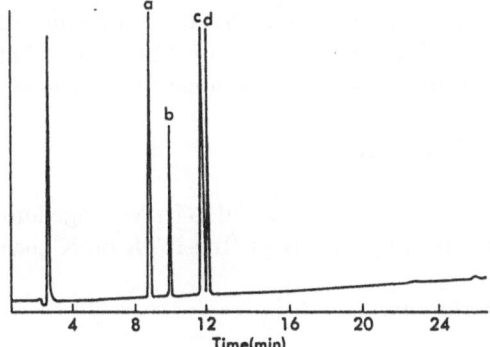

Fig. 3. HPLC profile of the mixture of IbTX, Biot-IbTX, hybrid peptide and ChTX. a: ChTX; b: hybrid peptide; c:IbTX; d:Biot-IbTX. Column: YMC-Pak A-302 (150×4.6 mm); Eluant:10-60% MeCN in 0.1% TFA (25min); Flow rate: 1.0 ml/min; Temperature: 50°C; Detection: 220nm.

From these experiments, the disulfide structure of syntheitc IbTX was confirmed to be

271

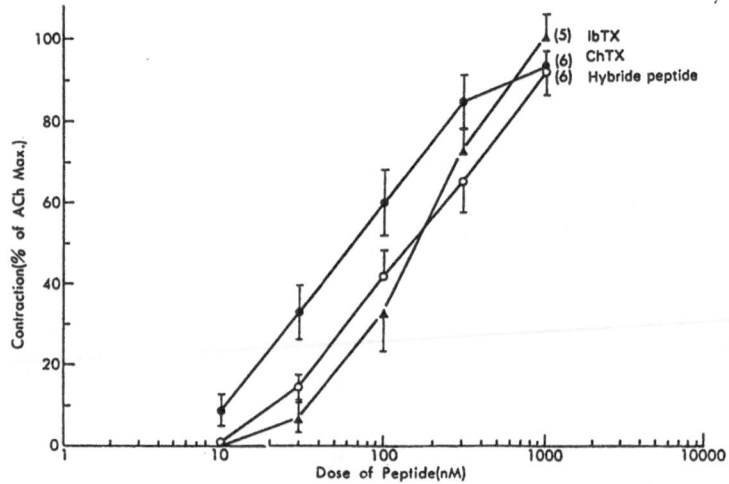

Fig. 4. *Effect of synthetic peptide on the contractility in guinea pig ileum.*

the same as that of ChTX [4]. The other synthetic peptides were also found to have the same disulfide strucutre. The biological activities of syntheitc peptides were measured using various smooth muscle preparations in guinea pigs and rats. As shown in Fig.4, the contractile activity in guinea pig ileum of IbTX was much less potent than ChTX, however, the hybrid peptide showed similar potencies to those of IbTX, although only the positions 21-24 in ChTX were replaced by those of IbTX. On the other hand, the potency of biot-IbTX was 1/100 of that of IbTX. This was further confirmed by the fact that only the higher concentration (1 μM) of Biot-IbTX was effective on the bindings to Ca^{2+}- activated K^+ channels in flog saccular hair cells (data not shown). From these results, we concluded that the N-terminus of IbTX is essential not only for biological activities *in vitro* but also for blocking K^+ channels, and the region of 21-24 in ChTX and IbTX might be important for K^+ channel binding selectivity.

Acknowlegement

The authors are grateful to Dr.M. Hagedorn of the University of Oregon for measuring the binding activity of Biot-IbTX on K^+channels.

References

1. Galvez, A., Gimenez-Gallego., G., Reuben, J.P., Roy-Contancin, L., Feigenbaum, P., Kaczorowski, G.J. and Garcia, M.L., J. Biol. Chem., 265(1990)11083.
2. Gimenez-Gallego, G., Navia, M.A., Reuben, J.P., Katz, G.M., Kaczorowski, G.J. and Garcia, M.L., Proc. Natl. Acad. Sci., USA, 85(1988)3329.
3. Kimura, T., Takai, M., Masui, Y., Morikawa, T. and Sakakibara, S., Biopolymers, 20(1981)1823.
4. Lambert, P., Kuroda, H., Chino, N., Watanabe, T.X., Kimura, T. and Sakakibara, S., Biochem. Biophys. Res. Commun., 170(1990)684.
5. Kuroda, H., Chen, Y-N., Kimura, T. and Sakakibara, S., Int. J. Peptide Protein Res., 40(1992)294.

Partially protected peptide thioesters as building blocks for protein synthesis

Saburo Aimoto and Hironobu Hojo
Institute for Protein Research, Osaka University, 3-2 Yamadaoka, Suita, Osaka 565, Japan

Introduction

Protein production by a chemical method is free from the restrictions which exist in preparation by recombinant DNA technology. In chemical preparation, the yield of a product is not affected by its physiological activity and the constituent of a protein is not limited to L-form proteinous amino acids. Synthetic proteins, which are site-specifically labeled by stable isotopes, consist of D-form amino acids or contain nonproteinous amino acids and so on, are expected to play very important roles in understanding the nature of protein and in promoting research in the field of protein engineering.

In most cases, however, proteins are prepared not by chemical methods but by recombinant DNA technology. Since both solid-phase and solution peptide synthesis have intrinsic methodological limitations, it is hard to adopt these processes routinely for a protein synthesis. Thus, a new method must be developed to realize rapid synthesis of a highly pure protein.

In 1991, we reported a method for protein synthesis in which partially protected peptide thioesters were used as building blocks. c-Myb(142-193) amide[1] and [Met[69]-methyl-d3]-HU-type DNA-binding protein [2] were synthesized by this method.

In this paper, we describe the thioester building block method for the synthesis of ^{2}H-, ^{13}C- and ^{15}N-labeled HU-type DNA-binding protein (HBs) of *Bacillus stearothermophilus*, a polypeptide containing 90 amino acids (Fig.1), as an example of the recent advances of this method.

Results and Discussion

We examined several types of thioester linker moieties connecting a resin and a peptide. Boc-Gly-SC(CH$_3$)$_2$CH$_2$CO-Nle-MBHA resin was used for peptide thioester preparation in this synthesis, since it gave a better yield than the resin containing the primary thioester moiety without a Nle or β-Ala spacer group.

An α-amino-protected peptide thioester was prepared by a solid-phase method employing Boc chemistry according to the procedure described previously [1]. Purified Troc-peptide thioesters were obtained after HF treatment followed by reversed-phase HPLC purification. The side-chain amino groups of these peptide thioesters were protected by Boc groups using Boc-ONSu. The partially protected peptide segments obtained were used for segment coupling. The yields are summarized in Fig. 2.

273

```
1                                    10                           ↓
Met-Asn-Lys-Thr-Glu-Leu-Ile-Asn-Ala-Val-Ala-Glu-Thr-Ser-Gly-
                20                                          30
Leu-Ser-Lys-Lys-Asp-Ala-Thr-Lys-Ala-Val-Asp-Ala-Val-Phe-Asp-
                                 ↓ 40
Ser-Ile-Thr-Glu-Ala-Leu-Arg-Lys-Gly-Asp-Lys-Val-Gln-Leu-Ile-
          *           50                         *           * ↓
Gly-Phe-Gly-Asn-Phe-Glu-Val-Arg-Glu-Arg-Ala-Ala-Arg-Lys-Gly-
    *                                *  70
Arg-Asn-Pro-Gln-Thr-Gly-Glu-Glu-Met-Glu-Ile-Pro-Ala-Ser-Lys-
                  *                                        90
Val-Pro-Ala-Phe-Lys-Pro-Gly-Lys-Ala-Leu-Lys-Asp-Ala-Val-Lys
```

Fig. 1. Amino acid sequence of HU-type DNA-binding protein of Bacillus stearothermophilus. *indicates the amino acids labeled with ^2H, ^{13}C or ^{15}N; $Phe^{47}(2-^{13}C)$, $Ala^{56}(1-^{13}C)$, $Gly^{60}(2-^{13}C)$, $Arg^{61}(^{15}N^w)$, $Met^{69}(methyl-d3)$, $Lys^{80}(^{15}N^\varepsilon)$. The arrows indicate the sites of segment coupling.

Yield

Boc-Met-Asn-Lys(Boc)-Thr-Glu-Leu-Ile-Asn-Ala-Val-Ala-Glu-Thr-Ser-Gly-
SC(CH₃)₂CH₂CO-Nle-NH₂

Boc-[Lys(Boc)³]-HBs(1-15)-SC(CH₃)₂CH₂CO-Nle-NH₂   (1)   22%

Troc-Leu-Ser-Lys(Boc)-Lys(Boc)-Asp-Ala-Thr-Lys(Boc)-Ala-Val-Asp-Ala-Val-Phe-Asp-
Ser-Ile-Thr-Glu-Ala-Leu-Arg-Lys(Boc)-Gly-SC(CH₃)₂CH₂CO-Nle-NH₂

Troc-[Lys(Boc)¹⁸,¹⁹,²³,³⁸]-HBs(16-39)-SC(CH₃)₂CH₂CO-Nel-NH₂   (2)   22%

Troc-Asp-Lys(Boc)-Val-Gln-Leu-Ile-Gly-Phe-Gly-Asn-Phe-Glu-Val-Arg-Glu-Arg-Ala-
Ala-Arg-Lys(Boc)-Gly-SC(CH₃)₂CH₂CO-Nle-NH₂

Troc-[Lys(Boc)⁴¹,⁵⁹]-HBs(40-60)-SC(CH₃)₂CH₂CO-Nle-NH₂   (3)   41%

Arg-Asn-Pro-Gln-Thr-Gly-Glu-Glu-Met-Glu-Ile-Pro-Ala-Ser-Lys(Boc)-Val-Pro-Ala-Phe-
Lys(Boc)-Pro-Gly-Lys(Boc)-Ala-Leu-Lys(Boc)-Asp-Ala-Val-Lys(Boc)

[Lys(Boc)⁷⁵,⁸⁰,⁸³,⁸⁶,⁹⁰]-HBs(61-90)   (4)   20%

Fig. 2. Partially protected peptide segments prepared for segment coupling.

Segment coupling was carried out according to the scheme shown in Fig. 3. As a typical example, thioester component 3 (15 μmol) and amine component 4 (15 μmol) were mixed in dimethyl sulfoxide (1 ml). To the reaction mixture, HONSu (174 μmol) and silver nitrate (51 µmol) were added. The coupling reaction completed within 1 day at room temperature in the dark. The progress of the reaction was easily monitored by reversed-phase HPLC after treatment of the reaction mixture with trifluoroacetic acid. Troc group was removed by zinc dust treatment in 80% aq. acetic acid under nitrogen. The reaction mixture was dialyzed against distilled water after the removal of zinc dust and freeze-dried to give a peptide containing peptide 5. According to the same procedure, peptide 2 and 1 were successively condensed to obtain peptide 7.

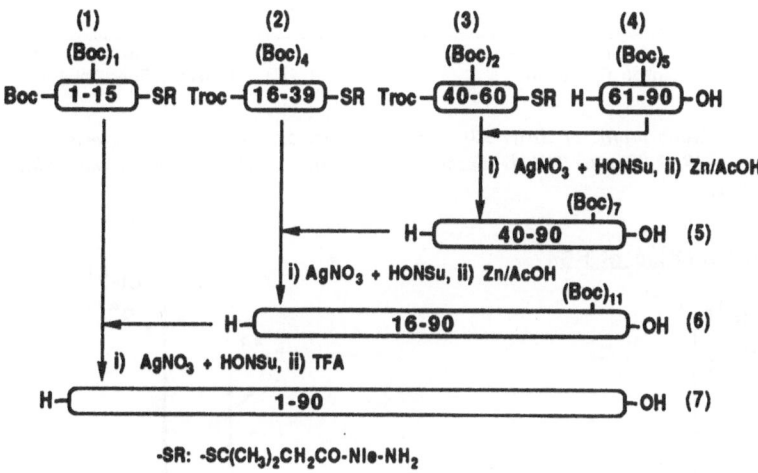

Fig. 3. Synthetic route of HBs(1-90).

The HPLC elution profile of crude HBs(1-90) is shown in Fig. 4. HBs(1-90) was purified by reversed-phase HPLC after trifluoroacetic acid treatment. Judging from the aromatic proton of phenylalanine residues, the ^{1}H NMR spectrum of purified HBs(1-90) in water solution showed that the protein was unfolded. However, HBs(1-90) assumed the native conformation, when dissolved in a 50 mM sodium phosphate buffer (pH 7.0). Additional purification was performed by ion-exchange chromatography using Pharmacia Mono S HR 5/5. The product corresponding to the main peak in Fig. 5 was isolated, desalted by HPLC using a C_4 column and freeze-dried to give 14 mg of the final product in a yield of 7.6% based on peptide 4.

The product thus obtained showed the same ^{1}H NMR spectrum as that of the native HBs except methyl-d3 of Met[69]. Also ^{13}C and ^{15}N signals were clearly observed by ^{13}C and ^{15}N NMR spectroscopy using 4 mg of this sample as shown in Fig. 6. This multi-labeled HBs(1-90) will give us information about the mode of interaction between HBs and DNA.

Tertiary alkyl type thioester gave good yield on segment preparation and converted rapidly to an active ester by silver ions in the presence of HONSu. Removal of the Troc group took a shorter time than that of iNoc group, but was accompanied by side reactions. On the other hand, iNoc group did not suffer any side reactions, but took a

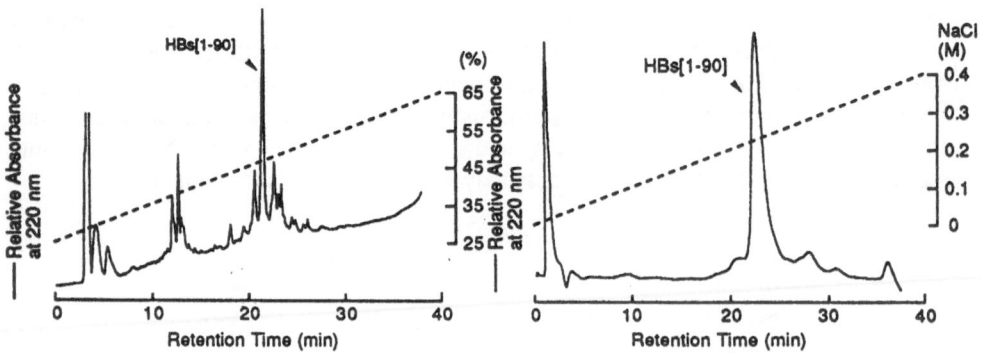

Fig. 4. HPLC elution profile of HBs(1-90) of a crude product of HBs(1-90) after TFA treatment of reaction mixture.

Fig. 5. Ion-exchange chromatogram of HPLC-purified HBs(1-90) by Pharmacia Mono S HR 5/5.

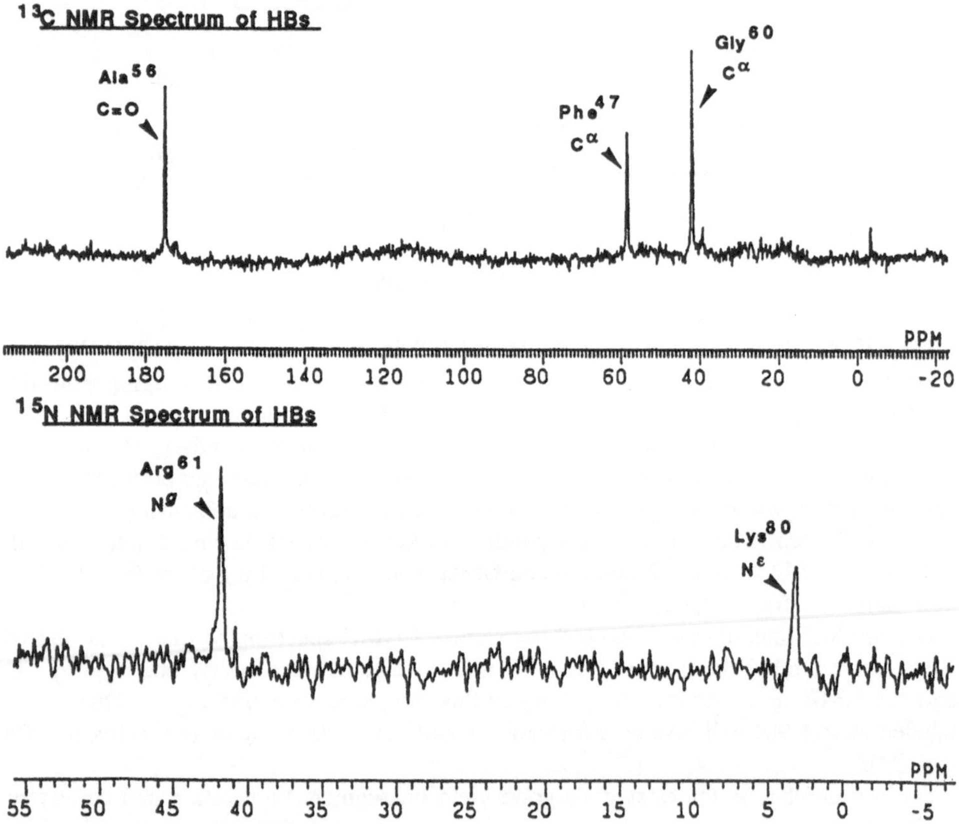

^{13}C NMR Spectrum of HBs

Ala^{56}
C=O

Phe^{47}
C$^\alpha$

Gly^{60}
C$^\alpha$

PPM

200 180 160 140 120 100 80 60 40 20 0 -20

^{15}N NMR Spectrum of HBs

Arg^{61}
N^g

Lys^{80}
N$^\varepsilon$

PPM

55 50 45 40 35 30 25 20 15 10 5 0 -5

Fig. 6. ^{13}C and ^{15}N signals obtained by ^{13}C and ^{15}N spectroscopy, respectively.

longer time to be removed. More consideration is required concerning the selection of the terminal amino protecting group.

To sum up, in this HBs(1-90) synthesis, the final product was eluted as a separated peak on reversed-phase HPLC, which suggests that a polypeptide, with neither standard material nor a specific feature such as enzymatic activity, can be synthesized using this method.

Acknowledgement

This research was partly supported by Grants-in-Aid for Scientific Research Nos. 62540407, 02263101 and 03740287 from the Ministry of Education, Science and Culture, Japan.

References

1.   Hojo, H. and Aimoto, S., Bull. Chem. Soc. Jpn., 6(1991)111.
2.   Hojo, H. and Aimoto, S., Bull. Chem. Soc. Jpn., 65(1992)3055.

Kinetic investigation of the chromogenic peptide substrates for chymotrypsin

Ruo-heng Zhang, Xue Ge, Xiao-jie Xu and You-qi Tang

Department of Chemistry, Peking University, Beijing 100871, China

Introduction

As one of important proteinase, activity assay of chymotrypsin is needed in its purification and application. The substrate containing para-nitroaniline was reported to improve the stability and sensitivity. In this communication, a series of chromogenic peptide amide substrate containing para-nitroaniline was synthesized and kinetic stydy on the specificity influenced by amino acid sequence was made to elucidate the best chymotrypsin substrate.

Results and Discussion

Phenylalanine is acrylated with phosphorus pentachloride and then, reacts with para-nitroaniline in dioxane/pyridine to give Phe-pNA in the yield of 38%. The peptide substrates were synthesized by the DCC/HOBt solution method, recrystallized with ethyl acetate/petroleum ether and analyzed by HPLC. The physical and chemical data are summarized in Table 1.

Table 1 *Physical and chemical data of synthetic peptides*

Peptides	m.p (°C)	Rf	$[\alpha]_D^{20}$	C	H	N	C	H	N
				Calcd %			Found %		
Boc-P-A-F-pNA	174–6	0.86	−10.1	60.75	6.31	12.65	60.99	6.92	12.50
Boc-P-L-F-pNA	202–4	0.78	−14.6	62.51	6.94	11.75	62.83	7.18	11.41
Boc-P-A-L-pNA	158–60	0.69	−36.5	57.70	7.18	13.48	57.57	7.51	12.85
Boc-A-A-F-pNA	202–4	0.86	17.1	59.19	6.31	13.37	58.37	6.36	12.89
Boc-F-N-F-pNA	>240	0.81	−55.7	61.29	5.92	12.99	60.91	5.93	12.73

λ max of substrates is about 325 nm and that of pNA is 385 nm. After chymotrypsin was added, the λ max of substrates changes from 325 to 385 nm, It is shown by UV spectrum and HPLC that the peptide amide is fully hydrolyzed by chymotrypsin to release chromophore pNA. The kcat/Km at different experimental conditions, λmeasure=410 nm or 385 nm and T=26 °C or 37 °C, are almost the same. The kinetic data of substrate are listed in Table 2. The substrate P-A-L-pNA can not be hydrolyzed

Table 2 *Kinetic data of substrate hydrolyzed by chymotrypsin*

Substrate	Km (mM)	kcat (S^{-1})	kcat/Km(S^{-1}M^{-1})
P-A-F-pNA	0.340	0.216	635.3
P-L-F-pNA	0.549	0.434	790.5
A-A-F-pNA	0.386	0.830	2150.0
F-N-F-pNA	1.520	0.648	426.4
P-A-L-pNA	–	–	–

by chymotrypsin. Though the other four substrates can be used in the activity measurement of chymotrypsin, Ala-Ala-Phe-pNA is best chromogenic substate.

Acknowledgement

We thank the National Natural Science Foundation of China for financial support.

References

1. Stakey, P.M. and Barrett, A.J., Biochem. J., 155(1976)273.
2. Zerner, B., Bond, R.P.M. and Bender, M.L., J. Am. Chem. Soc., 86(1964)3674.
3. Erlanger, B.F., Edel, F. and Cooper, A.E., Arch. Biochem. Biophys., 115(1966)206.
4. Bauster, J.H. and Wolfbeis, O.S., Analytical Biochemistry, 171(1988)393.

Synthesis and kinetics of fluorogenic substrates containing 7-amino-4-methyl-coumarin

Ruo-heng Zhang[a], Shao-juan Jia[a], Xiao-jie Xu[a], You-qi Tang[a], Guang-wei Yuang[b] and Yi-sheng Ni[b]

[a]Department of Chemistry and [b]Department of Biology, Peking University, Beijing 100871, China

Introduction

The fluorogenic substrates have being used for the highly sensitive determination of chymotrypsin. The sensitivity is due to the fact that the leaving group, some fluorogenic amine, is highly fluororesent. In this communication, we wish to report the design, synthesis and kinetic study of a series peptide amide substrates containing 7-amino-4-methylcoumarin(AMC). By calculating the enzyme's kcat and Km values for the substrates, the influence of amino acid sequence on the specificity of substrates was studied. Therefore, two best fluorogenic substrates for chymotrypsin, Suc-Ala-Ala-Phe-MC and Suc-Ala-Pro-Phe-MC, were found.

Results and Discussion

Amino acids at P_2, P_3 position of peptide substrates were choosed to be alanine, proline, glycine and valine. These substrates were synthesized by mixed anhydride and purified by recrystallization with ethyl acetate/petroleum ether. Their physical and chemical data are summarized in Table 1. To increase the solubility, Suc-P_3-P_2-P_1-MC instead of Boc-P_3-P_2-P_1-MC were used in the kinetic measurement.

These substrates when digested by chymotrypsin release AMC. Though both AMC and the peptide amide substrates are fluoresent, their excitation and emission maximum are different, the λex and λem of AMC are 345 nm and 445 nm. The enzymatic release of

Table 1 *Analytical data of the substrates*

Substrate	m.p. (°C)	Rf[a]	$[\alpha]^{20}_D$	$[m+1]$[b]	C% Calcd%	H%	N%	C% Found%	H%	N%
Boc-A-A-F-MC	222-4	0.35	15.5	565	63.82	6.43	9.92	63.99	6.45	9.85
Boc-A-P-F-MC	174-6	0.29	−36.3	591	65.02	6.48	9.48	64.86	6.47	9.32
Boc-A-G-F-MC	132-6	0.33	20.1	551	63.26	6.22	10.17	62.98	6.23	9.69
Boc-G-A-F-MC	137-43	0.33	31.7	551	63.26	6.22	10.17	61.59	6.28	9.71
Boc-P-V-F-MC	175-8	0.30	− 6.4	619	66.00	6.84	9.06	65.54	6.85	8.84
Boc-G-P-F-MC	191-4	0.25	− 0.4	577	64.57	6.29	9.72	63.84	6.46	9.24

a: $CH_2Cl_2:CH_3OH:HOAc =90:10:1$, b: Suc-P_3-P_2-P_1-MC.

Table 2 *Kinetic constants of substrates hydrolyzed by chymotrypsin*

Substrate	Km(μM)	kcat(s^{-1})	kcat/Km(S^{-1},M^{-1})
Suc-Ala-Ala-Phe-MC	459	10.80	2.35×10^{-4}
Suc-Ala-Pro-Phe-MC	197	4.27	2.17×10^{-4}
Suc-Ala-Gly-Phe-MC	395	0.821	2.08×10^{-3}
Suc-Gly-Ala-Phe-MC	1409	2.21	1.57×10^{-3}
Suc-Pro-Val-Phe-MC	1134	0.132	116
Suc-Gly-Pro-Phe-MC	337	3.05	9.04×10^{-3}

the fluorophore AMC was measuered at λex=380 nm and λem=460 nm to decrease the interference of the peptide amide substrates. Dealing with the data by Michaelis and Menten Equation, the relation of AMC concentration and its corresponding fluorescence intensity was obtained by making a standard curve under the same experimental condition. The kinetic data of the substrates are listed in Table 2.

From the above results, we conclude that the substrates with alanine or proline at P$_2$ position are more specific to chymotrypsin than those with valine or glycine at the same position.

Acknowledgement

We thank the National Natural Science Foundation of China for financial support.

Session VIII
Synthetic methodologies

Chairs: Yun-hua Ye
Peking University
Beijing, China

and

Yoshiaki Kiso
Kyoto Pharmaceutical University
Yamashina-ku, Kyoto, Japan

Improved synthesis of N-methyl arginine, N-methyl lysine and the ψCH_2NH surrogates for solid phase peptide synthesis

James P. Tam, Jane C. Spetzler and C. Rao
Department of Microbiology and Immunology, Vanderbilt University,
Nashville, TN 37232, U.S.A.

Introduction

N-Methylated amide and reduced peptide bond (ψCH_2NH) are useful modifications in the design of peptide analogs to provide backbone conformation restrictions and increased stability towards peptidase. These modifications have been successfully used in the design of metabolically stable hormone agonists and antagonists as well as protease inhibitors. Although the methodology to their synthesis using N-methyl amino acids [1] or amino aldehyde/NaCNB$_3$ [2] is well established for most amino acids, their synthesis for lysine and arginine with these modifications are complicated. In this paper, we describe an improved synthesis of N-methyl lysine (MeLys) and N-methyl arginine (MeArg) using a common synthetic route and Arg-containing peptides containing ψCH_2NH in solid phase peptide synthesis.

Results and Discussion

To study the structure-functions of a tetrapeptide Arg-Ser-Arg-Lys, we systematically modified each amide bond in two series of seven analogs either by N-methylation or with ψCH_2NH. These analogs would be conveniently accomplished by the solid phase approach using the Boc-benzyl chemistry. To obtain the N-methylated amide, we required MeArg and MeLys as synthons. Our strategy was to develop a common route for their synthesis because MeArg could be obtained by guanidination from N-methyl ornithine (MeOrn) [3]. To simplify our synthetic scheme, we planned to convert MeOrn to MeArg on the solid support during the solid phase synthesis. For this purpose, MeLys or MeOrn was derivatized with a protecting group scheme compatible with the Boc-benzyl chemistry and suitable for the guanidination of MeOrn on the solid support. The choice of such a protecting group scheme for the side chain amine of Lys or Orn to fit the dual purpose is somewhat limited because to methylate selectively the N^{α}-amine of Lys or Orn, its side chain amine would have to be diprotected. The phthalyl (Pht) protecting group removable by hydrazinolysis which would not affect the stability of the side chain benzyl protecting groups appeared to be a reasonable choice [4]. Boc-Lys(Pht) and Boc-Orn(Pht) were prepared in excellent yields from their respective amino acids (Fig. 1). Initial attempts to N-methylate directly Boc-Lys(Pht) or Boc-Orn(Pht) by NaH/MeI were unsuccessful due to the unusual base lability of the Pht group towards NaH. Similar attempts using Ag$_2$O/MeI also failed because NaOH or LiOH was required for the subsequent saponification of the ester resulted from the Ag$_2$O mediated reaction. However, Ag$_2$O/MeI did not affect the integrity of the phthalyl group during the N-methylation and if the use of the base saponification could be avoided, it was still the reagent of choice. The use of basic saponification was removed in our scheme by adding two extra steps, i.e. protecting the carboxylic acid with the benzyl ester prior to the methylation reaction and then deprotected the benzyl ester by hydrogenation. The successful synthesis of Boc-MeLys(Pht) and Boc-MeOrn(Pht) is shown in Fig. 1.

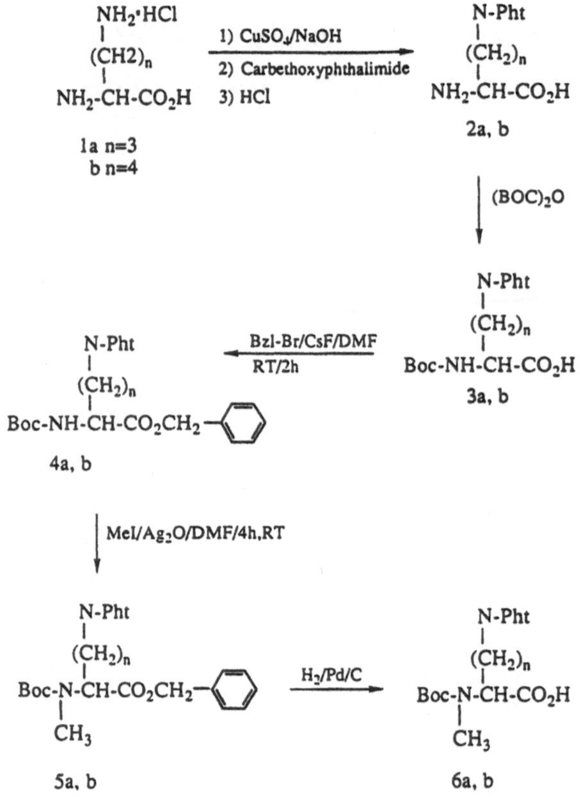

The use of Boc-MeLys(Pht) in the synthesis of four N-methylated tetrapeptide analog series proceeded smoothly and

Fig.1. Synthesis of Boc-MeOrn(Pht)OH and Boc-MeLys(Pht)-OH.

the Pht protecting group was removed by hydrazinolysis prior to the HF cleavage. The conversion of MeOrn(Pht) to MeArg [5] was mediated on a solid support during the synthesis as shown on Fig. 2. The nitroguanidine reagent, N^{im}-nitro-S-methyliso-thiourea, was superior than S-methylisothiourea and was conveniently synthesized by nitrating the S-methylisothiourea using a mixture of 70% HNO$_3$ and 95% H$_2$SO$_4$ (1:3, v/v) for 10 min at $-10°C$. The conversion yield of MeOrn to MeArg(NO$_2$) was about 97% when the guanidination procedure was repeated two times on the solid support.

In the second series, the ψCH$_2$NH was prepared by reductive amination of Boc-amino acid aldehyde by NaCNBH$_3$ on the solid support according to Coy and his coworkers [2]. While no particularly serious problem was encountered for the reduction of the O,N-dimethylhydroxamates of Boc-Lys(ClZ) and Boc-Ser(Bzl) by LiAlH$_4$ to their corresponding aldehydes, great difficulties were encountered with the Boc-Arg(Tos) derivative whose aldehyde was unstable to the reductive procedure. Substituting the Boc-Arg(NO$_2$) derivative or the milder reducing agent DIBAL gave some improvement. In addition, we found that the ψCH$_2$NH bond was prone to branching and protection by carbobenzoxyl chloride was necessary after the reductive amination step and prior to further elongation on the solid support. All peptide analogs were purified by C$_{18}$ RP HPLC and their integrity confirmed by several criteria including high resolution mass

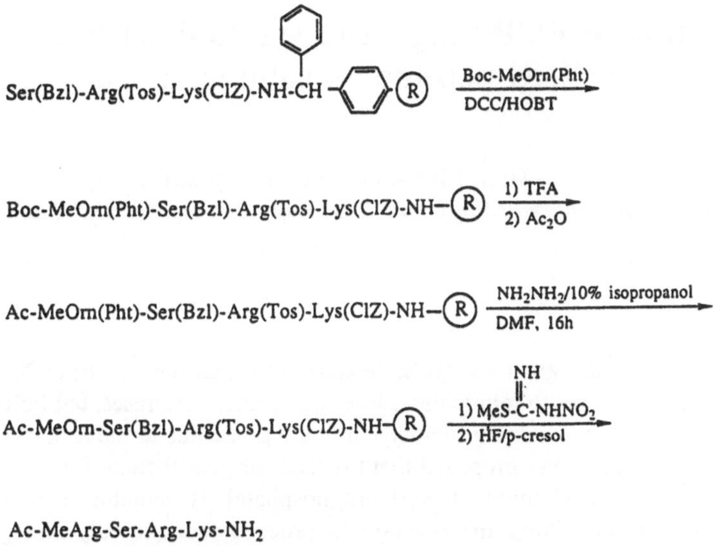

Ser(Bzl)-Arg(Tos)-Lys(ClZ)-NH-CH⟨⟩(R) $\xrightarrow[\text{DCC/HOBT}]{\text{Boc-MeOrn(Pht)}}$

Boc-MeOrn(Pht)-Ser(Bzl)-Arg(Tos)-Lys(ClZ)-NH—(R) $\xrightarrow[\text{2) Ac}_2\text{O}]{\text{1) TFA}}$

Ac-MeOrn(Pht)-Ser(Bzl)-Arg(Tos)-Lys(ClZ)-NH—(R) $\xrightarrow[\text{DMF, 16h}]{\text{NH}_2\text{NH}_2/10\% \text{ isopropanol}}$

Ac-MeOrn-Ser(Bzl)-Arg(Tos)-Lys(ClZ)-NH—(R) $\xrightarrow[\text{2) HF/p-cresol}]{\text{1) MeS-C-NHNO}_2}$

$$\overset{\text{NH}}{\underset{\|}{}}$$

Ac-MeArg-Ser-Arg-Lys-NH$_2$

Fig. 2. Synthesis of a tetrapeptide Ac-MeArg-Ser-Arg-Lys-NH$_2$ using BocMeOrn(Pht) as a synthon.

spectrometry.

In summary, problems encountered for the synthesis of the tetrapeptide analogs containing MeArg, MeLys and the ψCH$_2$NH surrogates could be improved by using suitable intermediate synthon such as Boc-MeOrn(Pht) for MeArg and the precautionary step in protecting the ψCH$_2$NH during the solid phase synthesis.

Acknowledgement

This work was in part supported by a grant from the U.S. PHS CA 36544.

References

1. Olsen, R.K., J. Org. Chem., 35(1970)1912.
2. Coy, D.H., Hocart, S.J. and Sasaki, Y., Tetrahedron, 44(1988)835.
3. Bodanszky, M., Ondetti, M.A., Birkhimer, C.A. and Thomas, P.L., J. Am. Chem. Soc., 86(1964)4452.
4. Ragnarsson, U. and Grehn, L., Acc. Chem. Res., 24(1991)285.
5. Tian, Z. and Roeske, R.W., Int. J. Peptide Protein Res., 37(1991)425.

Application of BOP reagent in synthesis of head-to-tail cyclized peptide on solid supports

Zong-jin Han, Zhen-kai Ding and Qi-kai Zhang

Institute of Pharmacology and Toxicology, Beijing 100850, China

Introduction

H.H. Seltzman [1] designed six cyclic hexapeptides that derived from X-ray data of numerous serine esterases to mimic the active site of serine esterases, but failed to obtain them using a variety of activating methods in liquid phase due to the lack of cyclization. In recent reports[2,3], it was proposed that the BOP reagent [benzotriazo1-1-yl-oxy-tris-(dimethyl-amino) phosphonium hexafluorophosphate] is suitable for solid phase cyclization because it allows the reaction to proceed more rapidly and gives purer products than DCC/HOBT method. In all these syntheses, the solid phase procedure was used with ring closure of the peptides still attached to the solid sopports. These observations prompted us to choose one of the six cyclic hexapeptides designed by Seltzman, cyclo-[His-Gly-Asp-Ser-Gly-Asp], to serve as a model peptide to evaluate the use of the BOP reagent for head-to-tail solid phase cyclization.

Results and Discussion

The strategy used for the synthesis of the cyclic hexapeptide is outlined in Fig.1. A classical N-Boc/Benzyl protection scheme was used except the first Asp. The Asp was combined with Mrrifield resin as N-t-butyloxycarbonyl-O^{α}-9-fluorenylmethyl-Asp (Boc-Asp-OFM) since the FM group remains stable during acidic treatment for removal of the Boc group. Excellent results of preparing Boc-Asp-O^{α}FM were obtained using a procedure of Bolin [4]: the commercially available Boc-Asp(OBzl)-OH was esterified with DDC and 9-fluorenylmethaol in methylene chloride using 4-dimethylaminopyridine (DMAP) as a catalyst followed by hydrogenation in MeOH with 10% Pd/C catalyst. Our experimental results revealed that the O^{α}FM group was stable in 0.8% DIEA/CH_2Cl_2 for 24 h, trace amounts of deprotection were observed after 24 h in 1% DIEA/CH_2Cl_2. The loading reaction of Boc-Asp-OFM Cs-salts to Merrifield resin proceeded with more difficult than that of Boc-Asp(OBzl)-OH, but completed with high yield and without noticeable loss of FM group at 40°C for 20 h by the presence of 1 equiv. NaI and 1 equiv. 18-Crown-6 in DMA solution. The following five protected amino acids were incorporated stepwise by DCC/HOBT method. The cyclization was studied using DCC and BOP coupling reagents after the N-terminal Boc and C-terminal FM group being removed by 50% TFA/CH_2Cl_2 and 20% piperidine/DMF respectively. The cyclization with 6 equive. DCC proceeded sluggishly and was less than 20% complete in 5 days as measured by the quantitative ninhydrine method. However, the cyclization with 3 equiv. BOP in DMF containing 6 equiv. DIEA was complete in 6 h as determined by

Table 1 *Head-to-tail cyclization of His-Gly-Ser-Asp-Gly-Asp*

Activation	Time	% Cyclization	
		Ninhydrin	HPLC
DCC[6 equiv]	2 days	9%	n.d.
DMF, 25°C	3 days	15%	n.d.
	5 days	19%	n.d.
BOP [3 equiv]	2h	84%	n.d.
DIEA [6 equiv]	4h	96%	95%
DMF, 25°C	6h	99.5%	99%

quantitative ninhydrin analysis (Table1). This result was also confirmed by analytical HPLC of the crude products following HF cleavage of the peptide-resin. The head-to-tail cyclo[His-Gly-Asp-Ser-Gly-Asp] was purified by semi-preparative RP-HPLC and the structure confirmed by amino acid analysis and FAB mass spectrometry. Based on our experimental results, it was concluded that the BOP reagent really resulted in more rapid cyclization and is the reagent of choice for solid phase head-to-tail cyclization reaction.

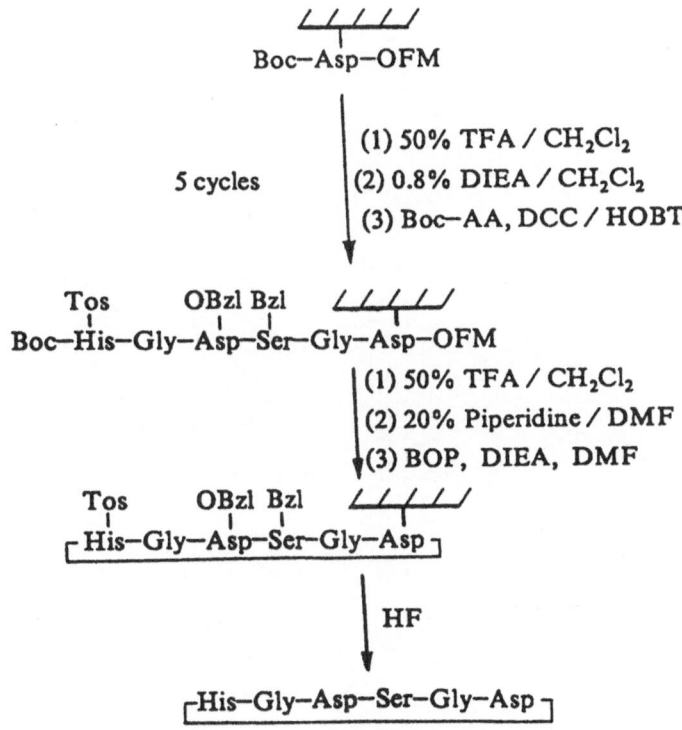

Fig. 1. *The synthetic scheme of cyclo[His-Gly-Asp-Ser-Gly-Asp].*

References

1. Seltzman, H.H., AD-A204 765.
2. Felix, A.M., Wang, C-T., Heimer, E.P. and Fournier, A., Int. J. Peptide Protein Res. 31(1988)231, 32(1988)441.
3. Plaue, S., Int. J. Peptide Protein Res., 35(1990)510.
4. Bolin, D.R., Wang, C-T. and Felix, A.M., Organic Preparations and Procedures INT., 21(1989)67.

Urethane n-carboxy anhydrides (UNCA's) and "allyl" side-chain protection for efficient peptide synthesis

A. Loffet[a], H.X. Zhang[a], P.A. Swain[b], N.J. Krotzer[c], W.D. Fuller[c] and M. Goodman[b]

[a]Propeptide BP 12, F-91710 Vert le Petit, France
[b]Department of Chemistry, UCSD, La Jolla, CA 92093, U.S.A.
[c]BioResearch 11189 Sorrento Valley Rd, San Diego, CA 92121, U.S.A.

Introduction

Urethane N-carboxy-anhydrides (UNCA'S) have been introduced by W.Fuller and M.Goodman [1] as powerful new acylating reagents for use in peptide synthesis. The synthetic utility of UNCAs has been demonstrated in their high yielding reactions with hindered α-amino-acids.

Allyl type side-chain protecting groups have been proposed by us [2] and D.Hudson [3] for trifunctional amino-acids. These may serve to prevent undesirable side reactions which arise from carbocation formation during the final deprotection of peptides.

The combination of UNCA's and allyl-type protecting groups allows new strategies of peptide synthesis. We now describe herein different approaches to the synthesis of LH-RH derivatives.

Results and Discussion

Synthesis of [D-Trp]⁶-LH-RH by continuous flow solid-phase methodology

The solid-phase method employed Polyhipe PA500 resin and was carried out with sequential coupling of 3 equivalents of Fmoc protected amino-acid derivatives in DMF and deprotection using 20% piperidine/DMF. The peptide was cleaved from the resin with NH₃/CH₃OH and the side-chains deprotected with TFA/thioanisole/EDT/anisole. After purification by HPLC the pure material was obtained in 66% yield. Analytical HPLC, capillary electrophoresis(CZE) and 500 MHz 1H-NMR were comparable with an authentic sample of the peptide.

Synthesis of [D-Trp]⁶-LH-RH in solution

The same target molecule was synthesized stepwise in solution using a 3 equivalent excess of the UNCA's with 0.1-5 equivalents of NMM in DMF. The synthesis is described in scheme 2. The yield of crude product was 63% and that of purified material was 58%. Again the purified peptide was comparable to that of the authentic sample by HPLC, CZE and NMR.

Synthesis of (4-10)-LH-RH and LH-RH

The peptides were assembled on a MBHA resin according to Scheme 3.

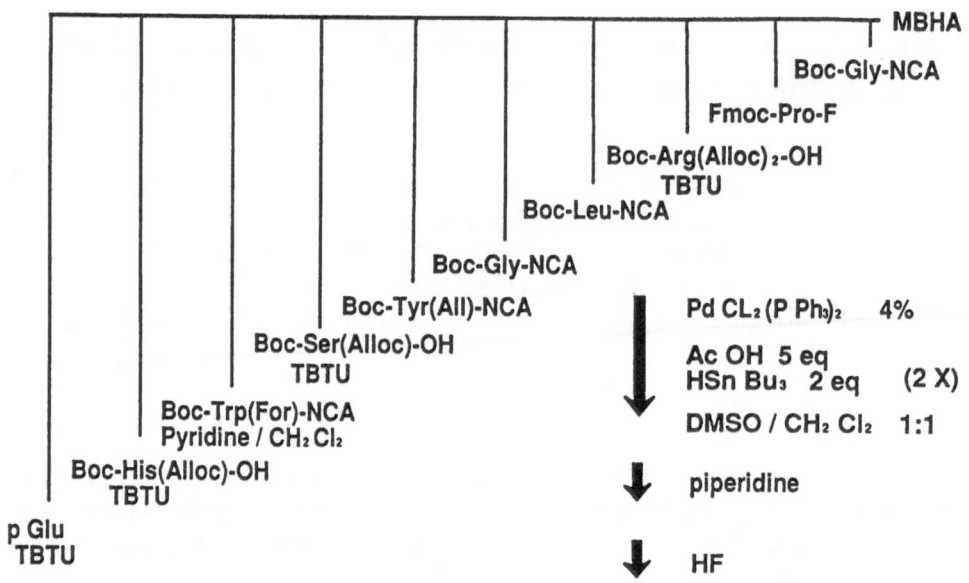

Scheme 1. Synthesis of [D-Trp]⁶-LH-RH by conntinuous flow SPPS.

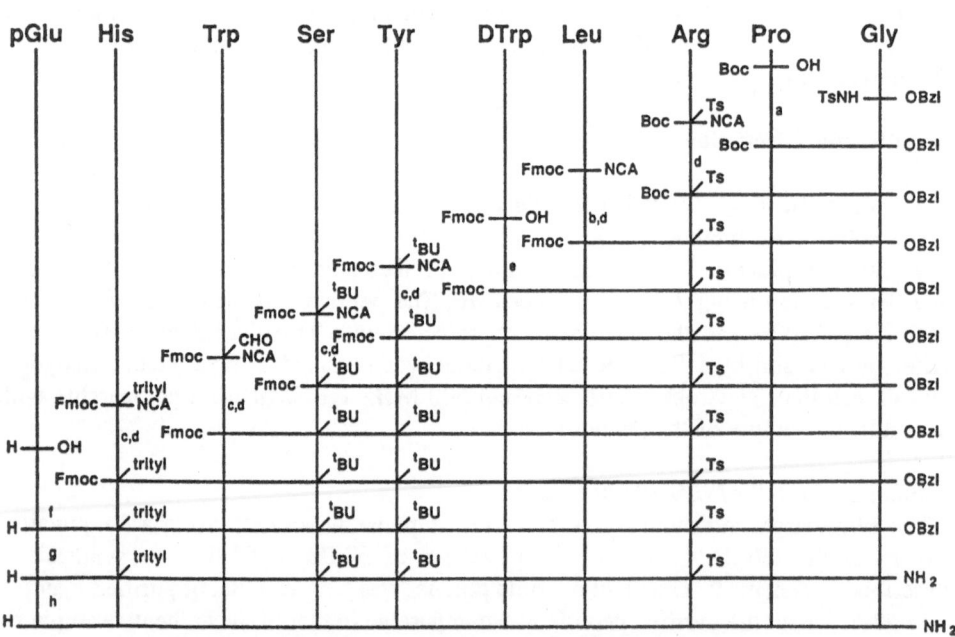

Reagents: a) HOBt/EDC.HCl (84%) or BOP (92%) ; b) TFA (100%) ; c) Piperidine/DMF ; d) NMM/DMF ; e) BOP/DMF/NMM ;
f) BOP/DMF/NMM ; g) NH₃/MeOH (91%) ; h) HF (77%) .

Scheme 2. Synthesis of [D-Trp]⁶-LH-RH in solution.

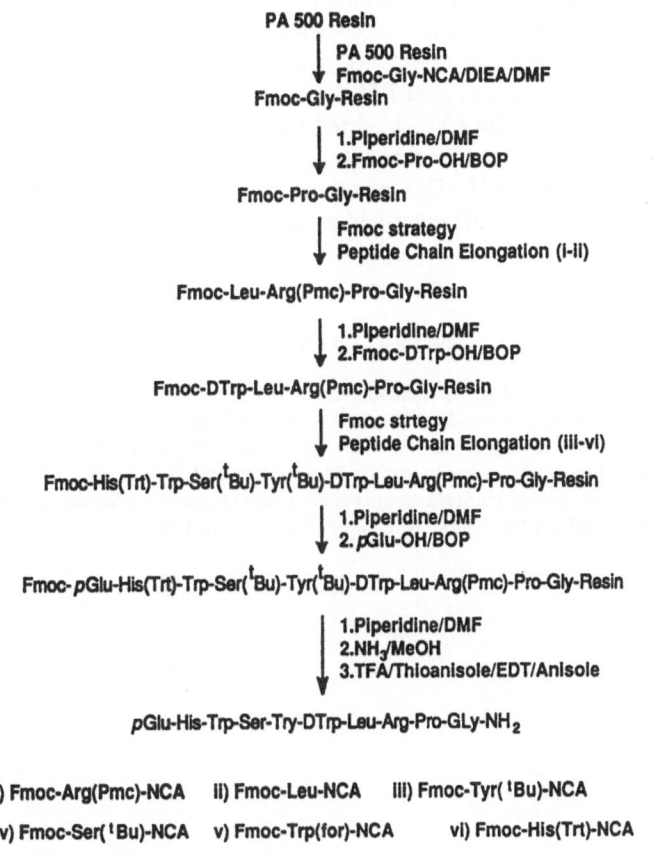

PA 500 Resin

↓ PA 500 Resin
Fmoc-Gly-NCA/DIEA/DMF

Fmoc-Gly-Resin

↓ 1.Piperidine/DMF
2.Fmoc-Pro-OH/BOP

Fmoc-Pro-Gly-Resin

↓ Fmoc strategy
Peptide Chain Elongation (i-ii)

Fmoc-Leu-Arg(Pmc)-Pro-Gly-Resin

↓ 1.Piperidine/DMF
2.Fmoc-DTrp-OH/BOP

Fmoc-DTrp-Leu-Arg(Pmc)-Pro-Gly-Resin

↓ Fmoc strtegy
Peptide Chain Elongation (iii-vi)

Fmoc-His(Trt)-Trp-Ser(^{t}Bu)-Tyr(^{t}Bu)-DTrp-Leu-Arg(Pmc)-Pro-Gly-Resin

↓ 1.Piperidine/DMF
2.*p*Glu-OH/BOP

Fmoc-*p*Glu-His(Trt)-Trp-Ser(^{t}Bu)-Tyr(^{t}Bu)-DTrp-Leu-Arg(Pmc)-Pro-Gly-Resin

↓ 1.Piperidine/DMF
2.NH₃/MeOH
3.TFA/Thioanisole/EDT/Anisole

pGlu-His-Trp-Ser-Try-DTrp-Leu-Arg-Pro-GLy-NH₂

i) Fmoc-Arg(Pmc)-NCA ii) Fmoc-Leu-NCA iii) Fmoc-Tyr(^{t}Bu)-NCA

iv) Fmoc-Ser(^{t}Bu)-NCA v) Fmoc-Trp(for)-NCA vi) Fmoc-His(Trt)-NCA

Scheme 3. Synthesis of (4-10)-LH-RH and LH-RH.

The polymer was divided in two parts after the coupling of the serine derivative.
The following coupling conditions were used:

1) Coupling of Boc-amino-acid-NCA: 2 equivalents of Boc-AA-NCA were stirred with the resin (1 equivalent of free NH₂ groups) in DMF for 15 min.: Kaiser test was negative after this time.

2) Coupling with TBTU: 2 equivalents of Boc-amino-acids and 2 equivalents of TBTU in DMF were stirred with the resin at pH 9 for 30 min.; Kaiser test was negative after this time.

3) Coupling of Fmoc-Pro-F: 3 equivalents of Fmoc-Pro-F and 3 equivalents of DIEA in CH₂Cl₂ were stirred with the resin for 30 min.

4) Coupling of Boc-Trp(For)-NCA to serine: after cleavage of the Boc group, the resin was neutralized with pyriding/CH₂Cl₂ (20%); 3 equivalents of Boc-Trp(For)-NCA were stirred in this mixture for 3 h.

The deprotection of the side-chains was carried out as follows: to the peptide-resin mixture swollen in DMSO/CH₂Cl₂(1:1) was added sequentially 4% PdCl₂(PPh₃)₂, acetic acid (5 equivalents /allyl group) and HSnBu$_3$ (2 equivalents/allyl group); the mixture was stirred for 15 min., filtered and the process repeated.

After thorough washings, the resin was treated with piperidine to eliminate the formyl group from the indole nucleus (LH-RH only).

The resin was washed with CH$_2$Cl$_2$, dried and finally cleaved with HF/anisole.

HPLC analysis of the crude products showed a common impurity in both the heptapeptide and the LH-RH. This common impurity is the oxazolidinone derivative which is formed by O--N shift even under our neutral coupling conditions.

References

1. Fuller, W.D., Cohen, M.P., Shabankarech, M., Blair, R., Goodman, M., Naider, F.R., J. Am. Chem. Soc., 112(1990)7414.
2. Loffet, A. and Zhang, H.X., In Innovations and Perspectives in Solid Phase Synthesis-Peptides, Polypeptides and Oligonucleotides-1992 (Epton, R. Ed), Intercept, UK, p. 77.
3. a) Lyttle, M.H. and Hudson, D., In Smith, J.A. and Rivier, J.E. (Eds.) Peptides: Chemistry and Biology (Proceedings of the 12th American Peptide Symposium), ESCOM, Leiden, 1992, p. 583.
    b) Hudson, D. and coll., In Innovations and Perspectives in Solid Phase Synthesis-Peptides, Polypeptides and Oligonucleotides-1992 (Epton, R. Ed), Intercept, UK, p. 135.

Attachment of obstinate amino acid to the chloromethyl resin

De-xin Wang, Gui-shen Lu and Min-tong Guo
*Institute of Materia Medica, Chinese Academy of Medical Sciences,
Beijing 10050, China*

Introduction

Peptide chemists are often bothered with asparagine and glutamine, because it is difficult to avoid troubles like dehydration and poor solubility caused by amide side-chain group of these two amino acids in peptide synthesis. Bases such as Cs_2CO_3 and KOH were useless, and TEA was reluctant to induce the esterification of Boc-Asn by the chloromethyl resin in our hands [1] although they are efficient reagents for other amino acids. In the present study, we tried to find out the optimum condition for the preparation of Boc-Asn-OCH$_2$-polystyrene(I), which is the key intermediate for the synthesis of hF-GRP(gonadotropin releasing peptide) [2] and its analogs.

Results and Discussion

Five organic bases, DEA (diethylamine), TEA (triethylamine), DCHA (dicyclohexyl-amine), Triton B (N-benzyltrimethylammonium hydroxide) and TMA-OH (tetramethyl-ammonium hydroxide), were compared to see how they would affect the esterification of Boc-Asn by the chloromethyl resin under the same condition: Boc-Asn/base/chloro-methyl resin (3:2.9:1, by mole) in DMF at 60°C for 20h, 40h and 60h. The yield of I was evaluated by ninhydrin quantitative test [3] after de-Boc treatment. The results (Table 1) showed that a moderate yield (54.8% and 68.1%) of I was obtained in the case of DCHA and Triton B, and that the potency of these organic bases to induce this reaction seemed out of accord with their basicity: TEA<DEA<DCHA<TritonB and TMA-OH.

Three kinds of potassium halide were explored as the catalyst to assist Cs_2CO_3 for the preparation of I. Very different yields of I were found from different KX: 64.4%(KF), 29.1%(KBr) and 96.5%(KI) under the same condition: KX/Cs_2CO_3/Boc-Asn/chloro-

Table 1 *The yields(%) of Boc-Asn-OCH$_2$-resin*

Time	TEA	DCHA	DEA	Triton B	TMA-OH
20h	26.80	9.01	5.26	47.10	5.94
40h	33.19	46.50	7.77	64.80	7.30
60h	33.50	54.80	9.32	68.10	7.19

methyl resin (0.15:1.5:3:1, by mole) in DMF at 60°C for 20h. The excellent yield of I with the catalysis of KI indicated that the halide exchange might be the dominant way in the first step of the whole reaction. Boc-Asn was esterified immedietely via benzylic cations because iodide is the best leaving group among the halides.

From the results of present study, the application of KI-Cs$_2$CO$_3$ to induce the esterification of Boc-Asn by the chloromethyl resin was proved to be an efficient and feasible way in solid phase peptide synthesis.

Acknowledgement

This work was supported by the grant of the National Natural Science Foundation, No. 39070951.

References

1.   Wang, D.X. and Lu, S.G., Chin. Chem. Lett., 2(1991)289.
2.   Li, C.H., Ramasharma, K., Yamashiro, D. and Chung, D., Proc. Natl. Acad. Sci. USA, 84(1987)959.
3.   Savin, V.K., Anal. Biochem, 117(1981)147.

Novel cyclic organophosphorus compounds as coupling reagents for peptide synthdesis

Chong-xu Fan, Xiao-lin Hao and Yun-hua Ye

Department of Chemistry, Peking University, Beijing 100871, China

Introduction

N-Diphenylphosphoryl benzoxazolone (DPBO) and N-diethylphosphoryl benzoxazolone (DEPBO) as coupling reagents were reported by our department previously[1,2]. Recently we developed two novel cyclic organophosphorus compounds, N-(2-oxo-1,3,2-dioxaphosphorinanyl)-benzoxazolone (I,DOPBO) and 3-[O-(2-oxo-1,3,2-dioxaphosphorinanyl)-oxy]-2,3-dihydro-1,2,3-benzotriazin-4-one (II, DOPBT), as coupling reagents for peptide synthesis. A series of depeptides were prepared by using these two coupling reagents in one-pot method. The racemization caused by them was detected.

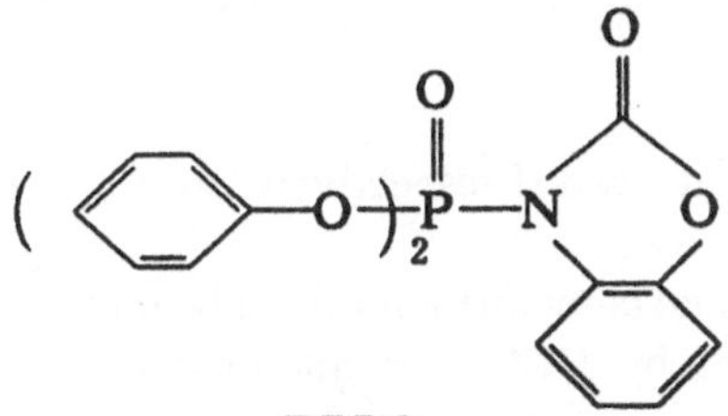

DPBO

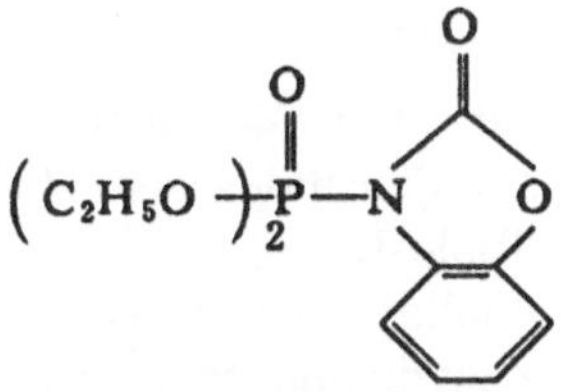

DEPBO

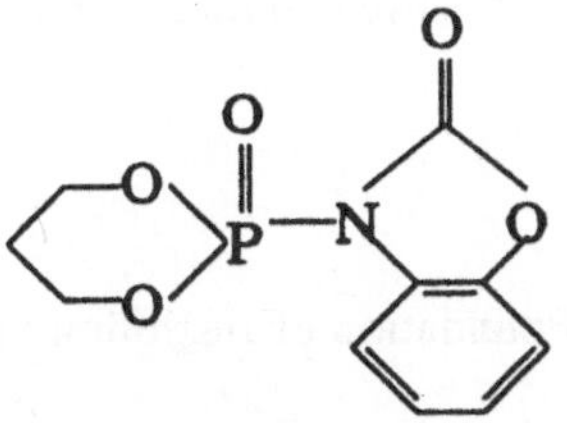

DOPBO

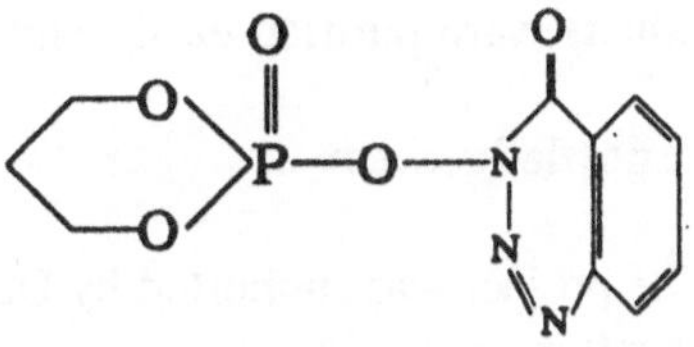

DOPBT

Results and Discussion

DOPBO and DOPBT were prepared conveniently by mixing equivalent 2-oxo-1,3,2-dioxaphosphorinanyl chloride, benzoxazolone or 3-hydroxy-2,3-dihydro-1,2,3-benzo-triazin-4-one and triethylamine in tetrahydrofuran at room temperature. Both DOPBO and DOPBT are stable colourless crystals and non-hygroscopic(I, m.p. 114-116°C, yield 65%; II, m.p. 157°C dec., yield 72%). They can be kept at room temperature for several

C.X. Fan, X.L. Hao and Y.H. Ye

Table 1 *Yields and physical data of dipeptides synthesized by DOPBT and DOPBO*

Peptide	DOPBT			DOPBO		
	Yield%	m.p.[a](°C)	$[\alpha]^{20}_D$(c,sol.)[b]	Yield%	m.p.[a](°C)	$[\alpha]^{20}_D$(c,sol.)[b]
Z-AlaTyrOMe	90	125-126	−6.4(1,MeOH)	79	122-124	−6.0(1,MeOH)
BocAlaPheOMe	84	88-89	−20.2(1,MeOH)	72	84-87	−
BocPhePheOMe	87	122-124	−12.9(1,MeOH)	−	−	−
Z-MetGlyOEt	81	90-92	−18.5(1,EtOH)	67	92-94	−19.0(1,EtOH)
Z-AlaPheOMe	83	101-103	−12.7(1,EtOH)	72	97-100	−
Z-AsnPheOMe	76	199-200	−5.1(1,DMF)	58	195-198	−4.3(1,DMF)
Z-AlaGlyOEt	72	101-102	−24.0(1,EtOH)	66	100-101	−22.1(1,EtOH)
Z-AlaSerOMe	68	137-138	−20.0(1,MeOH)	53	137-139	−19.5(1,MeOH)
BocTrpGlyOEt	73	118-121	−14.7(1,EtOH)	50	116-119	−12.7(1,EtOH)
Bz-LeuGlyOEt[c]	81	136-140	−31.7(1,EtOH)	80	135-140	−30.6(1,EtOH)

[a] m.p. were measured by Yanaco macto melting point apparatus and the thermometer was not corrected.
[b] $[a]^{20}_D$ were measured by Perkin-Elmer 241-MC. Values of $[a]^{20}_D$ and m.p. were in agreement with reported values.
[c] Bz-LeuGlyOEt was a crude product.

months without any decomposition.

A series of dipeptides were obtained using reagent I and II respectively. Results were given in Table 1.

In the coupling reaction, the hydroxy group in amino component need not be protected, and the dehydration of asparagine easily caused by DCC was not observed. The racemization caused by DOPBO and DOPBT was less then 7% and 6% respectively, which was detected by Young test[3]. The results show that DOPBT is more efficient than DPBO, DEPBO and DOPBO. It was prepared conveniently and reacted rapidly. The products were purified easily and in good yields.

Acknowledgement

The project was supported by Doctoral Programme Foundation of Instituion of Higher Education.

References

1. Li, G., Li, C.X. and Xing, Q.Y., Org. Chem. (in Chinese), 9(1989)395.
2. Zhang, D.Y. and Ye, Y.H., in Peptide: Biology and Chemistry: Proceedings of the Chinese Petide Symposium 1990 Ed. By Du Yu-cang et al, science Press: Beijing, (1991) pp.235.
3. Williams, M.W. and Young, G.T., J. Chem. Soc., (1963) 881.

A novel Fmoc-based anchorage for the synthesis of protected peptide on solid phase

Wei Lin[a], Lan Chen[b], Yin-zeng Liu[c] and Ching-I Niu[a]

[a]*Shanghai Institute of Biochemistry, Chinese Academy of Sciences, Shanghai, China*
[b]*Shanghai Institute of Materia Medica, Chinese Academy of Sciences, Shanghai, China*
[c]*Peninsula Lab. Inc., Belmont, CA 94002, U.S.A.*

Introduction

Although the Merrifield procedure of solid phase peptide synthesis has been extremely useful for the preparation of intermediate sized peptide, assembling the large peptides (>50 amino acids) step by step usually encountered trouble, and renders their purification more difficult. The synthesis of protected fragments by the solid phase methodology and subsquent coupling therefore represented an attractive alternative to the preparation of large peptides. Recently, an anchorage compound based on Fmoc, designed by Liu et al[1], 9-(hydroxymethyl)-2-fluoreneacetic acid (AFm), had been used successfully in the synthesis of protected peptides. During the synthesis of AFm, the key step, Clemmensen reaction, proceeded with difficulty and a low yield, and could not be easily scaled up. Therefore, a new compound, 9-(hydroxymethyl)-2-fluorenebutyric acid (BFm), was proposed and synthesized in this paper. The BFm was used to prepare a modeling protected fragment corresponding to the sequence of rat TGF-α(1-7) (Fig.1).

```
          Bzl Tos           Cl-Z
           |   |             |
Boc-Val-Val-Ser-His-Phe-Asn-Lys-OH
```

Fig. 1. The sequence of rat TGF-α(1-7).

Results and Discussion

A new anchoring compound 5 was synthesized in five steps in overall yield of 44.7% (Scheme 1). In the process of synthesizing the compound, the reductive reaction was studied, modified and successfully carried out in nearly quantitative yield[2]. Thereby, BFm could be produced in a large amount. The structure of the new compound was verified by mass, ^{1}H n.m.r spectra and elementary analysis. The coupling of BFm to BHA resin to give BFm-BHA resin was mediated with DCC. Subsquent conversion to 9-(Boc-aminoacyl oxymethyl)-2-fluorenebutyramido-BHA resin was achieved by DCC/DMAP coupling with a variety of N^α-Boc amino acid. The modeling peptide was released from the resin by the β-elimination reaction with 15% piperidine in DMF. The cleavage yield of protected peptide was nearly quantitative. After simple purification, we

299

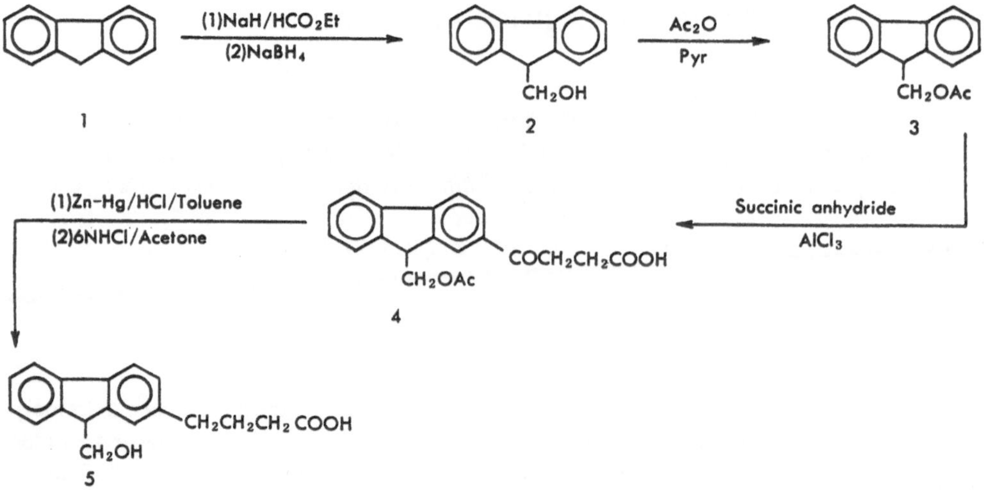

Scheme 1. Synthesis of 9-(hydroxymethyl)-2-fluorenebutyric acid.

could get the pure product with a yield of 46%. The elementary analysis, amino acid analysis and FAB-mass spectra were in accord with the expected structure.

The results demonstrated that the new bifunctional compound had the same properties as AFm in synthesis of protected peptides. But it had a much higher synthetic yield than AFm(44.7% vs. 22.1%), and could be produced in a large amount. Mutter synthesized a related compound 9-(hydroxymethyl)-fluorene-4-carboxylic acid as anchoring ligand for a polymer-supported peptide synthesis. Although the 9-(hydroxymethyl)-fluorene-4-carboxylic acid system was reported to be stable to neutralization with DIEA, we believed that the newly synthesized bifunctional reagent, by virtue of the additional methylene group between the fluorene nucleus and the carboxyl group, should be more stable to β-elimination. This may render the new resin more stable to premature anchor-bond cleavage.

Additional studies are projected using the base labile resin for the synthesis of other protected pepide fragments.

References

1. Liu, Y., Ding, S., Chu, J. and Felix, A., Int J. Peptide Protein Res., 35(1990)95.
2. Martin, E., in Org. Synthesis, Coll. Vol 2, Blatt, A.H.(Ed), John Wiley and Sons, New York, 1943, pp.499.

Preparation and applications of xanthenylamide (XAL) handles for mild Fmoc solid-phase synthesis of C-terminal peptide amides

Yong-xin Han[a], Mark C. Munson[a], Fernando Albericio[b] and George Barany[a]

[a]Department of Chemistry, University of Minnesota, Minneapolis, MN 55455, U.S.A.
[b]Millipore Corporation, 75A Wiggins Avenue, Bedford, MA 01730, U.S.A.

Introduction

Our laboratory reported on the 5-(9-(9-fluorenylmethyloxycarbonyl) aminoxanthen-2-oxy)valeric acid (XAL) handle [**1** in scheme 1] at the 21st European and 12th American Peptide Symposia[1,2]. The related (9-(9-fluorenylmethyloxycarbonyl) aminoxanthen-3-oxy)acetic acid handle [**2** in Scheme 2] is described in a patent application by Pessi and co-workers[3]. Tryptophan-containing peptides cleaved from XAL supports suffer less alkylation, by comparison to appropriate controls with PAL[2,4,5]. We report here additional improvements in the synthesis of XAL handles and their applications to peptide synthesis.

Results and Discussion

Borohydride reduction of 5-(9-oxoxanthen-2-oxy) valeric acid **3** to 5-(9-hydroxyxanthen-2-oxy) valeric acid **4** is very sensitive to reaction and workup conditions [Scheme 1]. At low pH(<8), the over-reduced by-product **5** invariably forms, while at high pH, reaction is incomplete. Earlier quench procedures using acetone or acetic acid were replaced by a new procedure involving careful addition of solid sodium bicarbonate. The resultant NMR and TLC-pure xanthenyl alcohol was reacted further with Fmoc-amide in acetic acid to provide title product **1**, admixed with a small amount (~5%) of acetate **6**. Alternatively, when the reduction mixture was quenched with solid ammonium bicarbonate, amine **7** was obtained directly. Further base-catalyzed reaction of **7** with Fmoc succinimide gave handle **1** in excellent yield [Scheme 1]. Corresponding chemistry in the 3-alkoxy series was more difficult [Scheme 2]. Borohydride reduction of xanthone **8** gave **9** as the principal product. However, sodium amalgam gave the desired xanthydrol, which was taken further to handle **2** according to the procedure of the Italian workers[3].

The usefulness of XAL is shown by two new examples. First, the nonapeptide oxytocin (one disulfide) was assembled starting with handle **1**, using standard Fmoc/DIPCDI chemistry and the novel 2,4,6-trimethoxybenzyl (Tmob) protecting group for cysteine[6]. The free peptide in the dithiol form was obtained directly in ~85% purity upon mild acid cleavage with TFA-CH_2Cl_2-Et_3SiH-anisole-H_2O (7:90:1:1:0.5), 25°C, 30 min. With higher TFA concentrations , undesired oxidation to oxytocin (50% or more) occurred during

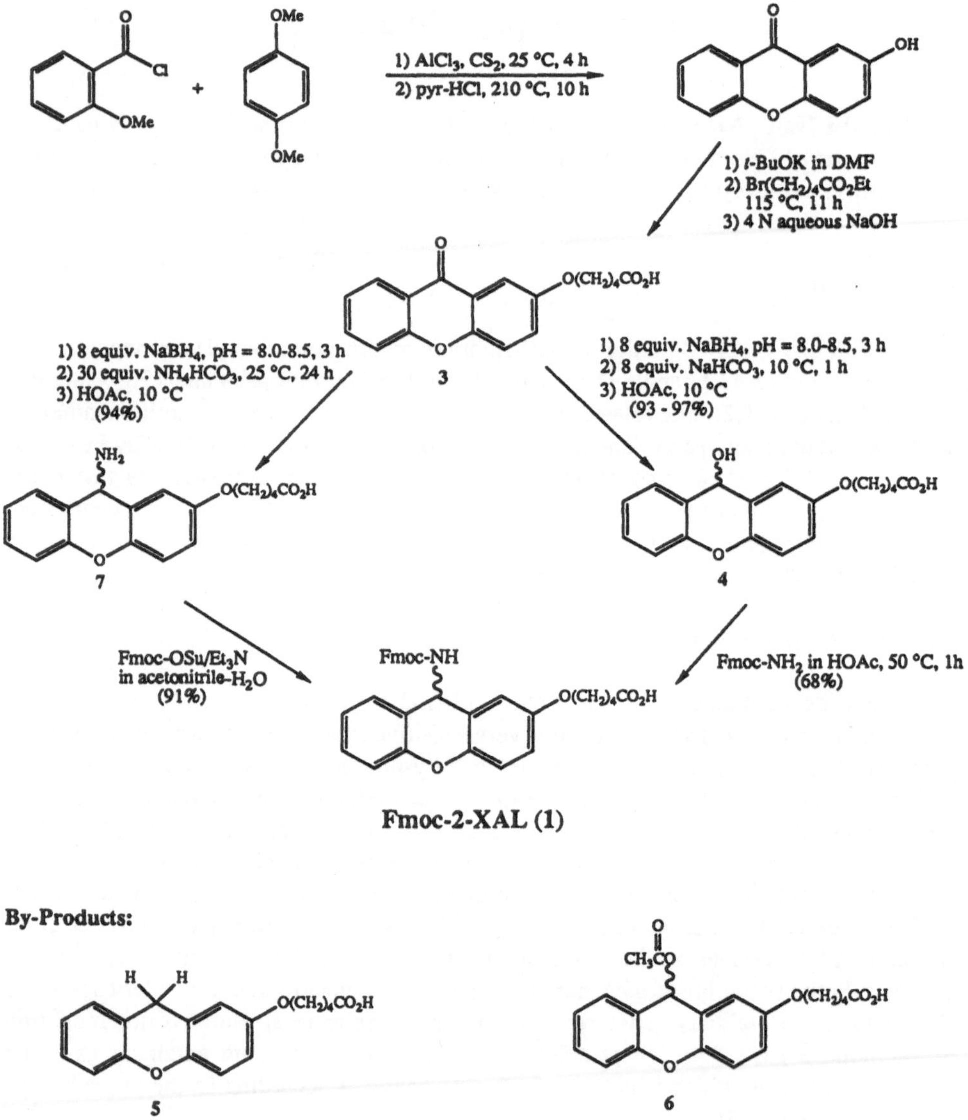
OMe
OMe
OMe
1) AlCl3, CS2, 25 °C, 4 h
2) pyr-HCl, 210 °C, 10 h
OH
1) t-BuOK in DMF
2) Br(CH2)4CO2Et
115 °C, 11 h
3) 4 N aqueous NaOH
O(CH2)4CO2H
3
1) 8 equiv. NaBH4, pH = 8.0-8.5, 3 h
2) 30 equiv. NH4HCO3, 25 °C, 24 h
3) HOAc, 10 °C
(94%)
1) 8 equiv. NaBH4, pH = 8.0-8.5, 3 h
2) 8 equiv. NaHCO3, 10 °C, 1 h
3) HOAc, 10 °C
(93 - 97%)
NH2
7
OH
4
Fmoc-OSu/Et3N
in acetonitrile-H2O
(91%)
Fmoc-NH
Fmoc-NH2 in HOAc, 50 °C, 1h
(68%)
Fmoc-2-XAL (1)
By-Products:
H
H
5
CH3CO
6

Scheme 1

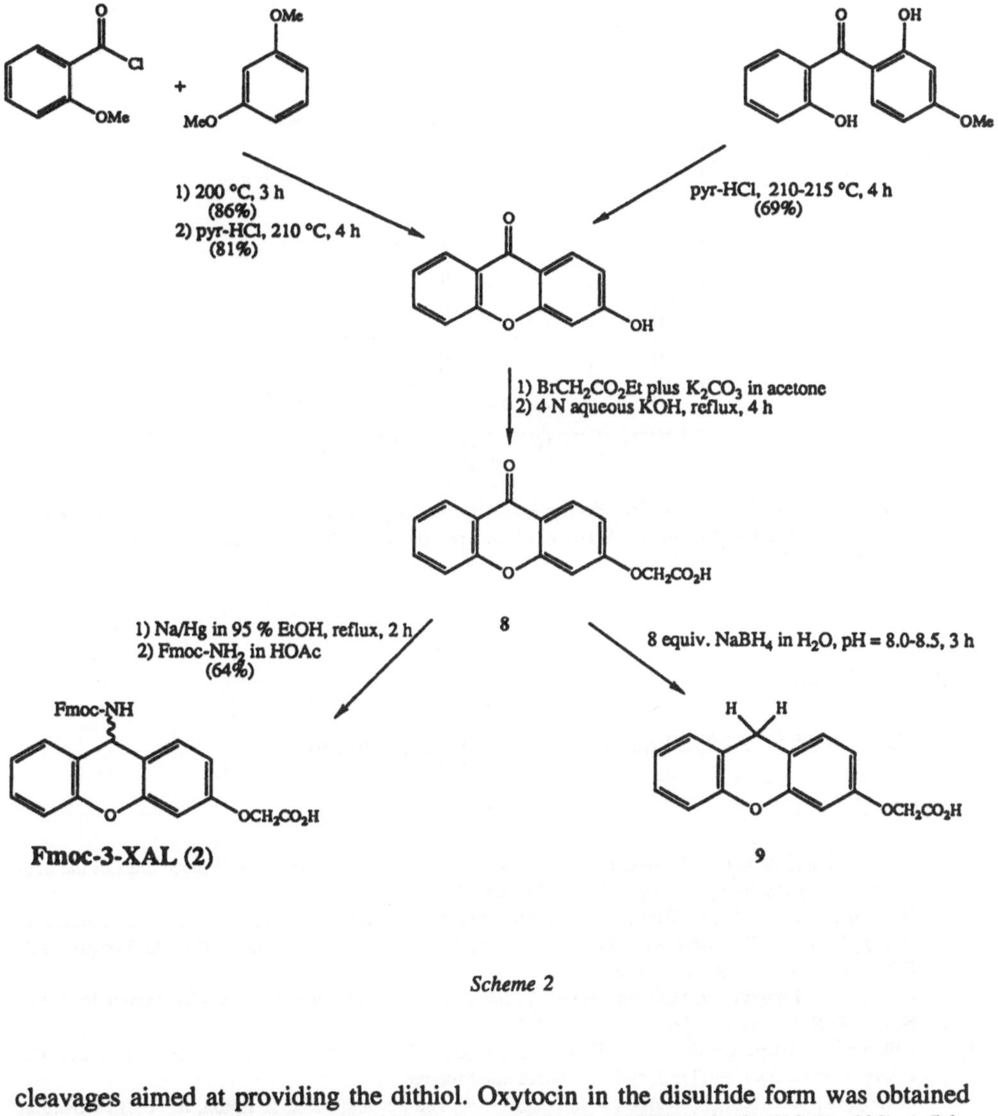

Scheme 2

cleavages aimed at providing the dithiol. Oxytocin in the disulfide form was obtained by resin-bound cyclization with 1.2 equiv. Tl(tfa)$_3$ in DMF-anisole (19:1), 0°C, 1.5 h, followed by acid cleavage. Final absolute yields of the monomeric oxidized peptide were in the range of 40-45%, and HPLC purities were >85%. For both reduced and oxidized oxytocins, cleavage of the XAL anchor was nearly quantitative (>95%). As a second illustration, handle **2** was used as the starting point for an Fmoc synthesis of the octapeptide CCK-8 sulfate. The synthetic protocol matched our earlier report[2], except that the two aspartate residues were protected as allyl esters and deprotected upon completion of chain assembly by use of Pd(PPh$_3$)$_4$ in CHCl$_3$-HOAc-NMM (20:1:0.5), 25°C, 2 h[7]. Subsequent cleavage by TFA-CH$_2$Cl$_2$-H$_2$O (1:18:1), 25°C, 15min gave the desired sulfated peptide in 71% yield and excellent purity (Fig.1).

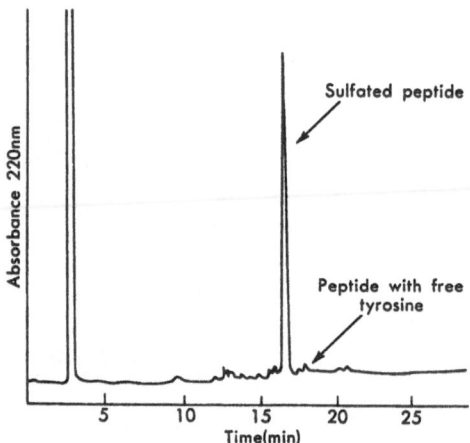

Fig. 1. HPLC of H-Asp-Tyr(SO$_3$H)-Met-Gly-Trp-Met-Asp-Phe-NH$_2$, synthesized as described in text, directly after cleavage. HPLC on C-18 column, linear gradient over 20 min of 0.1% TFA in CH$_3$CN-0.1% aqueous TFA from 1:9 to 2:3, flow rate 1.0 ml/min.

Acknowledgement

We thank NIH GM 42722 and 43552 for financial support.

References

1. Barany, G. and Albericio, F., In Giralt, E. and Andreu, D. (Eds.) Peptides 1990 (Proceedings of the 21st European Peptide Symposium), ESCOM, Leiden, 1991, pp. 139–142.
2. Bontems, R.J., Hegyes, P., Bontems, S.L., Albericio, F. and Barany G., In Smith, J.A. and Rivier, J.E. (Eds.) Peptides: Chemistry and Biology (Proceedings of the 12th American Peptide Symposium), ESCOM, Leiden, 1992, pp. 601–602.
3. Caciagli, V., Longobardi, M.G. and Pessi, A., Italian Patent Application No. 19697A, March 16, 1990, SCLAVO SpA, Siena, Italy.
4. Albericio, F., Kneib-Cordonier, N., Biancalana, S., Gera, L., Masada, R.I., Hudson, D. and Barany, G., J. Org. Chem., 55(1990)3730, and references cited therein.
5. Barany, G., Solé, N.A., Van Abel, R.J., Albericio, F. and Selsted, M.E., In Epton, R. (Ed.) Innovation and Perspectives in Solid Phase Synthesis and Related Technologies: Peptides, Polypeptides and Oligonucleotides 1992, INTERCEPT, Andover, England, 1992, pp. 29–38.
6. Munson, M.C., García-Echeverría, C., Albericio, F. and Barany, G., J. Org. Chem. 57(1992)3013.
7. Albericio, F., Barany, G., Fields, G.B., Hudson, D., Kates, S.A., Lyttle, M.H. and Solé, N.A., In Schneider, C.H. and Eberle, A.N. (Eds.) Peptides 1992 (Proceedings of the 22nd European Peptide Symposium), ESCOM, Leiden, 1993, pp. 191–193.

Enzymatic synthesis of peptides using racemic amino acids as carboxyl components

Gui-ling Tian, Dong-cheng Dai, Hong Wang and Yun-hua Ye

Department of Chemistry, Peking University, Beijing 100871, China

Introduction

Nowadays many bioactive peptide were modified by unnatural amino acids which were usually prepared by chemical method in racemic forms. Most of the synthetic amino acids need to be resolved before using them. We tried to condense the Z- or Boc-protected DL-amino acids (or esters) and GlyNHNHPh in the presence of papain at pH=5(or 8.2) or α-chymotrypsin at pH=10 to form expected optically active protected dipeptides. The reactions were carried out in an aqueous-organic mixed solvent system.

Results and Discussion

In the model reactions, Z-DL-AlaOH(Ia), Z-DL-AlaOCH$_3$(Ib),Boc-DL-AlaOH(Ic), Z-DL-TyrOEt(Id) and Boc-DL-TyrOEt(Ie) as carboxyl components were reacted with GlyNHNHPh (II) in the presence of different enzymes and the expected optically active protected peptides were obtained. Ia and Ic were coupled by papain with II in water-methanol at pH=5, 38°C for 72 h to form Z-L-Ala·GlyNHNHPh (IIIa) and Boc-L-Ala-GlyNHNHph (IIIc) respectively. Ib as an ester reacted with II by the same enzyme and in the same solvent system at 38°C and pH 8.2 for 72 h to form III a in a higher yield. Z-DL-TyrOEt(Id) and Boc-DL-TyrOEt (Ie) were condensed with II in the presence of α-chymotrypsin and water-dichloromethane (1/4.5,v/v) (Id) or water-DMF (1/5,v/v) (Ie) at pH=10 and room temperature for about 48 h to form Z-L-Tyr·GlyNHNHPh (IIId) and Boc-L-Tyr·GlyNHNHPh (IIIe) respectively. The results are given in Table 1.

The results indicate that both N-Z- or N-Boc- protected racemic amino acids or esters could form peptide bond containing only L-configuration throng the stereo-selectivity of enzymatic catalyzed reactions. The N-protected D-configuration amino acids or esters did not participate in the condensation reactions under the same conditions, so there is no need to resolve the racemic amino acids before peptide bond formation. The Z-DL-AlaOCH$_3$ as a acyl donor was better than Z-DL-AlaOH. All the dipeptides were identified by MS, IR, elemental analysis and they were identical with those synthesized by chemical coupling reagents DEPBO[la], DOPBO and DOPBT[lb]. The physical constants of the above peptides were identical with those reported in the literature[2,3]. The Z-Ala·GlyNHNHPh was oxidized with water solution of ferric chloride (30%) to deprotect the C-terminal phenylhydrazide to obtain Z-Ala·GlyOH in 60% yield, mp: 125-126°C, $[\alpha]_D^{20}$: −15.6° (c 1, EtOH) [lit[4]: mp: 125-128°C, $[\alpha]_D^{25}$:−16° (EtOH)].

Our experiments also showed that the speed for peptide bond formation catalyzed by papain was increased under nitrogen. Further studies are in progress.

Table 1 *The physical properties of dipeptides using racemic acids as carboxyl components by enzymatic synthesis*

Enzyme	Acyl donor	Nucle-ophile	Product	%yield[a]	$[\alpha]_D^{20}$ [b] (c 0.5,DMF)
papain	Z-DL-AlaOH	II	Z-L-Ala·GlyNHNHPh	55	+3.9°
papain	Z-DL-AlaOCH₃	II	Z-L-Ala·GlyNHNHPh	80	+4.5°
papain	Boc-DL-AlaOCH₃	II	Boc-L-Ala·GlyNHNHPh	60	−7.6°
α-chymotrypsin	Z-DL-TyrOEt	II	Z-L-Tyr·GlyNHNHPh	70	−20.6°
α-chymotrypsin	Boc-DL-TyrOEt	II	Boc-L-Tyr·GlyNHNHPh	60	−11.5°

[a] The yields were based on the L-configuration.
[b] $[\alpha]_D^{20}$ were measured by Perkin Elmer 241-MC.

Acknowledgement

This project was supported by the Doctoral Programme Foundation of the Ministry of Higher Education.

References

1. a. Zhang, D.Y. and Ye, Y.H., In peptides: Biology and Chemistry: Proceedings of the Chinese Peptide Symposium 1990, Ed. by Du Yu-cang et al, Science Press: Beijing, 1991, p. 235.
    b. Fan, C.X., Hao, X.L. and Ye, Y.H., In Du, Y.C., Tam, J.P. and Zhang, Y.S. (Eds.) Peptides: Biology and Chemistry, ESCOM, Leiden, 1993, pp. 297–298.
2. Sakina, K., Int. J. Peptide Protein Res., 31(1988)245.
3. Kullmann, W., J.Bio. Chem., 255(1980)8234.
4. Schroder, Von E., Ann. Chem., 679(1964)207.

306

Exploratory studies on sequence-dependence in solid phase peptide synthesis

Zheng-ying Chen, Xue-jun Fan, Xiao-jie Dong and Xian-kai Ma

Institute of Basic Medical Science, AMMS, Beijing 100850, China

Introduction

In solid phase peptide synthesis, sequence-dependent coupling inefficiencies continue to be a problem which bothers the peptide chemists. This paper tries to explore sequence-dependence in peptide synthesis and hope to find some generality so as to predict potential difficulties during the peptids synthesis. Using Fmoc solid phase strategy on a Biolynx 4170 Automatic Peptide Synthesiser, we have synthesized peptides of various length, among them we selected 19 peptides for analysis. These peptides ranged in size from 11 to 52 residues, mostly exceeding 15 residues.

Biolynx Synthesiser possesses the benefits of UV on-line monitoring especially the deprotection traces are most useful aspect. It provides information about the release of Fmoc group into solution, so we can identify sequences in which peptide primary structure may affect the efficiency of peptide bond formation according to the peak area and shape of the deprotection traces. In other words, the onset of such unpredictable anomalies is reflected in the deprotection peaks monitored during the synthesis (Fig. 1).

If the peak area decreases, the last amino acid addition may not be complete. If the peak shape flattened the deprotection may be difficult, suggesting that some form of steric hindrance is affecting the reaction rate.

Results and Discussion

Basing on the comparisons of deprotection traces of mentioned 19 peptides, following points were found:

1. 14 out of 19 peptides (73%) contained steric hindrance sequences (usually 3-6 residues). So we have good reason to say that in synthesis of large peptides, sequence-dependence is a general phenomenon.

2. The steric hindrance sequences of most peptides (13/14,93%) appeared in 8-20 residues from the C-terminus and only one (1/14,7%) in the 28-36 redidues. Most probably the secondary structure, formed during the elongation of peptide chain, creates a steric hindrance which slows down the coupling rate of subsequent amino acid.

3. Usually the steric hindrance sequences appeared after a segment of several consecutive hydrophobic amino acids or some amino acids with a large side chain protecting group. These phenomena lead us to think that when several consecutive hydrophobic amino acids appeared in the nascent growing peptide chain, secondary structure, like that in the natural peptide chain, might form. For certain amino acids with a large side chain protecting group, the steric hindrance may stem from their own

307

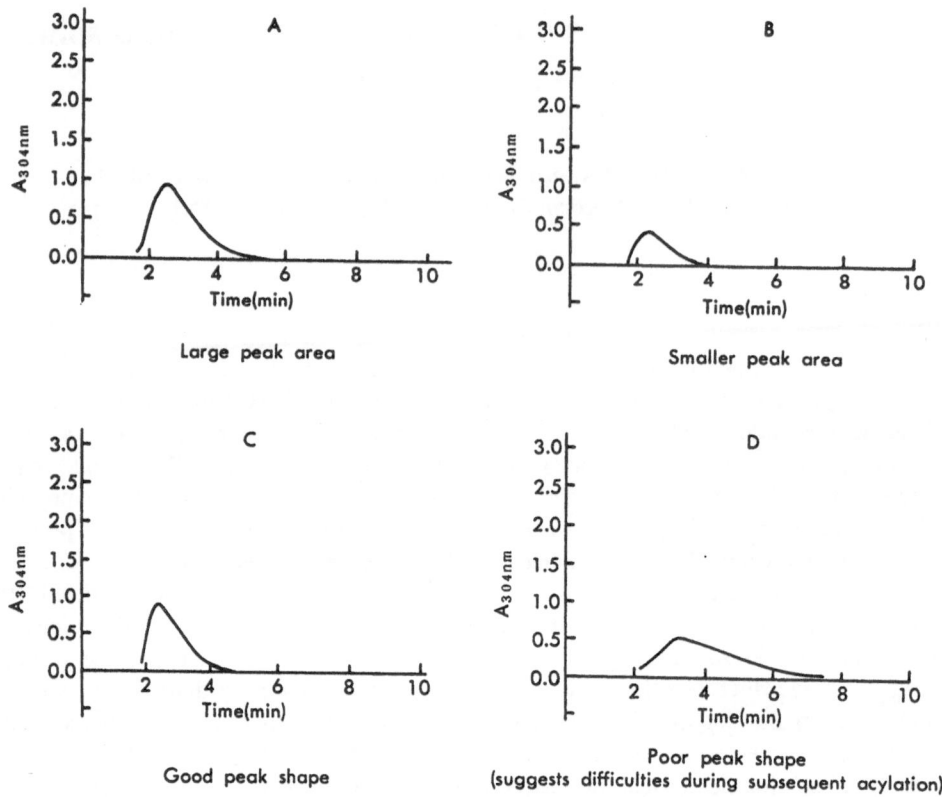

Fig. 1. *Some shapes of deprotection traces.*

structures.

4. According to the data from these 19 peptides: during the synthesis, when Arg, Val, Thr, Ile and Trp acted as carboxyl component, the coupling reaction became difficult. When Gly, Pro, Ala, Arg, Asn and Gln acted as amino component, the same problem occurred. The coupling reaction became more difficult when the following paris of amino acids were encountered: Arg-Pro, Arg-Arg, Arg-Gly, Arg-Glu, Thr-Lys, Trp-Gly, Ile-Ile and Val-Pro etc.

It would be very useful to predict when the peptide bond formation are going to be difficult. We could then take more care to keep the coupling yields high. This would reduce the number of deletion peptides.

Acknowledgement

The project supported by National Natural Science Foundation of China (3870675).

Carboxyl catalyzed phosphoryl group migration in phosphorylated glutathion

Hou-jun Yang, Jian Liu, Zeng-jia Yang and Yu-fen Zhao

Department of Chemistry, Tsinghua University, Beijing 100084, China

Introduction

Previously, we have successfully synthesized a series of phosphrylated amino acids using various dialkyl phosphites as the phosphorylating reagent[1], this method has also been employed to phosphorylate oligopeptides. In phosphorylating serine, we found that the phosphoryl group migrated from amino group to hydroxy group but not migrated from amino group to thiol group in phosphorylating cysteine[2]. When we tried to phosphorylate glutathion(reduced form), we found a minor peak at 23 ppm in ^{31}P-NMR spectra which may arise from thiol phosphorylation in addition to a major peak at 6.55 ppm from amino phosphorylation. In fact, many phosphorylated peptides display an important biological activity, and some of them contain one or several cysteines. Here, we studied the mechanism of phosphoryl migration in phosphorylated glutathion.

Results and Discussion

The amino acids and oligopeptides were phosphorylated with 1 mole diiso-propylphosphite (DIPP) and CCl_4 in a solution of ethanol and water. It was previously reported that N-DIPP serine and threonine were liable to have their diisopropyloxy phosphoryl group migrated from amino to hydroxy group[3]. It is thought that N-DIPP Cys might have similar behavior. Nevertheless, ^{31}P-NMR tracing experiment shows no N to S migration.

When glutathion (reduced form) was phosphorylated with the method described above, we found a minor peak at 23.82 ppm in ^{31}P-NMR which might arise from the phosphorylation of SH. In order to identify this peak, we phosphorylated glutathion with two mole of diisopropyloxy chlorophosphate and got the same peaks at 6.83 ppm of N phosphoryl group and 23.82 ppm of S phosphoryl group. Is SH of glutathion phosphorylated directly or from the migration of N phosphoryl? To answer this question, we repeated the reaction of cysteine phosphorylation with this method and got no S phosphorylated product. It was obvious that the thiol group can not be phosphorylated directly in the phosphrylation of glutathion. Then we tried to synthesis N-(diisopropyloxyphosphoryl) glutathion and purify it until a single peak at 6.82 ppm was observed in ^{31}P-NMR. When the purified N-DIPP glutathion was examined by ^{31}P-NMR tracing experiment, an observable peak was found at 23.82 ppm with relative intensity of 2.4 over 20 h, and the relative intensity became 17.8 after 44 h, then it remained constant. Now we can conclude that the phosphoryl migration does occur to N-

Table 1 ^{31}P-NMR and MS data

Compound	mp (°C)	^{31}P-NMR(ppm)	FAB-MS(m/z)
DIPPCysOH	Oil	6.10	MH⁺ 286
DIPPGluCysGlyOH	135-7	6.27(DMSO)	MH⁺ 472
DIPPGluCysGlyOH DIPPS	Oil	28.2, 6.83 (D₂O)	NH⁺ 636

phosphorylated glutathion, and the peak at ~ 23 ppm in ^{31}P-NMR comes from phosphorylated thiol group. Like the phosphoryl migration in serine and threonine, it is reasonable to postulate that the phosphoryl group migrates to the thiol group by a mixed phosphoryl-carboxyl anhydride intermediate (Scheme 1).

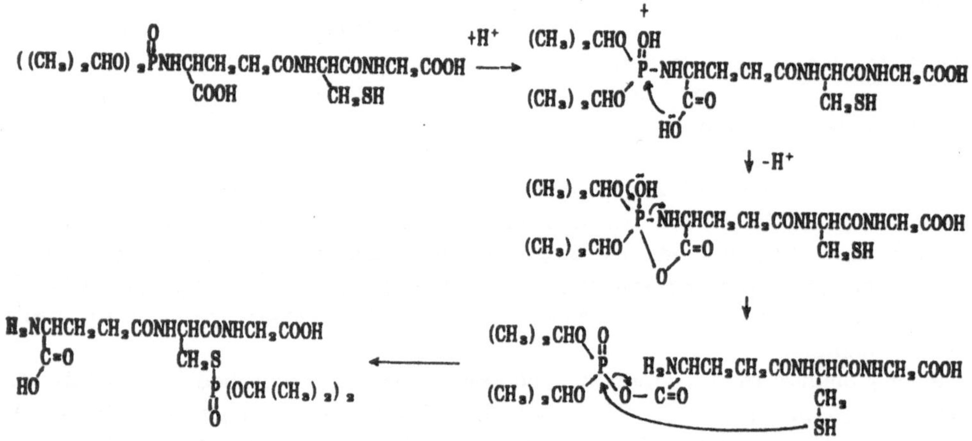

Scheme 1.

Acknowledgement

Thanks to the Fund Foundation of the Institute of Science of Tsinghua University.

References

1. Zeng, J.N., Xue, C.B., Chen, Q.W. and Zhao, Y.F., Bioorg. Chem., 17(1989)434.
2. Li, Y.C. and Zhao, Y.F., Phophorus, Sulfur and Silicon, 60(1991)233.
3. Xue, C.B. Yin, Y.W. and Zhao, Y.F., Tetrah. Lett., 29(1988)1145.

Application of BBC reagent for peptide synthesis

Shao-qing Chen and Jie-cheng Xu
*Shanghai Institute of Organic Chemistry,Chinese Academy of Sciences,
Shanghai 200032, China*

Introduction

In peptide chemistry, many of reagents have been used for peptide synthesis. So far, DCCI is still one of the most useful. Recently, phosphonium and carbonium-containing compounds such as BOP[1] and HBTU[2] have attracted high interest. A new carbonium compound, benzotriazolyloxy-bis(pyrrolidino)-carbonium hexafluorophosphate (BBC) has recently been introduced by us[3] and B. Castro et al[4] separately for peptide synthesis. Here we report some analogs and the application of BBC for peptide synthesis by solution or solid phase method, and especially for the synthesis of cyclic peptides.

Results and Discussion

BBC can be prepared by the reaction of carbonyl dipyrrolidine and phosgene or phosphorous oxychloride, or directly from pyrrolidine and phosgene.

By using BBC, the coupling reaction proceed very fast determined by TLC or HPLC using the model coupling of Z-Gly-L-Phe-OH and L-Val-OMe·HCl.

Several analogs of BBC were synthesized and used for the peptide synthesis. The results were shown in Table 1. It can be seen that BBC seemed to be the best. The racemization can be decreased by using the additive HOBt.

The facility for the use of this reagent was revealed by the direct use of TFA salt or DCHA salt but not CHA salt. Probably it can be explained from the reaction of PhCOOH and DCHA or CHA using BBC as coupling reagent. In the case of CHA, it gave the prospective amide; but in the case of DCHA, it didn't give amide but only-OBt ester, the same as the product of reaction of PhCOOH and BBC. This implied that the reaction intermediate might be the -OBt active ester. Because of the weak nucleophilicity and great steric hindrance of DCHA, it could not react further with this intermediate.

BBC was also used for the synthesis of amide and ester. The protected amino acid active esters suah as -ONp, -OPcp, -OPfp and -OSu esters were prepared.

Leu-enkephalin was synthesized via 3+2 approach by solution method. All the peptide bonds were coupled by BBC and the yield of each step was about 90% yield. The final cleavage was carried out with Low-TFMSA method and the crude product showed high purity (98.6%) in HPLC.

Due to the high coupling speed and short of insoluble by-product such as DCU, BBC can be successfully used in the solid phase peptide synthesis. Leu-enkephalin and another two peptides (Ile-Trp-Gly-Cys-Ser-Gly-Lys-Leu-Ile-Met and Ile-Trp-Gly-Met-Ser-Gly-Lys-Leu-Ile-Met), corrsponding to the aa sequence 600-609 of gp41 of HIV,

Table 1 *Comparison of racemization of using various reagents by HPLC method*

No.	Reagent	Structure	Yield(%)	D%
1	BBC	$(C_4H_8N)_2{}^+C\text{-OBt}\cdot PF_6{}^-$	100	3.4
2	BBC/HOBt		100	0.2
3	BC-Cl	$(C_4H_8N)_2{}^+C\text{-Cl}\cdot PF_6{}^-$	12.8	22.3
4	FBC	$(C_4H_8N)_2{}^+C\text{-OPfp}\cdot PF_6{}^-$	100	35.0
5	ABC	$(C_4H_8N)_2{}^+C\text{-}N_3\cdot PF_6{}^-$	79.4	·0.2
6	DBBC	$(C_4H_8N)_{2+}C\text{-ODHBt}\cdot PF_6{}^-$	65.5	0.2
7	HBTU	$(Me_2N)_2{}^+C\text{-OBt}\cdot PF_6{}^-$	100	4.5
8	FTU	$(Me_2N)_2{}^+C\text{-OPfp}\cdot PF_6{}^-$	100	27.9
9	FTU/HOBt		100	0.5
10	BOP	$(Me_2N)_3{}^+P\text{-OBt}\cdot PF_6{}^-$	100	4.8
11	FOP	$(Me_2N)_3{}^+P\text{-OPfp}\cdot PF_6{}^-$	93.3	37.6
12	DPPA	$(PhO)_2\overset{\displaystyle O}{\overset{\|}{P}}\text{-}N_3$	76.6	0.2
13	DCCI	$C_6H_{11}\text{-}N\text{=}C\text{=}N\text{-}C_6H_{11}$	51.1	16.2

were synthesized with BBC by solid phase synthesis. The coupling steps were accomplished within 0.5–1 h.

Although many of coupling reagents have been used for the synthesis of linear peptides, only few of them are suitable for cyclic peptides. We found that BBC is reliable for the synthesis of cyclic peptides with high yield and fast reaction speed even in very dilution of the reactants.

Cyclo-Leu-enkephalin was synthesized using BBC. TFA salt can be used directly in the cyclization without further treatment. By using slight excess BBC reagent, the cyclic peptide was obtained in 83.3% isolated yield.

The cyclic reaction was monitored directly and conveniently by HPLC method using H-Tyr(Bzl)-Gly-Gly-Phe-Leu-OH→cyclo-(Tyr(Bzl)-Gly-Gly-Phe-Leu) as a model. From the Table 2, it can be seen that BBC gave the better result than the other reagents. Using BBC, the cyclization reaction was accomplished within 0.5 h in almost quantitative yield even in low concentration, while using diphenyl phosphoryl azide (DPPA)[5], which is in common use for the cyclic peptide synthesis, the yield was only 56.5% even after 330

Table 2 *The coupling reaction speed using BBC and other reagent in synthesis of cyclo-(Tyr(Bzl)-Gly-Gly-Phe-Leu)*

Reagent	Concentration (M)	Time(min)	Yield (%)
BBC	2×10^{-3}	10	97.3
		30	100
BBC	1×10^{-3}	10	93.0
		30	100
ABC	1×10^{-3}	10	33.9
		30	37.1
		210	70.5
DPPA	1×10^{-3}	5	5.9
		30	11.1
		60	17.1
		330	56.5
DCCI	2×10^{-3}	15	0
		120	0

min. A BBC's analog (ABC) gave a little better result than DPPA. Although DCCI is the most popular peptide coupling reagent, it was not suitable for the cyclic peptide. Almost no cyclic product could be detected by HPLC after 2 h in the same condition.

BBC can successfully be used not only in solution method, but also in solid phase peptide synthesis, especially for the cyclic peptide synthesis.

References

1. Castro, B., Dormoy, J.R., Evin, G. and Selve, C., Tetrahedron Lett., (1975)1219.
2. Dourtoglou, V., Ziegler, J.C. and Gross, B., Tetrahedron Lett., (1978)1269.
3. Chen, S.Q., and Xu, J.C., Tetrahedron Lett., 33(1992)647.
4. Coste, J., Frerot, E., Jouin, P. and Castro, B., Tetrahedron Lett., 32(1991)1967. They referred this component as HBPyU and found that it showed low efficiency in the synthesis of some peptides containing N-methylamino acid.
5. Sioiri, T., Ninomiya, K. and Yamada, S., J. Am. Chem. Soc., 94(1972)6203.

Author index

Subject index